Foreword

Waterlogging and salinity are two of the major environmental problems associated with large irrigation systems. It is in this context that the Environment Department is very pleased to share the results of this case study for the Indus Basin irrigation system, which analyzes causes of waterlogging and salinity and suggests possible remedies. The methodology and the techniques developed for this case study can be considered as applicable generally to irrigation systems operating in semi-arid and arid environments.

Mohamed T. El-Ashry
Director
Environment Department

Acknowledgements

The authors have been inspired by, and have benefited from, discussions with Warren Fairchild, S.S. Kirmani, Peter S. Lee, Richard Meyers, Walter Ochs, Willem Vlotner, and Tae-Hee Yoon. Walter Ochs and Herve Plusquellec commented on an earlier draft. None of these should be held responsible for any errors or any views expressed in this study.

They are grateful to Anthony Brooke, John H. Duloy, Alexander Meeraus, and Gerald T. O'Mara for their earlier work on the Indus Basin Models.

Table of Contents

1	Introduction	1
	Historical Background	1
	Emerging Problems	2
	The Water Sector Investment Plan	4
	Objectives of this Study	5
2	The Indus Basin Model Revised	7
	Background	7
	Overview of the IBMR	8
	The Zone Models	13
	The Province Models	14
	Appendix: The Agro-Climatic Zones	15
3	Agricultural Performance 1988-2000	17
	Agricultural Production (1988)	17
	Resource Use (1988)	21
	Projections to 2000: Assumptions	24
	Projections for 2000: With No Further Investment	26
	Projections for 2000: With WSIP Investment	27
	Appendix: Surface Water System Inflows	32
4	Environmental Issues	37
	Environmental Priorities	37
	Waterlogging	37
	Salinity	43
5	Quantifying Waterlogging and Salinity Problems	47
	Modelling Salt and Groundwater Flow	47
	Water Balances	50
	A Model of Salt Distribution	51
	Dynamics of Salt Accumulation	65
6	Options for Groundwater Control	71
	Introduction	71
	Dimensions of the Policy Problem	72
	Sub-surface Drainage	75
	Water Allocations	77
	Canal Lining	80
	Watercourse Improvements	84
	Summary of Environmental Impacts	88

7	A Suggested Program	91
	Developmental Objectives	91
	The Need for Environmental Concern	91
	Elements of the Suggested Program	93
	Need for Future Work	95

Appendices:

    A.1: Data and Scenario Selection 101
    A.2: Model Specification 161
    A.3: Report Programs 171
    A.4: Groundwater Simulation Program 183

References

Acronyms and Abbreviations Used

ACZ	agro-climatic zone
ASP	Agricultural Statistics of Pakistan
CGA	canal-commanded area
CWM	Command Water Management
GCA	gross canal-commanded area
GW	groundwater
FGW	fresh groundwater area
IBM	Indus Basin Model
IBMR	Indus Basin Model Revised
IBRD	International Bank for Reconstruction and Development
ICM	integrated, comprehensive management
ISRP	Irrigation System Rehabilitation Project
LIP	Lower Indus Project
LBOD	Left-Bank Outfall Drain
MAF	million acre-feet
Mtons	million tons
NSP	net social product
OFWM	On-farm Water Management
PCA	Pakistan Census of Agriculture
RAP	Revised Action Programme for Irrigated Agriculture
SCARP	Salinity Control and Reclamation Project
SGW	saline groundwater area
SWM	surface water model
USAID	United States Agency for International Development
WAPDA	Water and Power Development Authority
WTD	water table depth
WSIP	Water Sector Investment Plan

CHAPTER 1
INTRODUCTION

Historical Background

From its birth as a nation in 1947 until very recently Pakistan has managed to achieve growth in agricultural production of 5% or more - substantially more than its robust population growth in excess of 3% per annum. Although not a "success story" by world standards, the record is impressive if one considers the ecological environment and paucity of infrastructure of forty years ago.

Because rainfall is highly variable and seasonal, averaging less than eight inches in much of country, 90% of agricultural production must rely on irrigation. Flood or inundation irrigation using the waters of the Indus river and its tributaries[1] has been practiced for perhaps thousands of years. Under British rule in the last century a network of canals and distributaries was begun to expand the irrigated area during the summer (kharif) growing season.

Development of the system accelerated after independence, fueled with financial and technical assistance from the World Bank/IDA and other donors. In 1960 the Indus Waters Treaty formally partitioned the Indus waters between Pakistan and India, with Pakistan acquiring rights to the Indus, Jhelum, and Chenab rivers. Shortly after, the Indus Basin Development Fund was established, in part to support the construction of two large dams, Mangla on the Jhelum, and Tarbela on the Indus. These dams, commissioned in 1967 and 1974 respectively, permitted both expansion of the irrigated area, and seasonal expansion to perennial irrigation where water supplies permitted.

Construction and modernization of the system progressed until today Pakistan possesses the world's largest contiguous surface distribution system, comprised of the Indus river and its major tributaries, three large storage reservoirs, 19 barrages/headworks, 43 main canals, 12 link canals, and about 89,000 watercourses. The total length of the canal system is about 36,000 miles of conveyance facilities. Watercourses, farm channels, and field ditches amount to more than one million miles in length. Inflow derived from glacier and snow melt, and rainfall largely outside the Indus plains averages 147 MAF/year.

In all, the canal commanded area (CCA) is 35 million acres. Most of the Indus plain is underlain by a highly transmissive, essentially unconfined aquifer. Although high salinities limit the area of useable groundwater to 25 million acres (about 60% of the total plain), by 1985/86 groundwater abstractions had reached 33 MAF/year, of which 75% was by private tubewells which now number over 250,000.

Most of the CCA is comprised of fine alluvial silt over a vast flat plain. Climate is arid to semi-arid. Temperatures permit year-round cultivation. Hence growing conditions are nearly ideal when water requirements are met.

---

[1] The Jhelum, Chenab, Ravi, Sutlej, Kabul, and Swat.

Emerging Problems

<u>Slowing Growth</u>.   The Fifth Plan (1978-83) targeted growth for 6.0%; 4.4% was achieved. The Sixth Plan (1983-88) target was for growth of 5.9% per year; 3.8% was achieved, but only because of a dramatic upsurge in cotton yields and area. Growth in production of wheat, rice, and sugarcane virtually ceased. In fact, production of all food crops fell below population growth for the decade of the 1980's (2.4% vs. 3.1%). Production of oilseeds, a major and growing item in the import bill, also stagnated. From the Water Sector Investment Planning Study (WSIP) projections, it can be calculated that agricultural growth to the year 2000 will be only 1.9% per year without additional investment in the water resources sector.[2]

That study also concluded that a 40% deficit in foodgrains, an 80% deficit in edible oils, and a 30% deficit in sugar will be facing Pakistan by 2000 if crop yields continue to grow at historical rates and no improvements in the irrigation system are forthcoming. This prospect, in a country which has mostly met its food and fiber needs heretofore is characterized by that report as "disastrous".

Prospects for future growth are not promising if one examines the sources of growth over the past decade. During that period crop output grew by 49%, but irrigated area grew by 14% and irrigation water supplies, by 23%. Yields of most crops stagnated with the exception of cotton which benefited from the introduction of new varieties and pesticides. As opportunities for expanding the irrigated area become increasingly scarce, and the costs of increasing the water supply ever higher, the prospects for future growth begin to look dim.

<u>Scarce Water</u>. Largely due to inadequate winter (rabi) water availability, cropping intensities are only 116%[3]. In the Sind, where much of the groundwater is saline and farmers cannot supplement surface supplies, the intensity is only 61%. Neither labor scarcity nor lack of markets can explain the low intensity. The main cause is scarce water during the rabi season and at the beginning and end of the kharif season (summer).

Of the 147 MAF of average total surface inflow to the system, 139 MAF pass the rim stations on the major rivers. Given the seasonal nature of the Himalayan runoff (the major source of Indus Basin waters), 85% of the total inflow occurs during the kharif season, and 15% during rabi. Thus rabi surface supplies must largely come from stored water. The available storage from Tarbela (8.86 MAF), Mangla (4.88 MAF), and Chasma (.44 MAF) reservoirs are obviously only a fraction of that needed for full land utilization. Opportunities exist for additional storage sites on the Indus, but only at great cost. Kalabagh Dam, considered periodically for the past 30 years, would today cost about US$ five billion (about 30% of which would allocable to irrigation; the remainder to power

---

[2]<u>National Investment Plan</u>, Federal Planning Cell (Lahore, 1990), Table S.1.

[3]100% implies one crop per year; 200% implies full cropping for both kharif and rabi seasons.

(65%) and flood control (5%)) and provide only 6.1 MAF of additional storage.

Opportunities for further exploiting groundwater use are severely limited and becoming ever more costly. About 21 million acres of the 35 total CCA are underlain by fresh groundwater, but most of this is already exploited by private farmers with tubewells where it has been economically feasible. Private investment in new tubewells has slowed to little above the replacement rate. Exploitation by public tubewell has actually declined, largely at the World Bank's urging, as management and maintenance problems eroded the benefits from their use. Aquifer mining, which implies that wells have to be ever deeper, and escalating energy costs have further impeded the continued development of groundwater resources. There is some evidence that infiltration from saline aquifers into fresh ones, exacerbated by inadequate drainage and poor water management, is deteriorating the quality of the fresh groundwater.

Given the constraints and costs associated with expanding both surface and groundwater supplies, it may not be unreasonable to conclude that Pakistan will have to do with the water supplies it now has for the foreseeable future.

<u>Deteriorating Infrastructure</u>. Partly due to age, and partly to poor or deferred maintenance, much of the physical system is in need renovation. The earthen bank canals leak and often breach during peak discharges. Losses average 40% of all water diverted between the canal head and the root zone. Many of the diversion structures are fifty years old, in need of repair, and are employed beyond their design capacities.[4] Two World Bank projects, Irrigation Systems Rehabilitation and Command Water Management, were begun in the past decade with the aims of renovating almost half of the canal system. Both projects have suffered from delays in implementation, with the result that the overall impact has been minimal to date.

The reservoirs are also suffering from a form of deterioration: siltation. Between 1980 and 1988 the three main reservoirs lost .644 MAF (4.3%) of their combined live storage capacity to siltation. Another .971 MAF (6.9%) is projected to be lost by 2000. Chotiari, the only new reservoir scheduled to come onstream by 2000, will barely offset the pace of siltation of the others since its live capacity will be 1.1 MAF.[5]

<u>Waterlogging and Salinity</u>. Perhaps of even greater potential devastation to the irrigation system than the scarcity of water and the deteriorating infrastructure are the twin menaces

---

[4] Prior to the inception of Mangla (1967), annual canal diversions averaged 83.5 MAF. During the post-Tarbela period (1976-88), average diversions increased by 21%, to 101 MAF with virtually no enlargement or improvement of the canals. These overruns resulted in encroachment of the designed freeboard, erosion of the channel banks, and deformation of the cross-sections. Combined with negligence of annual maintenance, the overruns have resulted in serious deterioration of much of the canal system.

[5] Chotiari will supply 2.177 acres in the Nara Canal area

of waterlogging and salinity. It has long been recognized that irrigation in environs like the Indus Basin (flat topography, poor natural drainage, porous soils, semi-arid climate with high evaporation) without adequate drainage will inevitably lead to rising water tables. And rising water tables bring salts to the surface, choking off the potential for cropping. The inevitable began to occur in the 1950's to a measurable extent. By the 1960's the problem required the attention of planners and donors and a series of SCARP (salinity control and reclamation projects) were initiated. Despite these efforts, today, fully 22% of the CCA is waterlogged[6] and 23% is "salt affected".[7] The WSIP estimates that average yield loss in waterlogged areas to dry-foot crops is in the order of 25%, with a smaller but not well defined loss due to salinity in the soils above affected areas.

It may be argued that Pakistan "mortgaged" its agricultural future by a policy of ignoring drainage needs while spreading available irrigation water over as wide an area as possible. It may also be argued that the strategy was the correct one: waterlogging and salinity may not appear until after 20 or 30 years of intense irrigation during which time the returns from agricultural production have been high.

Waterlogging and salinity problems are technically solvable. First, drainage must be provided, usually in the form of subsurface tiles or pipes installed at the proper depth which empty into a drain which sends the effluent to the sea, an evaporation pond, or, where feasible, back into the irrigation system for downstream re-use. Second, the soils must be leached or flushed of salts through application of large quantities of water. Third, the soil may need to be treated with gypsum and other chemicals to restore structure and chemical balance.

The hitch is that drainage is expensive. From recent Bank projects involving drainage, costs may be estimated at about US$ 1200 per acre. This amounts to a potential cost of about US$ nine billion for the approximately eight million acres of CCA affected - almost the total value of the agricultural sector's output in one year! Clearly, priorities for drainage must be carefully considered, which is one of the objectives of the Sectoral Environmental Assessment/National Drainage Program Concept study currently being executed by the Bank.

The Water Sector Investment Plan (WSIP)

Against this backdrop of probable severe shortfalls in production, a myriad of high priority investment needs, and a shortage of both domestic and foreign financial resources, the Water Sector Investment Plan study commenced in 1988. Amongst planning procedural

---

[6]To the extent that the water table is within five feet of the surface during at least part of the year. WSIP estimates that one-third of the CCA is currently in this category.

[7]"Salt affected" implies salinity levels sufficiently high to impede plant growth.
Source: <u>Report of the National Commission on Agriculture</u>, Government of Pakistan (Rawalpindi, 1988), p. 295.

and institutional objectives the WSIP undertook the highly challenging tasks of:

a) cataloging and updating the literally hundreds of ongoing and proposed projects of the Federal Government and the governments of the Balochistan, Punjab, NWFP, and Sind provinces.

b) devising evaluation criteria which would permit logical ranking of the potential projects.

c) defining alternative national plans for water sector investments in light of the sectoral objectives, the project evaluations, budgetary limitations, and the need for balance among the provinces.

Three innovative methodologies were employed by the WSIP. Net benefit/investment (N/K) ratios were calculated for each potential project taking into account not only economic costs and benefits but preferences for less developed areas and small farmers, and adverse environmental impacts. A large-scale mathematical programming model of the Basin's agricultural activities and water resource constraints (the Indus Basin Model Revised, or "IBMR") was employed to help estimate the benefits, and ensure consistency among projects in any plan. Finally, a "plan generator" was developed which used mixed-integer programming techniques to assist in project scheduling and to ensure adherence to financial and other macro-economic constraints.[8]

Despite uncertainties over the availability of funds from external sources, unresolved inter-provincial issues, and the absence of a national decision on additional storage (Kalabagh or Basha dams), the WSIP was able to recommend "minimum" and "maximum" plans.[9] The plans contain 41 and 48 specific projects, and were estimated to cost about US$3.75 and US$5.39 billions respectively during the decade of the 90's, and considerably more over their entire lives. The "minimum" plan is characterized as "expected", and its components will be described in Chapter 3. For now, it is worth noting that the total cost estimate is about half or less of the cost of providing drainage to the waterlogged parts of the Basin.

Objectives of this Study

Our intention in this study is not to criticize the WSIP plan. Indeed, we view it as a profound and timely step forward in a highly precarious and delicate planning situation. Our concern is with the implications for the environment of the Indus Basin, particularly with regard to the waterlogging and salinity problem. At the time of this writing, the WSIP plans are being considered, and the Sectoral Environmental Assessment/National Drainage Program Concept (NDP) study is being launched. Therefore it is timely to review the

---

[8] "National Investment Plan", chapter 6.

[9] "National Investment Plan", chapter 8.

present status of waterlogging and salinity, and what it is likely to be in the year 2000 if investments are limited to those in the WSIP plan.

Our primary tool of analysis will be the above-mentioned IBMR. Although this model was used by the WSIP to analyze the implications for agricultural production of the various plan components, it has far wider capabilities. To wit, it tabulates recharges to and withdrawals from groundwater for fifteen spatial divisions of the Basin,[10] and therefore can compute water balances and the implications for drainage needs. The impact of individual projects on the watertable can thus be projected, as well as the implications for water availability. Furthermore, because the surface water distribution system is explicitly contained in the model and endogenously operated within political as well as physical constraints, the impact of alternative water management policies for groundwater issues can be estimated. As we shall see, how water is allocated within the system is a major determinant of sector performance and has enormous implications for the design of programs to deal with waterlogging and salinity.

Chapter 2 provides an overview of the IBMR for those readers who are not familiar with its capabilities and limitations.

Chapter 3 reviews the present situation and that projected to prevail in 2000 as simulated by the IBMR with and without the WSIP "expected" plan.

Chapter 4 examines the environmental implications of the present situation and that projected for 2000. The causes of increasing waterlogging are traced to the various sources of groundwater recharge and discharge for the fifteen subareas of the Basin.

Chapter 5 examines the sources of salinity buildup in groundwater and soils. It offers a simple method for predicting the effects of alternative water management policies on salinity buildup, and projects the salinity status to the year 2000.

Chapter 6 tests the effectiveness of alternative projects to deal with waterlogging and salinity: drainage, water allocation policies, canal remodeling, and watercourse improvements (so-called On-Farm Water Management).

Chapter 7 concludes with a recommended package of investments plus associated water management policies which will both improve the performance of the WSIP program and reduce the deterioration of the Basin's soils and groundwater resources. Finally, deficiencies in the soil and groundwater data bases and agricultural research are noted, and a call is made for computerized management of the irrigation system.

---

[10]Nine agro-climatic zones, of which seven are divided into fresh and saline groundwater areas.

CHAPTER 2
THE INDUS BASIN MODEL REVISED

Background

The history of the Indus Basin Model (IBM) dates to 1976 when the World Bank's Development Research Center, in collaboration with the Water and Power Development Authority (WAPDA) of Pakistan, began developing the large scale linear programming model of Pakistan's irrigated agriculture.[1] The resulting model, completed in 1981, combined standard agricultural modelling techniques with a network model of the surface water storage and distribution network. Additional complexities introduced had to do with the existence of fresh and saline groundwater areas and the conjunctive use of surface and groundwater in the fresh areas.

In its complete version the model totaled about 8,000 equations. It proved valuable as a research tool in investigating water-related projects and agricultural policies where important externalities with respect to groundwater quality and the depth to the water table existed.[2] However, its sheer size prevented the researchers from obtaining more than a handful of solutions.

During the early 1980's various components of the model were used for the analysis of particular projects in Pakistan. These components were usually defined on a specific project area and took the applicable production activities and resource constraints from the large model. By altering the production technologies and/or resource availabilities according to the project's objective, the impact of the project on cropping patterns, resource use, and farmers incomes could be simulated. The LBOD, OFWM, and SCARP Transition are examples projects which benefited from the IBM.

In the late 1980's the model was completely re-worked and updated to analyze the proposed Kalabagh Dam Project. Because of the inter-connectedness of virtually the entire irrigation system, a tool which could simultaneously evaluate the benefits of additional water throughout the agro-climatic zones and among the major crops was desired. The resulting model, called the Indus Basin Model Revised (IBMR) contains about 2,500 equations yet retains all of the IBM's detail on the production side as well as the surface storage and distribution network.. An additional feature introduced for the Kalabagh work and essential to all project work is the ability to project the model forward in time. By varying those parameters for which future estimates are available (growth in demand, yields, growth in input use, effects of projects on the resource base),

---

[1]For a complete description of the model, including input data, validation, and a complete model description in GAMS (General Algebraic Modeling System), see chapters II-V.1 and Appendix A of Masood Ahmad, et. al., Guide to the Indus Basin Model Revised, World Bank/WSIP, 1990. Much of this chapter is adapted from that material.

[2]See John H. Duloy and Gerald T. O'Mara, "Issues of Efficiency and Interdependence in Water Resource Investments: Lessons from the Indus Basin of Pakistan", World Bank Staff Working Paper Number 665 (Washington D.C., 1984).

"snapshots" of agricultural performance may be made for future years. In the case of Kalabagh, this feature was used to simulate the agricultural economy of 2000 with and without the dam.[3]

The IBMR was again updated in 1988-89 for use by the WSIP. This version was also "streamlined" for solution in WAPDA. For the first time in its long history, the model was transferred to computers in Pakistan, and local analysts were trained in its use. Results reported in this study are based on this version.

Overview of the IBMR

Agricultural production and consumption is simulated within the above-mentioned nine agro-climatic zones. They are listed in Table 2.1 with their size (saline areas in parentheses), their component Districts, and their component canal commands, and are shown on the accompanying map, IBRD 21611. Although it is unfortunate that the zones do not correspond to strict aggregations of either Districts or canal commands, it must be recognized that homogeneity of agricultural activity combined with identifiable sources of water control are critical to keeping aggregation bias to a minimum.

Within each zone there are production possibilities for up to fifteen crops (wheat, Basmati and Irri rice, cotton, sugarcane, gur, maize, rape and mustard seeds, gram, onions, potatoes, chilies, orchard, and Kharif and Rabi fodders) and three livestock products (cow and buffalo milk, beef). A typical crop production possibility takes the form of an input-output vector (a column in a linear programming tableau) representing the cultivation of one acre of land in the crop. The vector contains elements for the primary crop output (the yield), and usually a by-product of animal feed. Various inputs required for the cultivation are also accounted for as elements in the vector or column: land, irrigation water, labor, bullock draft power, tractor services, fertilizer, seeds, and any other purchased inputs which may be required.

For all crops except orchard, two technology choices are given: bullock and semi-mechanized. The bullock technology relies primarily on manual labor and bullock power; the semi-mechanized one uses a combination of labor, bullocks, and tractors with the tractors used typically for plowing and sometimes harvesting. Depending on the relative costs of maintaining bullocks versus hiring or using tractors, the model will make the technology choice endogenously. For wheat, additional techniques are given differentiated by planting and harvesting dates, and water stress levels.

---

[3] See Masood Ahmad, Gary P. Kutcher, and Alexander Meeraus, "The Agricultural Impact of Kalabagh Dam (as Simulated by the Indus Basin Model Revised)", two volumes, World Bank Report Number 6884 PAK (Washington, D.C., 1987). Restricted circulation.

Table 2.1: Zone-District-Canal Mapping

Agro-climatic Zone	CCA (million acres)		Component Districts		Component Canals	
NWFP	.628 (no saline)		Peshawar Mardan		22-USW Lower Swat canal 23-LSW Upper Swat canal 24-WAR Warsak canal 25-KAB Kabul River canal	

Punjab zones:

Agro-climatic Zone	CCA (million acres)		Component Districts		Component Canals	
PRW	2.782 (no saline)		Lahore Sheikhupura Gujranwala Sialkot	.50 .32 .26	02-CBD Central Bari Doab 03-RAY Raya canal 04-UC Upper Chenab canal 05-MR Marala Ravi canal 11-JHA Jhang canal 12-GUG Gugera Branch canal	
PMW	2.413 (.574 saline)	.3 .55 .6 .25	Sarghoda Mianwali Muzaffarg. D.I. Khan	 .25	26-THA Thal canal 27-PAH Paharpur/CRBC canal 28-MUZ Muzaffargargh canal	
PCW	11.245 (2.45 saline)	 .4	Multan Vehari Sahiwal Muzaffgarh D. G. Khan Bahalwalpur Rahimyarkhan Cholistan Kasur	 .50 .75	01-UD Upper Dipalpur 02-CBD Central Bari Doab 06-SAD Sadiqia 07-FOR Fordwah 08-PAK Upper Pakpattan plus 09-LD Lower Dipalpur 10-LBD Lower Bari Doab 15-BAH Bahawal canal 16-MAL Lower Mailsi+lower Pakpattan 17-SID Sidhnai canal 19-RAN Rangpur canal 20-PAN Panjnad canal 21-ABB Abbasia canal 28-MUZ Muzaffargargh canal 29-DGK Dera Ghazi Khan	
PSW	4.398 (1.333 saline)	.68 .74 .7	Jhang Faisalabad Gujrat Sargodha		11-JHA Jhang canal 12-GUG Gugera Branch 13-UJ Upper Jehlum canal 14-LJ Lower Jehlum canal 18-HAV Haveli	

Table 2.1: Zone-District-Canal Mapping (continued)

Agro-climatic Zone	CCA (million acres)	Component Districts		Component Canals	
Sind Zones:					
SCWN	3.594 (1.749 saline)		Sukkur Khairpur Nawabshah	.59 .20	33-GHO Ghotki 37-KW Khairpur West canal 38-KE Khairpur East canal 39-ROH Rohri canal 41-NAR Nara canal
SCWS	2.791 (2.381 saline)	.9 .6	Hyderabad Badin Tharparkar Sangar	.41 .80	39-ROH Rohri canal 41-NAR Nara canal
SRWN	4.395 (2.819 saline)		Dadu Larkana Jacobabad Shikarpur Nasirabad		31-P+D Pat + Desert canal 32-BEG Begari canal 34-NW North West canal 35-RIC Rice canal 36-DAD Dadu canal
SRWS	2.775 (all saline)	.1 .4	Hyderabad Badin Thatta Karachi		42-KAL Kalri canal 43-LCH Lined channel 44-FUL Fuleli canal 45-PIN Pinyari canal

The inputs or resources are either constrained in availability, charged a price, or both. Land is strictly constrained in availability by month. Surface water is both constrained by the canal diversions to the zone and priced at the canal water rate charged each crop/acre. Bullocks are not constrained (except for historical solutions or near-term scenarios in which the herd size can be safely predicted), but their maintenance is charged via feeding requirement constraints. In each crop season, a sufficient supply of nutrients (carbohydrates, protein and green fodder) must be met through a combination of fodder crops and straws and by-products of other crops. The same holds for the stock of cows and buffaloes which have different nutrient requirements and produce milk and meat instead of draft services. Tractor services, fertilizers, and seeds are charged their respective prices. In some versions these inputs are constrained in supply; in others, supply is assumed to be infinitely elastic. Farm-level investment options have been introduced in some versions for tractors and tubewells. Labor is available from farm household sources up to a limit. Beyond that, activities which hire off-farm labor may be activated depending on the return to additional labor and the wage rate.

Each crop requires water in different quantities and at different times. To be consistent with land, labor, bullock and tractor power, the chosen time dimension is monthly. The actual water requirements are reduced by the expected effective rainfall and that available from subirrigation.[4] Both sources differ by zone, and subirrigation differs by the fresh/saline distinction within each zone because the depths to water table differ. Mixing of subirrigation supplies with surface water is restricted to acceptable ratios in saline areas. The remaining crop requirements must be met from surface water, supplemented in fresh areas by private tubewell operations up to the limit of the installed capacity plus any endogenously-determined private investment in new tubewells. Where applicable, government tubewell operations are simulated to augment the surface supplies. Within a zone, the surface flows from the canal heads are reduced for canal losses and again for watercourse and field losses prior to reaching the crops. It is easy to see how the model is used, for example, to simulate a project such as irrigation system rehabilitation: the loss coefficients are reduced allowing more water to reach the crops and lowering recharges. The benefits are endogenously generated and reflected in a higher objective function.

Groundwater inflows and outflows are also tabulated so that the net impact on the water table, by zone, and separately for fresh and saline areas within a zone, is obtained from each solution. Inflows comprise recharge from rainfall, river reaches and seepage losses in the water delivery system. Outflows are public and private tubewell pumpage, river bank recharge, and evaporation from the groundwater.

The input data for the Indus Basin Model has been refined over the years. This is particularly true for the technical coefficients (i.e. the land, labor, draft power and water requirements), which are not readily obtainable from published sources or from surveys. The information base for these data starts with the 1976 XAES Survey of Irrigated Agriculture.[5] It was supplemented by the Agricultural and Livestock Censuses and by the Farm Re-survey undertaken in 1988 under the WSIP. The current data base also reflects the cumulative knowledge acquired during the above-mentioned project studies.

Other data are readily available from published sources. Statistics on cropped area, crop production, total fertilizer use, canal diversions, and river inflows are extracted from publications and transformed as needed for the model. Yields, and prices of outputs and inputs, are obtained from the most recent sources. Changes to parameters describing the endowment of physical resources are made on the basis of information obtained from projects. For the WSIP version, the data base was updated to 1987-88 using the 1988 survey, recent issues of <u>Agricultural Statistics of Pakistan</u>, and WAPDA-supplied irrigation data.

The nine zone models are linked together via the surface storage and distribution

---

[4]"Subirrigation" refers to transpiration from groundwater as a result of capillary action.

[5]Extended Agricultural Economics Survey, undertaken by WAPDA under the Revised Action Programme (RAP).

model (SWM).[6] The SWM uses the network theory concepts of nodes and arcs to simulate the flows throughout the system of Indus rivers and link canals. A node is a point where a water flow decision is made and/or a mass balance for each month is desired. Thirty-seven nodes represent the reservoirs and rim stations, barrages/headworks, confluences of rivers, and the terminus of the Arabian Sea. Twenty-four sinks (terminal nodes) represent diversions to irrigation canals or groups of canals. Fifty-one arcs represent river reaches and link canals between nodes.

Surface water enters the system at the rim stations. Flows along river reaches are simulated with losses and gains from river bank storage. Water may be stored for later release in the reservoirs or allowed to flow down river. At each control point (usually a barrage), decisions are simulated to allow diversions for use within the adjacent production zones or release for downstream use. Flows along main and link canals are subject to losses based on the permeability of their beds and integrity of their banks. The entire system of river reaches and main canals is contained in the model, using months as the time period of reference.

The diversion decisions, together with the reservoir operating rules, may be left flexible within the physical constraints of the storage and distribution system and the needs of power generation - in which case the model will distribute water optimally to maximize some function of agricultural production. For realistic cases, however, constraints are inserted which reflect existing policy on how the water is to be allocated. Typical allocation rules simulated are canal-wise diversions proportional to those of the post-Tarbela period, and the so-called Ad Hoc allocations adopted over the past few years. Of course, the IBMR is uniquely suited for simulating the impacts of changes in the allocation system.

After a model has been solved, power generation at Mangla and Tarbela reservoirs is computed. This is done using the model's determination of monthly storage levels and outflows, and is tabulated in the solution report. The value of power generated has no direct bearing on model's decisions with regard to storage or releases from the reservoirs. (Power considerations were taken into account by the engineers who set the operating rule curves for the reservoirs. The rule curves also ensure safe operating conditions for the reservoirs, and are exogenous to the model.)

The linear programming format of the IBMR permits specification of several alternative objective functions. However, economic theory permits only one objective function for competitive agricultural sectors: the maximization of the sum of consumers and producers surpluses. Use of this objective function ensures that farmers will choose cropping patterns and input usage which maximizes their incomes while at the same time equating consumer demands with product supplies via adjustment in market prices. Thus a solution to the IBMR gives not only a simulated water distribution pattern, production

---

[6] Originally developed by Kutcher in 1976 based on WAPDA's COMSYM simulation model of the Basin. It can be solved as a stand-alone LP model to test the feasibility of meeting projected water demands and to gauge the degree of flexibility available in the system.

by crop, technology, and zone, and the input use required for that production, but market-clearing prices and hence farm income as well. Furthermore, every solution is assured by the LP framework to be feasible with respect to the available resources.

Use of this objective function, called NSP (net social product) does not preclude tabulation of other aggregates of interest. Value added, employment, exports and imports of agricultural commodities are just some of those produced by a solution to the model. For most project analysis, value added measured in economic prices is required to compute benefits. This measure is produced automatically in all solutions of the IBMR.

The Zone Models

Given a set of water allocations from the full IBMR, COMSYM, or from hand computation, the zone models may be solved independently. Each zone model contains about 200 equations and is solvable on an IBM-type personal computer, and is useful for indicative policy analysis as well small, localized projects so long as assumptions of constant surface water diversions are valid and provided an aggregation to national totals of production and resource use is not required. Application of a given zone model requires an assumption about the market pertaining in the zone. At one extreme, one might assume that the zone is small, and trade among neighboring zones is free and costless. If so, output prices may be taken as given and unalterable by any prospective simulation of the model. This assumption obviously applies to any given farm. No single farmer is large enough to influence market prices, and presumably he may trade among neighboring farms at little cost. If so, he will probably specialize in those crops in which he has a comparative advantage, trade the excess over his family's consumption needs with neighboring farmers, and, if he is large enough, export some produce to the rest of the zone, country, or the world.

Thus, if prices are fixed, and the objective function is one involving some measure of farm income, then the zone model in effect simulates a representative farm of the zone. Only the scale will differ, and even this difference can be eliminated either through scaling down the resources given to the zone model or scaling down the reported outcome of the model's solution.

There are two difficulties with this approach. First, when prices are fixed, the solution usually shows a good deal of specialization in the cropping pattern, as was speculated above for the individual farmer. If one observes twelve crops being grown in non-negligible quantities in a zone, the fixed-price model might produce only four of them. In this case the farm model is not representative at all, and the results are not aggregable in any meaningful sense. Second, farm sizes usually make a substantial difference in output patterns and input use, so that if farm level results are required, a disaggregation of farm sizes or types is called for.

The objective function structure used in the zone models is similar to the sector NSP, but scaled down to the zone level. In this case, the objectives of farmers and consumers within the zone are modelled in. What happens within the delimitation of the zone model can affect prices. Of course, those effects must be limited because if prices rise too much goods will flow in from neighboring zones, or if prices drop much below those in neighboring zones, "exports" will occur.

However, it is well recognized that farm families typically produce most if not all of their food consumption needs. The 1988 Re-survey allowed us to estimate the proportions of the production consumed on the farm, and hence to estimate the actual on-farm consumption. Current versions of the zone models and the full IBMR require that the farms embedded in the zone models meet consumption needs first, and then market any excess.

To sum up the structure adopted for the zone models: for production beyond the minimum consumption requirement, downward-sloping demand functions are derived for each zone model so that changes in production leads to changes in zone-level prices. Scaling down the solution to a representative farm is still possible, and indeed, more realistic, because the representative farm will in fact be an average of the zone in all respects. Hence some results, such as cropping patterns and production, will be presented at the zone level, while other information, particularly the components of farm income and costs, will be given for a representative farm.

The Province Models

Just as the full IBMR is a collection of zone models, models of the three main provinces irrigated by Indus waters may be constructed by grouping the appropriate zone models. The NWFP zone model in fact corresponds to the areas irrigated by the Kabul and Swat canals, not the entire NWFP Province. A model of the Punjab consists of the zone models PCW, PRW, PMW, and PSW linked together. Because PMW includes areas irrigated by the Paharpur and CRB canals in NWFP, the Punjab model comprises part of the NWFP province. A model of the Sind consists of SCWN, SCWS, SRWN, and SRWS. Because some of the zone models do not correspond exactly to the province to which they are "attached", a one-to-one correspond with actual provincial boundaries is not possible, and this discrepancy must be kept in mind in data work - both in preparing the input data from provincial-level statistics, and in interpreting results at the level of a province. Thus, for example, results from the "Sind" model will include that part of Balochistan irrigated by the Pat and Desert feeders and Northwest canal and which lies in the zone model SRWN.

Just as zone models require assumptions about the canal water allocated to them in each month, so do the province models (because water is a scarce national resource which must be shared by the provinces, it would make no sense to solve a provincial model without an allocation first being assumed). However, there is some flexibility in how the allocation is made: one could assume that the allocations by canal continue to hold (in which case the solution to the province model would simply amount to the sum of the component models), or one could allow endogenous re-allocation among the canals but within the overall provincial constraints. Such flexibility, of course, would be subject to the physical capacities of the surface storage and distribution system within the province being studied. In the Kalabagh and WSIP studies using the IBMR, it was found that there was a good deal of room for efficiency gains from <u>within</u> province re-allocations because the marginal productivity of water varies greatly from zone to zone. We will need to return to this topic later in this study as we search for means of achieving cost-effective methods of preserving the environmental balance.

Chapter 2 Appendix:
The Agro-climatic Zones

The Indus Basin Irrigation system is divided into nine Agro-climatic zones following the RAP. Map IBRD 21611 shows these zones and a brief description follows.

NWFP. (North-West Frontier Province). The ACZ NWFP comprises only the canal commanded areas of Peshawar and Mardan districts. This ACZ receives most of its surface water from the Swat and Kabul rivers before they reach the Indus. Water shortages can largely be traced to limited canal capacities. Groundwater is usable throughout most of the area. Cropping is dominated by sugarcane, maize, and wheat.

PMW. (Punjab Mixed-Wheat, called "Thal Doab" in the RAP) contains nearly two million canal commanded acres, mostly on the left bank of the Indus below the Jinnah barrage but also including the Paharpur and Chasma Right Bank canal command areas in the NWFP Province. The topography is rough, soils are sandy and seepage is high, resulting in low cropping intensities and yields. The fresh groundwater and localized waterlogging in most of the ACZ imply that the potential for tubewell development is favorable.

PRW. (Punjab Rice-Wheat) contains about 2.8 million acres, virtually all of which is underlain by fresh groundwater. This has spurred intense private tubewell development. As a result, cropping intensities are among the highest in the Punjab, with Basmati rice being the dominant cash crop. Relatively high returns to farming combined with a (reported) shortage of labor has led to rapid mechanization; this zone has more tractors per acre than any other.

PSW. (Punjab Sugarcane-Wheat, called "Punjab Mixed Crop" in the RAP) lies between PMW and PRW, and contains about 4.4 million acres. Wheat and sugarcane are the principal crops. About one-third of the zone is saline, but farmers make extensive use of groundwater in the rest. Water shortages do exist, and are largely attributable to low watercourse efficiencies.

PCW. (Punjab Cotton-Wheat) is by far the largest ACZ in the Basin, comprising over 11 million acres on the left bank of the Indus between Sind Province, India, and the other Punjab ACZs. Cotton and wheat, the main crops, have some of the highest yields in Pakistan. About one fourth of the ACZ suffers from severe waterlogging and salinity. Groundwater is extensively used in the rest of the zone, but adequate water remains an overall constraint.

SCWN and SCWS. (Sind Cotton-Wheat North and South), each of which covers over three million acres, are disaggregations of the Sind Cotton-Wheat zone. Nearly half of the north, and most of the south is saline and/or waterlogged. Yields on areas remaining in use are favorable. Groundwater use is minimal, and surface water supplies are hampered by high losses, particularly at the watercourse level.

SRWN and SRWS. (Sind Rice-Wheat North and South) are the right and left bank delineations of the Sind Rice-Wheat zone. About two-thirds of the 4.4 million acres in the north are saline and all of the south is similarly classified. Soils are favorable for rice, the dominant crop. Because of the high water table (at saturation level in many areas) yields for other crops are poor, and cropping intensities, particularly in the south, are the lowest in the Basin. The XAES found that surface water supplies were inadequate, and other inputs, such as fertilizer, used sparingly.

CHAPTER 3
AGRICULTURAL PERFORMANCE 1988-2000

In this chapter we briefly examine the 1988 base case solution of the IBMR as it compares with the available data,[1] and project the model forward in time to the year 2000 planning horizon. Solutions for 2000 are presented for the scenario of no further investment in the water resources sector, and for the "expected" WSIP plan scenario as it pertains to investments maturing prior to 2000. The important assumptions about surface water inputs to the system are presented in an appendix to this chapter.

Agricultural Production (1988)

Table 3.1 reports the cropping patterns by province as projected from the 1980 Pakistan Census of Agriculture (PCA) and the recent Agricultural Statistics of Pakistan (ASP), and as simulated by the model. Some minor definitional problems are worth noting: actual areas of individual crops refer to areas under all types of irrigation, whereas the model covers only canal commanded areas. In the Sind, 2.8% of all irrigated land was irrigated from non-canal sources in 1980; in the Punjab, the figure was 18.75%. It is indeterminable, however, how much of this area is irrigated by tubewell and is in fact included in the model's CCA. The problem is much more serious in NWFP where most of production occurs on rainfed land or irrigated land outside of the canal system. Recall from the last chapter that "Sind" includes part of Balochistan irrigated by the Indus canals, and "Punjab" includes part of D.I. Khan district in NWFP. "Pakistan" includes both the above and the ACZ called "NWFP". In all definitions, only canal commanded, irrigated areas are included.

Major differences between the simulated area and the actual occur for cotton and maize (higher) and sugarcane (lower). The model appears to be capitalizing on the recent and impressive growth in cotton yields, largely at the expense of sugarcane, the yields of which have not kept pace. The markedly higher maize area, combined with the lower fodder areas (next paragraph) may imply a substitution of maize for fodder, the former having a substantial green by-product fed to livestock. The fodder areas reported undoubtedly include non-irrigated areas. It is extremely difficult to estimate the actual irrigated fodder areas, as well as what proportions of the livestock population are attached to irrigated areas. Hence the model's simulated fodder areas in total are significantly below the reported actual.

Seasonal and annual cropping intensities are reported at the bottom of Table 3.1. The model produces a lower intensity in Rabi in the Punjab, and a much higher Kharif intensity in NWFP. Otherwise, the intensities are remarkably close. We speculate that the reported intensity for the Punjab includes areas irrigated by private dams and tubewells situated outside the CCA, therefore not included in the model's coverage. The NWFP discrepancy is probably due to the above-mentioned data mapping difference.

---

[1] A more thorough comparison, including statistical validation tests of the performance of the model, was presented in the Guide, op. cit., pp. 95-102.

Table 3.2 reports the simulated vs. actual livestock populations. Recall that the model's structure implies that bullocks will be kept primarily for draft power, and secondarily for meat. The model chooses to maintain about 14% fewer bullocks in total than the estimated actual population, the biggest difference appearing in the Punjab. There are two possible explanations for this discrepancy. First, the model may be picking up the trend toward more intensive mechanization under which fewer bullocks are required. Second, the mapping of the bullocks to the canal-irrigated areas may be incorrect, i.e. we may be wrong in expecting to find 5.253 million bullocks in the model's domain as many are kept outside of the canal-irrigated areas.

The cow and buffalo populations are completely endogenous, depending only on the model's desire to produce fodders in lieu of other crops to obtain revenue from meat and milk sales. The "actual" figures in the table are adjusted from LS totals to reflect independent evidence that only half the cows are maintained on irrigated areas, and 70% of the buffaloes. The model still chooses about 35% fewer cows and about 41% fewer buffaloes. Again, this may be a mapping problem, or in fact a true bias against livestock. Perhaps the competition of cash crops for land is squeezing the livestock population, as evidenced by the low figures on simulated fodder area described above. Regardless, this topic requires further investigation.

Table 3.3 reports production and disposition of crop and livestock products as simulated by the IBMR. Production totals are straightforward summations from the hundreds of cropping activity levels described in the previous chapter. "On-farm consumption" are the totals from the simplistic consumption functions contained in the model. Production not consumed on the farms enters the price-endogenous demand functions, and "exports" are quantities exported to the outside. Production therefore equals the sum of "on-farm consumption", "marketed" commodities and "exports". Imports are an endogenous source of supply, alternative to domestic production, entering the domestic demand functions.

In addition to crop and livestock production, the IBMR tabulates hydropower generation from the endogenous releases at Tarbela and Mangla reservoirs. These outputs are given in Table 3.4. Because the IBMR optimizes over agricultural outputs only given power generation needs as defined by the constraining reservoir rule curves, the value of these hydro outputs do not directly affect the solution.

Table 3.4:
Simulated Hydro-Power Generation
(billion kilowat hours)

Mangla	5.671
Tarbela	11.040
Total	16.711

Table 3.1: 1987-88 Cropping Pattern
Model Simulation vs. Estimated Actual
(000s acres)

	NWFP Model	NWFP Actual	Punjab Model	Punjab Actual	Sind Model	Sind Actual	Pakistan Model	Pakistan Actual
Rice								
Basmati			2,188.7	2,001			2,188.7	2,001
IRRI			415.8	620	1,523.1	1,491	1,938.9	2,111
Cotton			7,027.0	4,782	1,526.1	1,555	8,553.1	6,337
Rabi Fodder	68.3	47	1,616.3	2,513	565.1	415	2,349.8	2,975
Gram	86.7	132	1,159.0	2,586	224.0	218	1,469.7	1,936
Maize	199.1	78	298.4	101			497.5	179
Mustard/Rape	35.1	52	364.9	385	214.2	225	614.2	662
Kharif Fodder	54.5	14	1,138.1	2,732	281.0	142	1,473.6	2,888
Sugarcane for Mill	173.4	143	1,361.0	1396	635.6	542	2,170.0	2,081
Wheat	224.8	224	11,521.9	12,321	4,076.1	2,616	15,822.7	15,161
Orchard	27.0	27	691.0	691	200.0	200	918.0	918
Potatoes	9.1	8	123.6	113	38.2	38	170.9	159
Onion	5.8	6	98.0	89	33.6	29	137.3	124
Chilli	11.2	9	145.0	123	46.8	41	203.0	173
Intensities								
Rabi	100.4	101.8	81.7	91.6	44.5	31.9	67.7	68.7
Kharif	74.1	43.2	63.7	59.7	30.7	29.0	51.1	47.5
Annual	174.4	144.9	145.4	151.4	75.2	60.9	118.8	116.2

Table 3.2: Livestock Population
Model Simulation vs. Estimated Actual
(000s animals)

	NWFP Model	NWFP Actual	Punjab Model	Punjab Actual	Sind Model	Sind Actual	Pakistan Model	Pakistan Actual
Bullocks	66.2	151.0	3,183.8	3,630.0	1,758.7	1,472.0	5,008.8	5,253.0
Cows	130.3	252.5	1,301.0	2,465.0	1,008.2	949.0	2,439.5	3,666.5
Buffaloes	121.3	255.2	4,313.5	7,858.4	1,619.8	1,581.6	6,054.6	9,695.2

Source for actual: Agricultural Statistics of Pakistan, 1987-88.

Table 3.3: Production, Consumption, and Exports from the 1988 IBMR Base Solution

Crop (000 tons) and Livestock (million tons/liters) Commodities

	NWFP	Punjab	Sind	Pakistan
Production				
Basmati		1,058.7		1,058.7
Irri		358.4	1,385.6	1,744.1
Cotton		5,890.4	954.1	6,844.5
Gram	14.4	196.5	57.9	268.8
Maize	106.9	159.7		266.6
Mustard and Rape	6.7	136.7	52.6	196.0
Sugar Cane				
Mill	2,974.7	20,044.5	11,511.3	34,530.4
Gur	49.5	409.3	202.0	660.8
Wheat	161.2	8,675.9	3,265.1	11,797.8
Orchard	85.9	2,513.9	680.0	3,279.8
Potatoes	35.4	471.1	131.2	637.7
Onions	31.5	487.3	129.1	647.9
Chili	5.0	95.5	24.9	125.4
Livestock Products				
Cow Milk	46.8	394.0	423.7	864.5
Buffalo Milk	73.0	2,699.0	1,390.8	4,162.8
Meat	4.8	137.6	70.5	212.8
On-Farm Consumption				
Basmati		144.0		144.0
Irri		54.7	312.2	366.9
Gram	10.1	118.2	25.2	153.5
Maize	18.9	22.6		41.5
Mustard and Rape	6.7	64.2	24.9	95.8
Sugar Cane - Gur	49.5	409.3	202.0	660.8
Wheat	110.0	6,130.0	1,851.9	8,092.2
Potatoes	13.4	191.6	57.4	262.4
Onions	11.3	183.5	46.7	241.5
Chili	2.2	37.1	10.0	49.3
Exports				
Basmati		250.0		250.0
Irri		213.7	781.0	994.7
Cotton		588.0	82.9	670.9
Wheat		1,182.2		1,182.2
Onions	1.4	3.3	5.4	10.1

Resource Use (1988)

The pattern of land use is shown in Table 3.5 which highlights the under-utilization of land in a slightly different manner than did the cropping intensity figures. Land is cropped fully (in at least one month of the year) in only the fresh areas of NWFP, the Punjab, and SCWS in Sind. Nearly 50% of CCA in the Sind is never cropped, as is nearly 30% for all of irrigated Pakistan. The model clearly confirms that arable land is not a constraint.

Casual observation has revealed that labor, if it is a constraint, is so only on a local and seasonal basis. The model produces a pattern of labor demand such that family labor can supply all of the needs in NWFP and the Punjab except for the fresh areas of PMW and PCW. Labor hire is used in most of the Sind, but at moderate levels. In no month or zone does the demand for non-farm labor approach the limits of availability. The model also confirms that labor availability is not a constraint on irrigated Pakistani agriculture.

The tractor stock as estimated is adequate to meet tillage requirements in all ACZ's. This result corresponds with the observed rapid buildup of tractor stocks in the past few years.

Although all fresh subareas make use of private tubewells in the solution, the available capacity is only binding in NWFP and SCWN. Thus the model indicates that private tubewell investment in new capacity has probably reached the exhaustion stage in canal-irrigated areas.

If the model has not revealed any of the above inputs or resources to be universally binding, it clearly does so with respect to surface irrigation water. With canal diversions set at post-Tarbela average levels, the supply of canal water is binding in most months in all canals and thus in all ACZ's. This means that more surface water can be usefully absorbed in all parts of the Basin served by the canal system. Furthermore, the model produces shadow values of additional water supplied up to nearly Rs 35 per acre foot measured in financial returns to farmers, 1988 prices.

Table 3.6 reports the water balance at the root zone by province and fresh/saline distinction. "Water requirements" are those needed to meet the model's simulated cropping pattern. Supplies come from rainfall, sub-irrigation, surface water (canal), and tubewells in fresh areas. "Slackwater" is the difference between total supply and the requirements and usually indicates a discrepancy between the canal diversions as dictated from use of historical diversions and the model's desires given the cropping pattern. Most of the slackwater appears during the peak flow kharif months and has little consequence apart from unnecessary groundwater recharge.

Table 3.5: Land Use, 1987-88 Base Case
(000s acres)

	CCA Fresh	CCA Saline	Area Cultivated Fresh	Area Cultivated Saline	Percent Cultivated Fresh	Percent Cultivated Saline	Labor Hire? Fresh	Labor Hire? Saline
NWFP	628.0	-	628.0	-	100.0	-	no	-
PMW	1,838.9	574.4	1,838.9	337.6	100.0	58.7	yes	no
PCW	8,795.5	2,449.8	8,795.5	1,117.7	100.0	45.6	yes	no
PSW	3,065.3	1,332.8	3,065.3	687.5	100.0	51.5	no	no
PRW	2,782.4	-	2,782.4	-	100.0	-	no	-
SCWN	1844.8	1,749.4	1576.0	849.7	85.4	48.6	yes	yes
SRWN	1576.0	2,819.0	1243.5	1,245.4	78.9	44.2	yes	no
SCWS	409.8	2,381.1	409.8	1,105.5	100.0	50.7	yes	no
SRWS	-	2,775.0	-	991.4	-	35.7	-	yes
	Total		Total		Total			
Punjab	20,839.0		17,093.4		82.0			
Sind	13,555.0		7,000.1		51.6			
Pakistan	35,022.0		24,608.8		70.3			

The heavy reliance on groundwater in fresh areas, for 42% of total requirements, goes far in explaining why fresh areas are able to crop virtually all of the CCA, while saline areas, restricted to the non-groundwater supplies, can cultivate less than half their CCA. Another way to view the water shortage is given at the bottom of Table 3.6. To crop all of the available CCA in each season (a 200% intensity) would require an additional 21.6 MAF in Punjab and 50.5 MAF in Sind of water *at the root zone*. If these requirements were to be met by surface supplies (virtually the only alternative for saline areas), canal diversions would need to be increased from the present 115 MAF to 366 MAF - an impossibility given the sources of surface water, as well as the capacities for storage and distribution. The reason such a large increase in diversions would be required is the "leaky" nature of the system: on average, only 41.3% of water diverted at the canal head reaches the root zone. We will return to this important topic in chapter 6 when we examine the benefits to canal lining, amongst other options.

Table 3.6: Water Balance at the Root Zone
(million acre feet)

	NWFP	Punjab	Sind	Pakistan
Fresh				
Water Requirements*	1.720	52.554	9.098	63.372
Water Supply				
Rain	.426	8.992	.521	9.939
Subirrigation	.100	1.452	.779	2.331
Canal Supply	1.257	18.764	5.680	25.701
Private Tubewells	.113	20.256	2.704	23.073
Government Tubewells		3.438	.170	3.608
Total Supply	1.897	52.901	9.854	64.653
Slack Water	.177	.348	.756	1.281
Saline				
Water Requirements*		4.855	13.025	17.880
Water Supply				
Rain		.851	.898	1.749
Subirrigation		.169	1.203	1.372
Canal Supply		5.288	16.632	21.920
Private Tubewells				
Government Tubewells				
Total Supply		6.308	18.732	25.040
Slack Water		1.454	5.707	7.160
Total				
Water Requirements*	1.720	57.408	22.123	81.252
Water Supply				
Rain	.426	9.843	1.419	11.688
Subirrigation	.100	1.621	1.982	3.703
Canal Supply	1.257	24.052	22.311	47.621
Private Tubewells	.113	20.256	2.704	23.073
Government Tubewells		3.438	.170	3.608
Total Supply	1.897	59.210	28.586	89.693
Slack Water	.177	1.802	6.463	8.441
Requirements for 200% intensity		79.0	72.6	151.6
(increase in diversions)		(21.6)	(50.5)	(72.1)

* to meet simulated cropping pattern

Projections to 2000: Assumptions

In this section we project the IBMR forward in time from the 1988 base to 2000. A solution for 2000 will provide us with a framework for a) predicting the results of the economic and environmental trends that are currently in force, and b) analyze the impacts of the WSIP and other investments from a consistent, future, base. It must be emphasized that this is not a forecasting device; we simply recognize that changes to the system are and will continue to take place, and to the extent that we can include them in the model's parameters, the more realistic our analysis. In turn we will describe demographic, economic, and agronomic trends, and then how future investment affects the model.

Population growth affects the model in two ways: through growth in the agricultural labor force, and through rightward shifts in the product demand functions. Population is projected to grow to about 145 million by 2000, with most of the increase attributed to the urban sector. Data from the PCA's of 1972 and 1980 showed the agricultural labor force growing by 1.24% in Punjab and 2.40% in Sind over that period. These rates are used to extrapolate the farm family labor force. The supply of non-farm labor is not a constraining factor in any of our 1988 solutions, so we assume that a perfectly elastic supply will be available in 2000.

Combining population growth with projected per capita income growth and income elasticities for the major consumables produces growth in product demand of: Irri rice, 4.0%; pulses, 4.7%; wheat, 4.1%; oilseeds, 3.4%; sugar, 5.5%, and milk, 6.3%. These rates are used to shift the demand functions. On farm consumption is assumed to grow at the same rates as agricultural labor force mentioned above. The limits on exports of cotton, Basmati, etc., set to reflect world market shares, are allowed to grow at 5% p.a., the assumed rate of growth of world trade in Pakistan's major exports.

The farm re-survey undertaken for WSIP was in part designed to provide updated input-output coefficients for the embedded farm models. The sample, however, was too small and unreliable to accomplish this. Statistically, the re-survey did not warrant changes to the elements of the I-O vectors covering labor, tractor, and bullock inputs. The re-survey did reveal, however, a fairly rapid tendency away from the purely bullock-based technologies toward a mix of tractor and bullock. Since the IBMR already allows this technology choice endogenously, no further changes were necessary.

Fertilizer requirements were updated from the re-survey, and calibrated for consistency with national production and import totals. Projections of future fertilizer use are available from the National Fertilizer Corporation and World Bank estimates. The Ministry of Food and Agriculture also provide recommended application rates and N/P ratios. At a 3% growth rate between 1988 and 2000, the resultant total use will be about 90% of the recommendations, and still lie within forecasts of industry capacity expansion. Assuming there will be little change in the cropped area, the fertilizer requirements in the I-O vectors are increased by 3% a year for the 2000 solutions, and this figure is also used for the monetary aggregate, which includes chemicals for plant protection.

Growth in crop yields is a major determinant of future performance. Trend analysis of yields as reported in the annual <u>Agricultural Statistics of Pakistan</u> reveals growth rates for the five years prior to our 1988 base as in the first column of Table 3.7. Growth in cotton yields have been outstanding, and sugarcane has been acceptable, but the others reported in the table are disappointing. Worse, crops not appearing in the table (but in the model) showed no statistically significant trend, or a negative one. For the latter group we assume no improvement to 2000. In the next chapter we will offer conjectures for the dismal performance of many crops experiencing low or negative yield growth. For the crops showing positive growth for the previous five year period, we extrapolate that growth to 2000. For cotton, whose growth can be explained by introduction of new American varieties and a massive effort at pest control, we assume that the growth will drop from 5% to 3%. Output of crop by-products fed to livestock are assumed to grow at the same rate as primary crop yields. Output of meat and milk are assumed to grow at 2.5% p.a., per animal. This growth is partly due to improved animal husbandry, and partly to the productivity arising from more specialization as farms become more highly mechanized and fewer animals are required to perform work-related tasks.

Table 3.7: Past and Projected Growth of Yields
(percent)

	1983-88 Trend	1988-2000 Assumed	Export Limits
Cotton	5.00	3.00	5.00
Sugarcane	2.29	2.29	N.A.
Wheat	0.41	0.41	5.00
Maize	0.73	0.73	N.A.
Onions	1.66	1.66	5.00
Basmati	0.00	0.00	5.00
Irri	0.00	0.00	5.00

In the first of the 2000 simulations reported below, that without the WSIP investments, only two changes to the irrigation system are taken into account. First, allowance is made for the continuing siltation of Tarbela, Mangla, and Chasma reservoirs. The expected changes in live storage capacity in MAF is as follows:

	1988	2000
Tarbela	8.861	8.357
Mangla	4.881	4.617
Chasma	.435	.231

Second, watercourse efficiencies (1-losses) increase between 2 and 8%, depending on the canal as a result of ongoing OFWM projects maturing and exogenous improvements undertaken privately by farmers.

Projections for 2000: with No Further Investment

Given the above assumptions, the 2000 solution without the WSIP investments is summarized in the second column of Table 3.8, with the corresponding items from the 1988 base reproduced in the first column.

The first block in the table contains measures of performance and income measured in constant 1988 rupee *economic* prices. The projection shows total output (value of production) growing at an annualized rate of 3.5%. This compares with 3.7% for the period 1980-88[2], a time of substantial investment and policy intervention in agriculture and water resources. Value added, a measure of keen interest because it is used to measure the benefits of any project/policy interventions, grows slightly higher, at 3.6%, largely because non-wage costs grew more slowly, at 3.2%. The latter component is probably understated: fertilizer is the largest component of cash costs, and our projections did not take into account the recent reductions in subsidies which will probably induce a lower rate of increase in this input.[3] The hired labor wagebill, although pitifully small and still reflecting the model's bias against hired labor as discussed above, does grow at a promising 4.8% rate. Farmers' real income, which includes a reservation wage attribution of half the hired labor rate, also increases sharply, at a 4.2% rate.

The overall performance, as measured by the value of production, is above the growth in population of 3.1%, but substantially below that called for by the NCA (5%). This could be expected, given this no-investment scenario. However, if we look at the crop-wise sources of this growth, we see that the overall performance in closely tied to our assumption about exogenous yield and export market growth. Cotton and rice make major contributions to growth, largely via the export route. Orchards are assumed to continue their recent rapid advance in area, while total production nearly doubles by 2000. Milk from both buffaloes and cows grows by 4.6 and 5.7% respectively, due to a combination of growth in demand and yield. Many of these sources of growth are traceable to assumptions, which if not met in reality, will certainly reduce the overall performance.

Of particular concern ought to be the projections for food grains, which grow by 0.6% in total, weighed down by stagnant growth in wheat production. The latter, a major concern for policy makers since wheat yields have shown little growth since the green revolution, suffers from the competition from the livestock sector. Substantial increase in fodder areas (not reported in the table) are required to increase milk production. Increases in rabi fodder compete directly with wheat for both land and water. Given tight rabi water supplies, it appears that policy makers will have to make the difficult choice between milk and bread.

---

[2] Reported in ASP 1988-89, p. 3.

[3] Although the economic costs of inputs as measured here excludes subsidies, the model's objective function uses financial prices (which include them) because farmers base their decisions on the prices they pay and receive.

Despite the growth in hired labor use, overall labor employment rises by only about 1.5% - far below the probable growth in the rural labor supply. This is due to two factors. First, mechanization has been heavily subsidized both through favorable pricing and the credit markets. By 2000, the stock of tractors is expected to be sufficient to perform all heavy tillage and threshing, and many intermediate operations. While there are arguments to the contrary, the pace of mechanization appear without doubt to be reducing the growth in agricultural employment. Second, employment is closely tied to land use, which grows by only 1.3% on an annual basis. Growth in kharif area (2.6%) is constrained by the availability of land, while growth in rabi area (only 0.3%) is constrained by rabi water. Overall intensity does show an improvement from 116% to 136%, but remains far below the 200% potential.

Fertilizer use grows about as expected by industry forecasts, and mirrors the growth in total output.

Canal diversions, measured at the canal head, decline 0.1% p.a. because of the accumulating reservoir siltation. Note that the proportionality assumption in allocations require a uniform percentage reduction in both provinces. NWFP is excluded from this requirement for the reasons discussed earlier. Groundwater usage, as measured by tubewell extractions, continues to grow at 1.8% p.a. in Punjab, but actually declines modestly in Punjab. Overall water availability measured at the root zone, rise by about 1.0% p.a. In Punjab, the increase is mostly due to the increased tubewell withdrawals, but partly to exogenous decrease in system losses brought about by continuing on-farm investments in watercourse and field efficiencies. In Sind, where both canal water and tubewell water decline, the 0.9% growth in water available at the root zone is partly explained by the latter factor, and partly from the increase in subirrigation from a continually rising average water table depth. This is both a blessing and a curse, as will be seen in the next chapter.

Projections for 2000: with WSIP Investment

The rightmost column of Table 3.8 report the same measures of performance, output, and input use for the 2000 case with the WSIP investments included. Although the full "expected" plan includes some 41 specific projects, many of these will not mature until well after our 2000 reference year. Accordingly, we have included only those which have a projected completion date of 1999 or earlier, listed in Table 3.9. Thus, *the 2000 projections reported here do not reflect the expected impact of the entire plan, only a subset of it.*

When implementing the Plan in the IBMR, many difficulties were incurred. First, most projects are multi-faceted in nature, often involving elements which are not represented in the system. The strengthening of the institutions involved is one prime example. Second, most projects are expected to improve yields in one way or

Table 3.8: 2000 Base Case Projections
(water allocations proportional to post-Tarbela period)

	1988 base	2000 w.o. WSIP		2000 with WSIP	
Run#	33	41		43	

Performance (millions of 1988 Rupees valued at economic prices):
and annualized growth rates from 1988 base

All irrigated agriculture:					
Value of production	130,724	198,280	3.5%	204,264	3.8%
Value added	93,608	143,475	3.6%	146,774	3.8%
Hired labor wagebill	2,450	4,277	4.8%	5,955	7.7%
Non-wage costs	34,666	50,528	3.2%	51,534	3.4%
Farm real income	74,717	122,296	4.2%	124,957	4.4%
Hydel output (BKWH)	16.711	16.775	0.03%	16.793	.04%
Punjab					
Value of production	96,598	145,092	3.4%	149,637	3.7%
Value added	69,362	105,508	3.6%	107,882	3.7%
Hired labor wagebill	1,814	2,419	2.4%	3,974	6.8%
Non-wage costs	25,423	37,165	3.2%	37,781	3.4%
Farm real income	55,627	90,468	4.1%	92,335	4.3%
Sind					
Value of production	31,324	49,451	3.9%	50,382	4.0%
Value added	22,559	35,533	3.9%	36,143	4.0%
Hired labor wagebill	636	1,858	9.3%	1,981	9.9%
Non-wage costs	8,129	12,061	3.3%	12,258	3.5%
Farm real income	17,977	29,979	4.4%	30,550	4.5%

Production (thousands of tons, all canal-irrigated areas)

Wheat	12,101.3	11,657.9	-0.3%	12,188.8	0.0%
Basmati	1,058.7	1,087.2	0.2%	1,103.8	0.3%
Irri	1,744.1	2,709.4	3.7%	2,725.5	3.8%
Maize	266.6	256.7	-0.3%	280.0	0.4%
Total Grains	15,170.7	15,711.2	0.3%	16,298.1	0.6%
Cotton	6,844.5	11,481.9	4.4%	11,763.4	4.6%
Gram	268.8	308.3	1.1%	308.1	1.1%
Mustard & Rape	196.0	212.4	0.7%	214.9	0.8%
Sugarcane (raw)	42,790.5	66,529.4	3.7%	67,634.1	3.9%
Orchards	3,279.8	6,416.0	5.8%	6,416.0	5.8%
Potatoes	637.7	736.4	1.2%	750.0	1.4%
Onions	647.9	818.5	2.0%	818.5	2.0%
Chilies	125.6	143.3	1.1%	144.3	1.2%
Cow milk	864.5	1,399.1	4.1%	1,491.2	4.6%
Buffalo milk	4,162.8	7,565.0	5.1%	8,118.7	5.7%
Meat	212.8	336.6	3.9%	366.0	4.6%

Table 3.8 (continued)

	1988 base	2000 w.o.	WSIP	2000 with	WSIP
Run#	33	41		43	

Use of selected inputs
Labor (million man-days)

Family	1,275	1,451	1.1%	1,453	1.1%
Hired	65	121	5.2%	171	%

Land (cropped area, thousands of acres)

Rabi	24,057	25,030	0.3%	26,073	0.7%
Kharif	16,647	22,580	2.6%	23,378	2.9%
Total	40,704	47,610	1.3%	49,451	1.6%
Intensity	1.16	1.36	1.3%	1.33	1.1%

Fertilizer use (thousand tons)

Nitrogen	1,406.9	2,283.6	4.1%	2,357.7	4.4%
Phosphate	435.3	687.2	3.9%	714.3	4.2%

Water Inputs (MAF)
Canal diversions (measured at canal head)

NWFP	3.04	3.31	0.7%	3.31	0.7%
Punjab	61.20	59.89	-0.2%	60.03	-0.2%
Sind	51.51	50.49	-0.2%	50.60	-0.1%
Pakistan	115.75	113.69	-0.1%	113.95	-0.1%

Tubewell extractions (measured at watercourse head)

NWFP	0.21	0.04	-	-	-
Punjab	43.57	53.77	1.8%	57.13	2.3%
Sind	5.94	5.77	-0.2%	5.58	-0.5%
Pakistan	49.71	59.58	1.5%	62.71	2.0%

Total water supply (measured at the root zone)

NWFP	1.90	2.07	0.7%	2.26	1.5%
Punjab	59.21	67.71	1.1%	70.31	1.4%
Sind	28.59	31.82	0.9%	32.02	0.9%
Pakistan	89.69	101.51	1.0%	104.64	1.3%

another, and the IBMR, in its current version, is not capable of endogenizing yield effects. This is most important for projects containing drainage, because lowering of water tables and salinity levels almost always brings about a substantial improvement in crop yields. Third, many of the projects were inadequately prepared with data required for simulation or evaluation missing. Given these crucial points, we cannot use the IBMR to evaluate the WSIP as the data in Table 3.8 might imply.

Table 3.9: Projects in the Minimum ("Expected") Plan
Which Will Impact the 2000 Solution

			funding years
NWFP			
	MS	Mardan Scarp	90-91
	CRBC3	Chasma Right Bank Canal Stage III	96-01
	OFWM3	On-Farm Water Mgt Phase III	90-94
	ISRP2	Irr. System Rehabilitation Phase II	94-98
	PC	Pehur Canal	95-98
Punjab			
	CWM	Command Water Management	90-92
	CRBC3	Chasma Right Bank Canal Stage III	99-04
	OFWM3	On-Farm Water Mgt Phase III	90-94
	ISRP2	Irr. System Rehabilitation Phase II	95-99
	GTC	Greater Thal Canal	94-00
	DBE	Dajal Branch Extension	94-99
	DGK	Dera Ghazi Khan SCARP	95-98
	PA	Panjnad-Abbasia SCARP	90
	LRR4D	Lower Rechna Remain. (4th Drainage)	90-92
Sind			
	NDSD2	North Dadu Surf. Drain. Phase II	90
	LBOD1	LBOD Stage I	90-93
	CWM	Command Water Management Phase I	90
	OFWM3	On-Farm Water Mgt Phase III	90-94
	ISRP2	Irr. System Rehabilitation Phase II	99-03
Balochistan			
	OFWM3	On-Farm Water Mgt Phase III	90-94
	ISRP2	Irr. System Rehabilitation Phase II	95-99

Instead, we use the model primarily to isolate the impact of the WSIP on the water resource system, and the benefits derived from that subsector, while retaining all of our previous assumptions. To this end, the following coefficients in the IBMR are modified:

Canal Command Area (CCA)
Canal Capacity
Canal Efficincies (loss rates)
Watercourse and Field Efficiencies
Reservoir Storage (Chotiari only; 1.1 MAF)

Despite the limited impact on the model's structure, the results from Table 3.8 show the Plan making a modest improvement in all performance measures. Total production is 3% higher in 2000 than without the plan, value added increases by 2.3%, and farm incomes, by 2.2%. The annualized growth rates for these measures also rise accordingly. Production impacts are mostly combined to wheat (4.6% higher) and milk and meat output (up 7.2%

and 8.7% respectively). Since no exogenous yield improvements are associated with the Plan (by our assumption), these increases imply a much improved rabi water availability.[4]

The employment effects of the Plan as projected appear to be sharply positive, with most of the gains appearing in hired labor. Fertilizer use also increases as a result of the higher cropping intensities in general, and an increase in the area under wheat in particular.

Canal diversions are not affected by the Plan, but tubewell withdrawals, which are endogenous, are. They show a slight drop in Sind, but a 6.2% rise in Punjab. In the "without WSIP" case, no investment in additional tubewells beyond our projections was necessary; with the Plan, farmers find it profitable to invest in tubewells in the Punjab for pumping capacity.

The increased efficiencies throughout the system brought about by the Plan investments results in water availability at the root zone increasing by 3.1%, with the gain benefiting all provinces. However, due to the tubewell investment in Punjab, that province shows the largest benefit.

Finally, we need to note once again the effect of the assumed water allocation policies. Prior to the settlement of the inter-provincial allocation issue, our allocation constraints require proportional allocation to all canals. When one canal hits a capacity constraint, the system must effectively "waste" any remaining water; otherwise a change in the allocation rules would be required. Obviously, this is a most inefficient way to operate the system, but one imposed by current political realities. In chapter 6, when we investigate policy measures which might have a favorable impact on the environmental problems, we will have occasion to relax the proportionality assumption.

---

[4]Increased rabi water availability does induce some _endogenous_ increase in wheat yields as the model is able to shift away from water-stressed wheat activities toward higher yielding activities.

Chapter 3 Appendix:
Surface Water System Inflows

Because all agricultural production in the IBMR requires irrigation water from some source, and most of the water requirements are derived from surface water flows, a critical assumption affecting the performance of the model involves water availability from surface inflows to the system.

According to the Indus Waters Treaty of 1960 between India and Pakistan, the waters of the Indus itself and the two "Western Tributaries" (the Jhelum and Chenab rivers), were allocated to Pakistan and the waters of the three "Eastern Rivers" (the Ravi, Beas, and Sutlej), to India. The major portion of Pakistan's surface water supplies come the from Indus, Jhelum, Chenab, and the Kabul/Swat rivers in NWFP province. Additional surface inflow is received from the Haro and Soan rivers, hill torrents into the Indus upstream from Taunsa barrage, and the upper reaches of the eastern rivers. Some flow still escapes the eastern rivers to Pakistan, but for purposes of water resource planning, these must be ignored.

For IBMR input, inflows are grouped into two categories, rim station inflows and tributary inflows. Rim stations are point where rivers enter Pakistan or key locations such as reservoir sites upstream of the irrigated area. Tributaries enter the river reaches laterally below the rim stations. Additional lateral inflow, largely from rainfall runoff and surface drains within the canal-irrigated areas are captured as a part of river gains (in some cases, losses, if lateral seepage of groundwater to river beds dominates) measured between the various nodes of the river system.

Unless otherwise noted, the use of the IBMR in this study employs 50% probability flows for the available historical period. Table 3A.1 and 3A.2 present the rim station and tributary inflows as used in the model. Of the minimum 147.04 MAF available with 50% probability, 138.69, or 94%, comes from rim stations, of which 80% is attributed to the four main rivers. As already noted, about 85% of inflows occur during the kharif season.

How the surface water is allocated to the various irrigated areas is captured by the canal diversion pattern. This pattern is established annually on the basis of political negotiations with much consideration of precedent. Table 3A.3 reports the average diversions by canal and regional canal groupings for the post-Tarbela period (1976-1988) which form the basis for most of the IBMR simulations. All base cases termed "proportional" constrain the endogenous canal allocations to be proportional to those in Table 3A.3. Exact duplication of these diversions is not always possible because, for example, reservoir siltation in future runs may make the historical diversions infeasible.

Table 3A.1: Rim Station Inflows, 50 Percent Probability (MAF)

River	Apr	May	Jun	Jul	Aug	Sep	Kharif
Indus	1.931	3.984	7.191	16.148	17.437	6.196	52.887
Jehlum	3.826	4.355	2.993	3.429	2.236	.846	17.685
Chenab	1.361	2.281	4.036	5.279	5.097	2.540	20.594
Ravi	.226	.161	.038	1.089	1.732	0.841	4.087
Sutlej	.031	.045	.052	.483	1.003	1.198	2.812
Haro	.016	.013	.014	.048	.204	.073	.368
Soan	.012	.007	.090	.225	.426	.068	.828
Swat	.483	.636	1.049	.863	.642	.346	4.019
Kabul	1.512	1.883	3.467	3.490	2.619	.975	13.946
Total	9.398	13.365	18.930	31.054	31.396	13.083	117.226

River	Oct	Nov	Dec	Jan	Feb	Mar	Rabi	Annual
Indus	2.297	1.279	1.188	.952	.973	1.468	8.157	61.044
Jehlum	.988	.521	.412	.460	.621	1.417	4.419	22.104
Chenab	.728	.487	.407	.480	.711	1.079	3.892	24.486
Ravi	.188	.134	.131	.090	.026	.455	1.024	5.111
Sutlej	.129	.009	.004	.060	.000	.000	.202	3.014
Haro	.037	.020	.016	.019	.019	.051	.162	.530
Soan	.010	.005	.020	.022	.028	.043	.128	.956
Swat	.195	.122	.088	.079	.068	.140	.692	4.711
Kabul	.523	.405	.354	.379	.440	.687	2.788	16.734
Total	5.095	2.982	2.620	2.541	2.886	5.340	21.464	138.690

Period of Record:
Indus   1937 - 1988
Jehlum, Chenab   1922 - 1988
Ravi, Sutlej   1966 - 1988
Others   1966 - 1976

Table 3A.2: Tributary Inflows, 50 Percent Probability (MAF)

River Reach	Apr	May	Jun	Jul	Aug	Sep	Kharif
Marala-Khanki	.004	.006	.023	.120	.240	.056	.449
Mangla-Rasul	.120	.180	.150	.170	.320	.210	1.150
Chasma-Taunsa	.051	.013	.021	.048	.021	.012	.166
Tarbela-Kalabagh	.041	.087	.182	.350	.602	.180	1.442
Balloki-Sidhnai	.000	.001	.009	.096	.206	.041	.353
Amandara-Munda	.036	.337	.637	.854	.544	.062	2.470
Maunda-Kabul	.015	.013	.005	.008	.007	.007	.054
Warsak-Nowshehra	.126	.102	.059	.070	.056	.056	.469
Total	.393	.739	1.086	1.716	1.996	.624	6.553

River Reach	Oct	Nov	Dec	Jan	Feb	Mar	Rabi	Annual
Marala-Khanki	.017	.007	.008	.007	.006	.007	.052	.501
Mangla-Rasul	.060	.000	.000	.000	.000	.000	.060	1.210
Chasma-Taunsa	.004	.005	.010	.006	.015	.026	.065	.231
Tarbela-Kalabagh	.100	.064	.038	.055	.052	.130	.439	1.881
Balloki-Sidhnai	.002	.003	.008	.005	.007	.006	.030	.383
Amandara-Munda	.133	.153	.150	.119	.119	.221	.894	3.364
Maunda-Kabul	.003	.003	.004	.004	.004	.009	.028	.082
Warsak-Nowshehra	.024	.027	.035	.035	.038	.070	.229	.698
Total	.343	.262	.253	.231	.241	.469	1.797	8.350

Period of Record: 1966 - 1976

Table 3A.3: Canal Diversions; 1976-1988 Average (MAF)

	Marala Canals	Mangla Canals	Upper Indus Canals	Lower Indus Canals Sind	Kabul Swat Canals NWFP	Punjab	Pakistan
January	.114	0.931	0.702	1.697	.072	1.748	3.517
February	.197	1.584	1.002	2.297	.121	2.784	5.202
March	.203	1.961	1.297	2.305	.178	3.461	5.944
April	.251	2.018	1.359	2.081	.215	3.629	5.925
May	.553	2.785	2.262	3.276	.252	5.600	9.129
June	.734	2.872	2.512	6.075	.252	6.117	12.444
July	.865	2.837	2.519	6.955	.238	6.221	13.413
August	.823	2.777	2.453	5.926	.210	6.053	12.190
September	.756	2.854	2.520	5.432	.234	6.131	11.797
October	.482	2.542	2.182	4.113	.243	5.206	9.562
November	.250	2.092	1.340	2.710	.209	3.681	6.600
December	.165	1.973	1.149	2.376	.191	3.287	5.853
Rabi	1.412	11.083	7.672	15.498	1.014	20.167	36.680
Kharif	3.982	16.144	13.625	29.745	1.402	33.751	64.899
Annual	5.394	27.227	21.297	45.244	2.416	53.918	101.578

Marala Canals : 01-UD, 02-CBD, 03-RAY, 04-UC, 05-MR

Mangla Canals : 06-SAD, 07-FOR, 08-PAK, 09-LD, 10-LBD
11-JHA, 12-GUG, 13-UJ, 14-LJ,

Upper Indus Canals : 15-BAH, 16-MAI, 17-SID, 18-HAV, 19-RAN,
20-PAN, 21-ABB, 26-THA, 27-PAH, 28-MUZ, 29-DGK

Lower Indus Canals : 31-P+D, 32-BEG, 33-GHO, 34-NW, 35-RIC, 36-DAD
37-KW, 38-KE, 39-ROH, 41-NAR, 42-KAL,
43-LCH, 44-FUL, 45-PIN

Kabul-Swat Canals : 22-USW, 23-LSW, 24-WAR, 25-KAB

Rabi - October through March: Kharif - April through September

CHAPTER 4
ENVIRONMENTAL ISSUES

Environmental Priorities

A recent World Bank report[1] reviewed a host of environmental problems in rural Pakistan:

- Threatened extinction of the coastal mangrove forests due to increased diversion of Indus River waters for irrigation with serious consequences for the port of Karachi, crustacean breeding grounds and migratory birds habitat;

- Deforestation for fuelwood leading to flash flooding and soil erosion, and shifting sand dunes which bury irrigated fields and rural infrastructures;

- Degradation of range lands due to over-grazing by cattle;

- Pollution of potable water supplies and river fish habitats from the increased use of agro-toxic chemicals.

But the most important, by far, is the degradation of soils due to waterlogging and salinity. Quoting earlier sourcework, the environmental study claims that 25% of irrigated lands are saline due to high watertables, of which 60% is unsuitable for cultivation, and another 30% suitable for selective cultivation only.

This study focuses on waterlogging and salinity problems because they affect both of the critical resources land and water. As population pressures mount relentlessly and the supply of both resources is virtually fixed, it is imperative that their qualities not be allowed to deteriorate further. The IBMR is a useful tool in addressing these problems because it tabulates, by source, recharge and discharge from groundwater, and watertable depths: the key variables in assessing the pace of waterlogging and salinity. In this chapter, we will investigate the status of these problems prior to using the IBMR to simulate interventions to control or, at least, forestall the disastrous consequences often predicted.

Waterlogging

The water table is the surface of the zone of water-saturated soils lying somewhere beneath the ground topography. We shall use WTD to refer to the depth of the water table, measured in feet. When there is no recharge to or discharge from groundwater, the WTD will tend toward a constant, or stable level. Even in the absence of irrigation, the WTD is constantly changing and tending toward a new equilibrium if there is any rainfall at all, and if there is any discharge or outflow. Figure 4.1 illustrates some of the factors involved in the WTD determination. Part of any rainfall is dispersed at the surface through

---

[1] Pakistan: Medium Term Policy Agenda for Improving Natural Resources and Environmental Management, World Bank, EMENA Region, 1991.

Figure 4.1: Factors Determining Groundwater Depth and Salinity

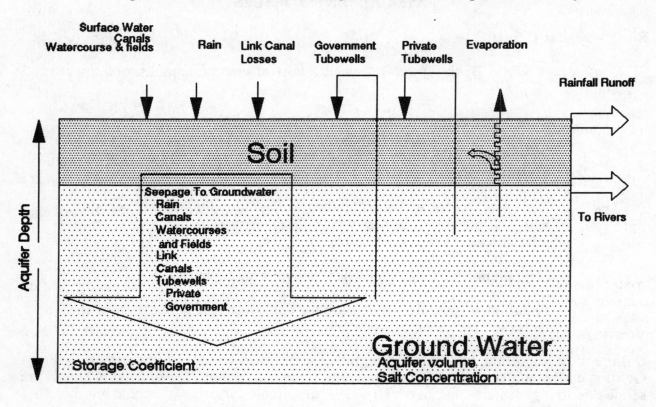

Figure 4.2: Evaporation and Watertable Depth Relationship

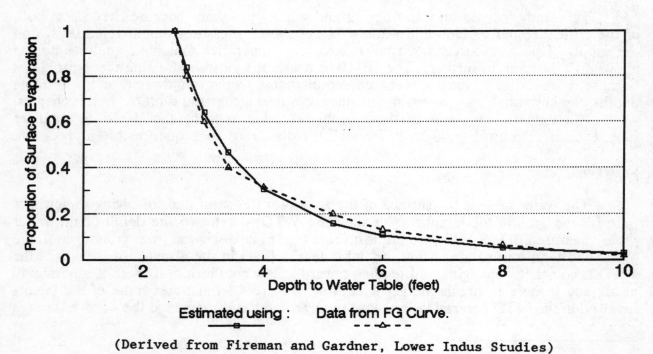

(Derived from Fireman and Gardner, Lower Indus Studies)

runoff to streams, rivers, etc. Runoff is facilitated in most of the Indus Basin through surface drainage which is necessary because of the flat topography. A surface drainage network of nearly 9,000 miles has been constructed within the Indus system, but much of it is in ill-repair and not capable of handling runoff of heavy rains. What rainfall does not runoff enters the soil through infiltration. After the soil is saturated, depending on the permeability of the soils, and the WTD, water seeps to the saturated zone, becomes a net addition to groundwater, and tend to raise the WTD. Similarly the rivers, link canals, irrigation canals, watercourses and water applied to the fields contribute to the recharge of the groundwater reservoir.

In most of the Indus Basin the underlying aquifer is believed to be unconfined and is thus synonymous with the "saturated zone" referred to above.[1] It varies in depth from about 300 meters in parts of Punjab to about 60 meters in the Sind. In the Punjab it exists in mostly sandy alluvium; in the Sind, the soils are dominated by less permeable clay and silt.[2]

Groundwater is naturally discharged, or exported, mainly in three ways. Evaporation from groundwater takes place continuously even at quite deep levels. The rate of evaporation from groundwater depends on climatic factors usually represented by pan evaporation (the rate of evaporation from a pan at the surface which depends primarily on climatic factors), the capillary conductivity and the moisture contents of the soil, and, of course, depth. The Indus Basin has very high potential evaporation, a significant amount of groundwater evaporates in the areas with high water table.

Groundwater is also discharged from a given area through horizontal movement or natural drainage to an adjacent area or to a nearby stream or river bed. Horizontal movements of water are exceedingly complex and difficult to measure. In most of Pakistan, we can assume that they are slow if not negligible, apart from areas close to the rivers. This is evident from viewing Figure 4.3 which shows that most waterlogged areas lie away from the river valleys and are bounded by mountains or desert. Poor natural drainage was not a problem prior to intensive irrigation; pan evaporation greatly exceeds rainfall in most of the Basin.

The third way in which groundwater is discharged is through uptake of vegetation for transpiration purposes. Transpiration varies according to many factors including the type of vegetation and its root structure, the intensity of cultivation, and the WTD. Transpiration from groundwater is termed "subirrigation" elsewhere in this study.

The method used in IBMR to estimate the evaporation and subirrigation is based on the results of the studies by Gardner and Fireman [1958]. They gave a relation for evaporation from soil columns having different water table

---

[1] A confined aquifer is one separated from the saturated zone (groundwater) by a relatively impermeable layer of, say, clay.

[2] W. Bakiewicz, "Supplementary Report: Groundwater", WSIP Volume III, p. S-1.

Figure 4.3: Areas of Saline and High Watertable

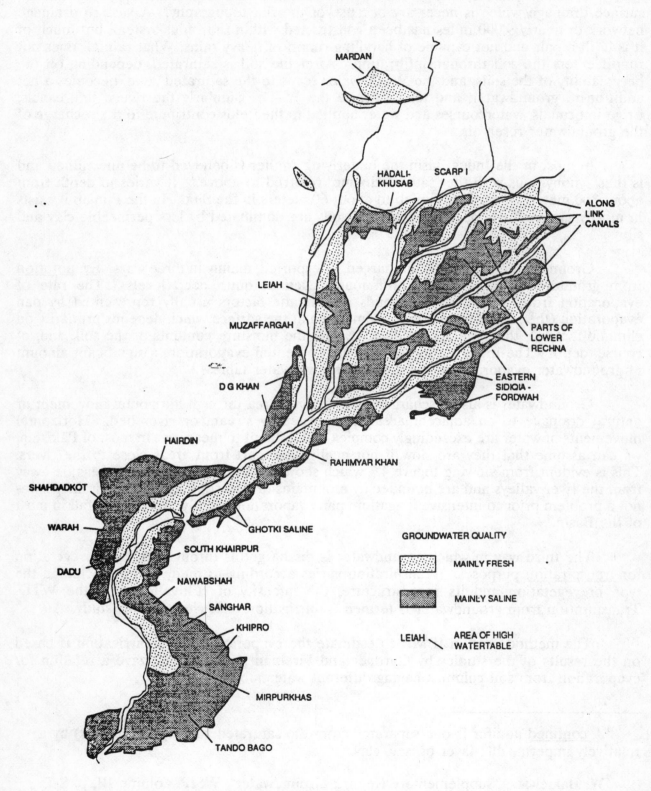

Source: Water Sector Investment Planning Study.

depths. In the Lower Indus Project[1], studies were done to estimate evaporation from different water table depths for an actual case in the region and results were compared with the Fireman-Gardner and various other methods.[2] LIP presented the results as a relation between depth to water table and proportion of free water surface evaporation, making the relation portable to other areas. For IBMR an exponential equation was fitted to the data values read from this curve. This relation and the original curve are shown in Figure 4.2. The groundwater discharge D_e via evaporation (feet) is estimated using the equation:

$$D_e = E * \frac{10.637}{WTD^{**2.558}} \qquad 4.1$$

where E is the pan evaporation (feet) and WTD is depth to the water table (feet).

<u>Waterlogging</u> refers to a WTD shallow enough to adversely affect plant growth. Although waterlogging is now recognized as an almost certain outcome of intensive irrigation in arid or semi-arid climes, as recently as the fruition of the Aswan High Dam agricultural planners were reportedly surprised at the speed of its manifestation.[3] From Figure 4.1 and the above discussion, it should be evident that any additional recharge given fixed natural drainage capacity will inevitably result in a rise in the water table. Irrigation typically involves dramatic increases in recharge. For example, in zone SCWS, seepage from canals, watercourses, and irrigated fields is estimated in the IBMR to be 1.971 feet of water per year. Seepage from rainfall is only .056. Little wonder, then, that in Khaipur district, "the water table rose by 10 cm per year between the early 1930's, when Sukkur Barrage was opened, and 1965".[4]

The most recent complete soil survey was undertaken by WAPDA during 1980-81, and the province-wide totals are given in Table 4.1 below. About 2.9 million acres (7% of GCA) had a WTD less than 3 feet, and another 6.2 million acres (15%) had a WTD between 3 and 6 feet. More recent investigations by WAPDA, reported by NCA, found 12.2 million acres in the "disastrous" range of less than 5 feet.[5]

Depending on the crop, a high water table induces adverse effects on yields. Table 4.2 shows estimates for Pakistan of the percentage of maximum yield obtained as the water

---

[1] Lower Indus Report, West Pakistan Water and Power Development Authority, 1965 (seven volumes).

[2] W. R. Gardner and Milton Fireman, "Laboratory Studies of Evaporation from Soil Columns in the Presence of a Water Table", <u>Soil Science</u>, Vol. 85, 1958.

[3] Ian Carruthers, "Protecting Irrigation Investment: The Drainage Factor", <u>Ceres</u> 106, p. 16.

[4] <u>Ibid</u>., p. 17.

[5] NCA, p. 291.

Table 4.1: Extent of Waterlogging (1981)

	Water Table Depth (feet)						Total GCA
	0 - 3	3 - 6	0 - 6	0 - 3	3 - 6	0 - 6	
	Percent of GCA			000s acres			
NWFP	6	12	18	91	182	273	1,517
Punjab	7	11	18	1,758	2,763	4,521	25,117
Sind	6	24	30	827	3,309	4,136	13,786
Baluchistan	1	6	7	9	52	61	873
Pakistan	7	15	22	2,891	6,194	9,084	41,293

Source: WAPDA, Soil Survey of Pakistan, 1981.

Table 4.2: Percentage of Yield Obtained at Different Water Table Depths

WTD (feet)	Cotton	Sugar	Wheat	Berseem
over 5.7	100	100	100	100
4.9 - 5.7	99	99	100	100
4.1 - 4.9	95	95	99	100
3.3 - 4.1	88	86	95	98
2.5 - 3.3	79	71	87	91
1.6 - 2.5	65	54	72	76
0.8 - 1.6	43	34	51	55
0.0 - 0.8	2	9	21	23

Source: Derived from WSIP Table 3.16, p. 3-49.
(Estimates from LBOD Project background studies.)

table rises from the neighborhood of five feet. It is clear that in the area of three feet, a high WTD exerts a substantial negative impact on all crops listed. At two feet, in all probability cultivation becomes uneconomic.

Although these estimates were not made with the carefully controlled trials their originators would have desired, evidence from Egypt tends to support them. The Drainage Research Institute found that cotton yields declined by 39% when the water table rose from about 3 feet to about 2 feet.[1]

Salinity

Salts occur naturally in most soils and in all sources of irrigation water. Salty soils show high contents of various kinds of salts and/or a high percentage of exchangeable sodium. Heavily salinized soils may even show effloresences or complete salt crusts, formed by such salts as gypsum ($CaSo_4$), common salt NaCl), soda (Na_2CO_3) or more complex salts. Salty soils come into being because either the parent material was salty or by irrigation with salty water or they were flooded by the sea water. The majority of the salty soils, however develop as a result of the upward capillary flow of water exceeding its downward movement. As a parameter for the soil salinity the electrical conductivity of the saturation extract (EC) is used, expressed in mmhos, mho being the reciprocal value of ohm. The classification of the saline and sodic soils is based on the EC and the exchangeable sodium percentage (ESP).

At high levels, plant growth is retarded both by soil salinity and salinity of the irrigation water. Salts affect crops through the increased osmotic pressure of the soil solution, which results in a reduction of a plant's capacity to withdraw water from soil. There may be effects due to the toxic ions, and unfavorable structure of the salty soils. Soil salinity buildup is a complicated process closely "interrelated with such factors as salinity of irrigation water, drainage, chemistry of soil water, and type of crop."[2] The processes are not entirely understood, and continue to be debated in the soil science literature. For our purposes, a simplistic explanation will suffice.

Even if irrigation water is relatively free of salts, repeated irrigation which induces a rise in the water table as described above will dissolve the salts in the soil and bring them toward the surface. Neither evaporation nor transpiration removes salts, so they continue to buildup in regions ever closer to the surface. When salts rise to the area of the root zones of crops, they begin to inhibit the uptake of nutrients productivity is affected, and soils are known to become "saline affected".

Many irrigation scheme failures in history due to salinity buildup have been

---

[1] Reported in H. J. Nijland and S. El Guindi, "Crop Yields, Watertable Depth, and Soil Salinity in the Nile Delta, Egypt", International Institute for Land Reclamation and Development Annual Report, 1983, pp. 19-21.

[2] R.J. Hanks, "Prediction of Crop Yield and Water Consumption under Saline Conditions", in I. Shainberg and J. Shalhevet (eds.), <u>Soil Salinity Under Irrigation</u>, New York: Springer-Verlag, 1984, p. 272.

documented. "FAO estimates that 50% of the world's irrigated land is salinized to the extent of affecting productivity. In Iran, Iraq, Egypt, and Pakistan more than 70 per cent of the farmland is so affected."[1] The visual evidence of salinity buildup can be startling. In parts of the southern Punjab in which SCARP VI is attempting to reclaim saline land, vast tracts are completely barren and give the appearance, from the air, of having received a recent snowfall. The types of salts present in the soil vary over the basin. In Sind the dominant salts are sodium chloride in Kotri canals and sodium sulphate elsewhere. In the Punjab the dominant salt is sodium bicarbonate.[2] Table 4.3 reports the amount and percentage of such soils in the Indus Basin as classified by WAPDA.

These data show that 8% of the irrigated area is "severely" (WAPDA's characterization) salt-affected, and another 6% is "moderately" affected. Most of these areas are in the Sind, where 28% of all irrigated land is affected.

What are the implications of these levels of soil salinity for crop yields? A seminal and still widely employed work by Maas and Hoffman[3] provides approximations of yield declines for soil salinities above threshold tolerance levels. Table 4.4 reports the basic parameters of threshold tolerance and percentage yield decline for unit increases above these levels for the major crops in the Indus Basin. From Table 4.3, it can be seen that 6% of the irrigated area experiences a minimum yield decline given in the third column of Table 4.4 (between 2% for cotton and 60% for rice), while 8% of the area experiences the minimum yield declines in column 4 (38% for cotton to 100% for rice). In the two rightmost columns we have calculated the average yield decline for the "saline" areas as defined for the IBMR, using salinity levels derived in Table 4.3, e.g., EC = 8.8 for the Sind-saline and EC = 7.5 for the Punjab-saline. Note that these salinity levels apply to about 78% and 21% of the Sind and Punjab, respectively. If these numbers are anywhere near being correct, soil salinity is "robbing" Pakistan of about 25% of its potential production of these major crops, or about US$ 2.5 billion per year!

---

[1] Carruthers, op. cit., p. 106. The data in Table 4.3 draws into the question the 70% figure, at least for Pakistan.

[2] WSIP : Main Report.

[3] E. V. Maas and G.J. Hoffman, "Crop Salt Tolerance - Current Assessment", Journal of the Irrigation and Drainage Division (June, 1977), pp. 115-134.

Table 4.3: Extent of Soil Salinity (1981)

	Electrical Conductivity (EC Millimohs/CM)						Total GCA
	>15	8-15	<8	>15	8-15	<8	
	Percent of GCA			000s acres			
NWFP	2	2	4	30	30	61	1517
Punjab	3	4	7	754	1005	1758	25117
Sind	18	10	28	2481	1379	3860	13786
Baluchistan	4	5	9	35	44	79	873
Pakistan	8	6	14	3303	2478	5781	41293

Derived Data: Average Salinity

	EC		PPM		Percent area	
	Fresh	Saline	Fresh	Saline	Fresh	Saline
Punjab	2.0	7.5	1280	4800	79	21
Sind	2.0	8.8	1280	5620	28	72

Source: WAPDA Soil Salinity Survey, 1981.

Table 4.4: Salinity-Yield Relationships

	Threshold tolerance (TT) EC	Percent Yield Decrease			Saline Areas	
		For unit increase above TT EC	EC=8	EC=15	Sind	Punjab
Cotton	7.7	5.2	2	38	6	0
Wheat	6.0	7.1	14	64	20	11
Rice	3.0	12.0	60	100	70	54
Sugarcane	1.7	5.9	37	78	42	34
Rabi fodder	1.5	5.7	37	77	42	34
Kharif fodder	1.8	7.4	46	98	52	42

Source: derived from Maas and Hoffman, op.cit. For Rabi and Kharif fodders we have used their data for berseem and maize forage.

CHAPTER 5
QUANTIFYING WATERLOGGING AND SALINITY PROBLEMS

Modelling Salt and Groundwater Flow

If rational policies and investments are to be devised to deal with the problems of waterlogging and salinity, a clear understanding of their underlying causes is a priority. In this section we attempt to unravel the causes by examining the water balances obtained from a combination of historical and model output data. The IBMR was modified to quantify water and salt flows to the groundwater aquifer and to compute water and salt balances for each zone. In addition spreadsheet techniques were used whenever found suitable. In this chapter we will describe the mathematical equations and methodology used to model groundwater and salt flows. The groundwater and salt balance for fresh and

---

Box 5.1 Notation used in the Groundwater and Salt flow equations.
(acre-feet for entire area; feet for per acre)

R_r recharge from river seepage

R_c recharge from canal seepage

R_w recharge from watercourses and fields

R_t recharge from tubewell operations (R_{gt}-government + R_{pt}-private)

R_l recharge from lateral flows from adjacent zones

R_p recharge from precipitation

TR   total recharge

D_e discharge from evaporation and subirrigation

D_t discharge from tubewell withdrawals (D_{gt}-government) + D_{pt}-private)

D_d discharge from subsurface drainage (subsurface drainage is provided in the saline areas only)

D_l discharge from lateral flows to adjacent zones

TD   total discharge

TS   total salt contents of the groundwater (tons)

ΔS change in volume of the groundwater aquifer

---

saline areas of each zone computed using this model will also be presented in this chapter. To reduce the potential for ambiguity in the relations and assumption, some definitions and notations used in the equations are given in Box 5.1.

Groundwater Balance

Recharge to the groundwater:

Saline areas: $TR = R_r + R_c + R_w + R_p + R_i$ 5.1
Fresh areas : $TR = R_r + R_c + R_w + R_p + R_i + R_t$ 5.2

Discharge from the groundwater:

Saline areas: $TD = D_e + D_d + D_x$ 5.3
Fresh areas : $TD = D_e + D_t + D_x$ 5.4

Water Balance:

$$\Delta S = TR - TD \quad\quad 5.5$$

where ΔS is the change in storage for the period.

Water table depth (feet):

$$WTD_t = WTD_{t-1} + (\Delta S/A^g)/c \quad\quad 5.6$$

where A^g is the gross area and c is the storage coefficient

Salt Content of the Groundwater

$$TS = TS_{t-1} + \Sigma s_i R \rho - \Sigma s_i D \rho \quad\quad 5.7$$

where s_i is the salt concentration of the various sources of recharge and discharge in PPM and α is conversion factor from PPM to tons/acre-feet.

Salt concentration the groundwater (PPM):

$$SGW = TS/V/\alpha \quad\quad 5.8$$

where V is the volume of the active aquifer computed as

$$V_t = (d - WTD_{t-1})c + \Delta S_t \quad\quad 5.9$$

where d is the depth of the active aquifer (feet).

All recharge and discharge data vary by month both in the data and in the model's usage. For reporting purposes, only the annual totals are usually given.

R_p is given exogenously, based on historical data. R_r is determined from the estimates of river losses and gains computed endogenously to route water through the river system.

R_c and R_w vary according to the mode of system operation simulated by the model and the seepage factors which vary by zone. The latter are affected by the state of system and can be reduced through rehabilitation. i.e., canal-lining projects serve to reduce seepage from canals and thus lower R_c; OFWM-type projects serve to lower R_w where they are implemented.

R_t, the sum of R_{gt} and R_{pt}, depends directly on the level of tubewell operations in fresh groundwater areas only. R_{pt} is determined endogenously within the model as a function of private irrigators' profit-maximizing behavior. R_{gt} is given exogenously in the model given prior decisions by the government operators, and has a much higher percentage of losses the private tubewell operations because government tubewells typically discharge at the level of the *mogha*[1] whereas private tubewell typically discharge at the farm.

D_e is function of the pan evaporation for the area and the WTD.

D_t is straightforward given the above explanation for recharges from tubewell operation, R_t.

D_d depends on the installed subsurface drainage, if any, in saline areas, and is assumed to equal:

$$D_d = TR - D_e - D_x \qquad 5.10$$

D_e in this case is estimated at the WTD with the installed drains. Comparison of the equations 5.3, 5.6 and 5.10 reveals that ΔS for areas with subsurface drainage will be zero as any net recharge will be exported as drainage effluent.

R_i and D_x, the subsurface inflows and outflows from neighboring areas depend on a host of physical and geological factors in addition to head (difference in WTD between adjacent zones). These flows must be ignored in the limited scope of this study, except that we will be able to indicate the likely direction and intensity of such flows from the changes in WTD.[2] Also we are interested in investigating scenarios which do not alter the WTD substantially.

ΔS, the difference between recharges and discharges in any given period, is of particular interest because, except in drained areas for which ΔS is zero by assumption (i.e., all net recharge is drained), it determines the change in WTD. A positive ΔS, typical in saline areas, indicates progressing waterlogging and probably salinity. A negative ΔS, typical

---

[1] the outlet from a distributary canal to a watercourse.

[2] Duloy and O'Mara, in formulating the original Indus Basin Model, attempted to endogenize these lateral flows, but at the expense of requiring a virtually unsolvable model. The results they did obtain, however, revealed that lateral flows among their 53 "polygons" in a solution year amounted to only about 2% of the total volume of groundwater flows, lateral and vertical.

in fresh groundwater areas, indicates a falling water table, a mining of the aquifer, and more costly tubewell operations.

Note briefly the paucity of instruments the government has available to control the various sources of recharge and discharge and hence ΔS. Policy instruments are limited to water allocations and taxes/subsidies on tubewell operations. The former can affect R_c and R_w to some extent by altering water allocation patterns among the various zones and groundwater types. The extent to which it may be effective as a groundwater control mechanism will be addressed in Chapter 6. Government tubewell operations in fresh areas, as noted previously, are being transferred to the private sector in an attempt to make them more efficient and more responsive to farmers' needs. Subsidies on private tubewell operations are probably in effect given the electricity rate charge system. Taxing them would be administratively difficult and costly in terms of national production objectives because the bulk occur in the most productive areas. Pakistani officials proclaim that no policy or mechanism exists to restrict private tubewell withdrawals.

Project options include canal and watercourse improvements/lining, and of course, drainage. These will be investigated in Chapter 6, both with respect to their potential impact on waterlogging and salinity, and on output.

Water Balances

Using the equations listed above the water balance of the groundwater aquifer both fresh and saline within each ACZ was computed. Table 5.1 reports annual estimates of surface water flows, seepage, net recharge, and implied changes in watertable depth for 1988 base case by zone and groundwater classification. The groundwater balance for Punjab and Sindh is presented in schematic form in Figure 5.1. These data are a compilation of IBMR input and output: rainfall, canal diversions, and government tubewell operations are input to the model as well as the WTD for 1987. River seepage, evaporation, private tubewell and link canal operations are endogenously produced by the model given canal diversions (fixed for this table) and rainfall. Seepage is determined in the model and in the table by the various loss coefficients.

The evaporation and the WTD are interrelated variables as shown by the equations 4.1, 5.5 and 5.6. They are computed by solving these equations iteratively. Given WTD_{t-1}, D_e is computed from equation 4.1, which is then used to compute WTD_t from equation 5.6. A new estimate of D_e is made using an average of WTD_t and WTD_{t-1}. Using the new D_e, WTD_t is computed again. This procedure is repeated until the values of WTD_t and D_e don't change.

The WTD varies considerably during the year, being higher towards the end of Kharif due to higher canal deliveries and rains. Table 5.1 gives the mean annual change in the WTD. Consider first the signs on net recharge per acre and change in WTD. They are positive in all saline areas and the fresh zones of northern Sind. They are negative in all fresh zones of the Punjab and in SCWS. Only PMW Fresh and all of SRWN are near equilibrium; all of the others are sharply out of balance and exhibit a rapidly rising or falling WTD: over 4 feet per year in places.

Most contamination of fresh areas occurs through lateral movement of saline groundwater, spurred by differing watertable depths. The higher the head (the difference in watertable depth between adjacent fresh and saline areas), the greater the pressure for lateral movement, and the more likely contamination will occur, in the absence of vertical hydrologic barriers. Too little is known about these barriers, or the characteristics of groundwater movement to make predictions. It is clear, however, that the widening differential in watertable depths (rapidly falling in fresh areas, rising to saturation levels in most saline areas) is a disaster waiting to happen. Moreover there are patches of saline areas within the fresh zones, due to which mixing is more likely. No time-series evidence on contamination of fresh groundwater areas is available (only the 1981 WAPDA Soil Survey tested groundwater quality on a basin-wide basis). There have been, however, recent reports of farmers complaining about the deteriorating quality of their tubewell water.

In fresh areas, there appear to be no significant pressures leading to waterlogging. Only in the northern Sind zones are average recharges positive. What natural drainage there is probably absorbs most of it. The reason, of course, is the large withdrawals of groundwater by tubewell, only a fraction of which returns as recharge. In the saline areas, water tables are rising, on average, in all zones, and at very rapid rates. Apart from natural drainage, which is unknown (except for that captured by rivers) but presumably small, the only outlet for positive recharge is evaporation. The additional recharge induces a rise in the watertable which in turn increases the evaporation. The watertable thus rises to the level where all the recharge is balanced by evaporation. As discussed in the following section this high rate of evaporation moves substantial quantities of salt to the surface.

From Figure 5.1, the culprits can be identified: seepage from canals and watercourses (including fields). In the Sind-saline zones, these seepages account for 92% of total recharge, since rainfall is so low. Obviously, these losses ought to be addressed if costly drainage is to be avoided or delayed, and we do so in the next chapter.

A Model of Salt Distribution

From the flow and seepage data in Table 5.1, and estimates of the salinity of the various components, it is possible to measure the salinity buildup over the course of a year. Table 5.2 presents such an effort, again by zone and fresh-saline distinction. At the top of each fresh-saline section, the estimated salinity concentration (in PPM) of the surface flows and groundwater are given. Surface flows include all canals and losses from link canals. Note that the PPM of surface water reaching the Sind is 25% higher than in the Punjab due to the effects of runoff and surface drainage effluent re-entering the rivers. Groundwater PPM's are the same as given in the IBMR input. Salt concentration of rainfall is assumed to be zero. Salt concentration of river losses (seepage to and from the river bed) is equal to groundwater PPM if the flow is from the zone to the river(negative) and equal to the surface PPM otherwise. River seepage varies by month, and often changes direction; the numbers presented here are net annual totals. Given the flow, and the PPM of that flow, the salt addition is computed on the basis of 1 MAF of water at PPM = 1000 contains 1.36 tons of salt. Note that tubewell operations and evaporation are not included in the

Table 5.1: Sources of Waterlogging Annual Flows (MAF)

Fresh Areas

	PMW	PCW	PSW	PRW	SCWN	SRWN	SCWS	Punjab	Sind
Gross area (million acres)	2.35	9.87	3.52	3.14	2.07	1.81	0.45	18.88	4.33
Surface Water flows									
Canal Diversions	3.85	24.84	6.85	6.06	6.14	4.44	1.40	41.60	11.98
Link canal losses	0.26	0.40	1.08	0.83	-	-	-	2.57	-
Rainfall Runoff	0.31	1.12	0.64	0.72	0.14	0.08	0.04	2.80	.25
Rainfall	1.74	6.37	3.65	4.09	0.77	0.43	0.21	15.84	1.42
Ground Water flows									
Pumpage									
Private Tubewells	1.89	18.35	4.30	7.10	1.50	0.46	1.56	31.64	3.52
Gov't Tubewells	1.24	0.86	3.13	1.52	0.30	0.02	0.04	6.76	.37
Seepage to Ground Water									
Rainfall	0.26	0.64	0.30	0.33	0.09	0.05	0.03	1.52	.17
Private Tubewells	0.53	4.18	0.98	1.57	0.42	0.11	0.43	7.26	.96
Gov't Tubewells	0.48	0.27	1.14	0.54	0.13	0.01	0.02	2.43	.16
Canals	0.86	4.89	1.36	1.00	0.93	0.54	0.02	8.11	1.66
Watercourses and Fields	1.09	6.21	1.83	1.64	2.10	1.15	0.49	10.77	3.75
Link canals	0.18	0.28	0.75	0.58	-	-	-	1.80	-
Rivers	-0.14	-0.32	-0.10	-0.70	0.30	0.13	0.03	-1.26	.47
Evaporation	0.55	0.86	0.40	0.18	1.49	1.28	0.15	1.99	2.92
Net Recharge	-0.41	-3.94	-1.57	-3.83	0.69	0.23	-0.56	-9.75	.36
Net Recharge/acre	0.18	-0.40	-0.45	-1.22	0.33	0.13	-1.24	-0.52	.08
Change in Water table (feet)	-0.70	-1.60	-1.79	-4.88	1.11	0.42	-4.13	-2.07	.28

Table 5.1 (continued)

Saline Areas

	PMW	PCW	PSW	SCWN	SRWN	SCWS	SRWS	Punjab	Sind
Gross area (million acres)	0.73	2.75	1.53	1.96	3.24	2.61	2.98	5.01	10.79
Surface Water flows									
Canal Diversions	1.52	7.78	3.02	6.32	8.97	8.03	9.95	12.32	33.26
Link canal losses	0.08	0.11	0.47	-	-	-	-	.66	-
Rainfall runoff	0.10	0.31	0.28	0.14	0.14	0.22	0.32	.69	.80
Rainfall	0.54	1.77	1.59	0.73	0.78	1.23	1.80	3.90	4.54
Seepage to Ground Water									
Rainfall	0.08	0.18	0.13	0.09	0.09	0.15	0.22	.39	.54
Canals	0.36	1.44	0.63	0.97	1.15	1.07	1.36	2.43	4.55
Watercourses and Fields	0.42	2.17	0.78	2.16	2.15	2.85	2.54	3.37	9.69
Link canals	0.06	0.08	0.33	-	-	-	-	.46	-
Rivers	-0.04	-0.09	-0.04	0.29	0.23	0.20	-	-0.18	.72
Evaporation	0.46	1.50	0.44	2.16	2.91	2.73	3.54	2.40	11.33
Drainage	-	.21	.04	.11	-	.13	-	.25	.24
Net recharge	0.41	2.06	1.35	1.24	0.72	1.40	0.57	3.82	3.93
Net recharge/acre	0.56	0.75	0.88	0.63	0.22	0.54	0.19	.76	.36
Change in Water table (feet)	2.24	3.00	3.53	2.11	0.74	1.80	0.64	3.05	1.21

Source: IBMR run WSISRG1A.

Figure 5.1a Water and Salt Inflows and Outflows to Groundwater Fresh and Saline Areas of Punjab

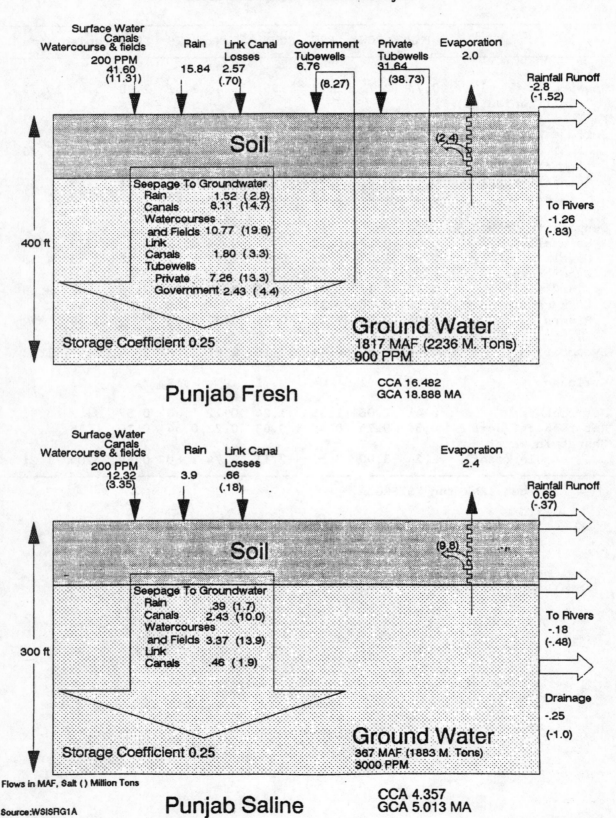

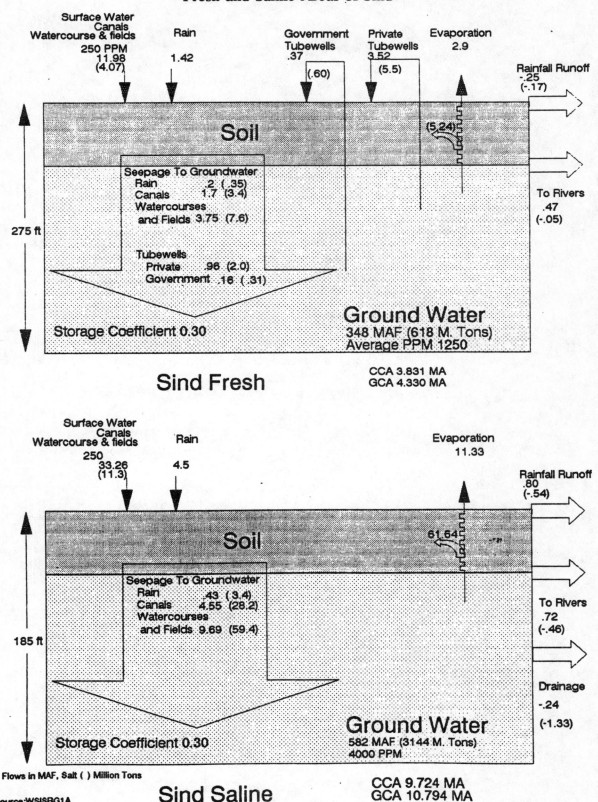

Figure 5.1b Water and Salt Inflows and Outflows to Groundwater
Fresh and Saline Areas of Sind

Table 5.2: Sources of Salinity Addition

Fresh Areas

	PMW	PCW	PSW	PRW	SCWN	SRWN	SCWS	Punjab	Sind
Salinity concentration of:									
Surface water	200	200	200	200	250	250	250	200	250
Groundwater	900	900	900	900	1200	1500	1000	-	-
Relevant flows (MAF):									
Canal Diversions	3.85	24.84	6.85	6.06	6.14	4.44	1.40	41.60	11.98
Link canal losses	0.26	0.40	1.08	0.83	-	-	-	2.57	-
Rainfall Runoff	0.31	1.12	0.64	0.72	0.14	0.08	0.04	2.80	.25
River Losses	-0.14	-0.32	-0.10	-0.70	0.30	0.13	0.03	-1.26	.47
Salt addition (million tons) from:									
Surface Water	1.12	6.86	2.15	1.87	2.09	1.51	0.48	12.01	4.07
Rain runoff	-.17	-.61	-.35	-.39	-.09	-.05	-.03	-1.52	-.17
River Losses	-.10	-.22	-.09	-.43	.04	.04	.01	-.83	.05
Net per zone	.85	6.04	1.71	1.05	2.04	1.50	.46	9.66	3.95
Net per acre	.36	.62	.48	.33	.96	.81	1.02	.51	.91

Saline Areas

	PMW	PCW	PSW	SCWN	SRWN	SCWS	SRWS	Punjab	Sind
Salinity concentration of:									
Surface water	200	200	200	200	250	250	250	200	250
Groundwater	3000	3000	3000	3000	4000	4000	4000	3000	4000
Relevant flows (MAF):									
Canal Diversions	1.52	7.78	3.02	6.32	8.97	8.03	9.95	12.32	33.26
Link canal losses	0.08	0.11	0.47	-	-	-	-	.66	-
Rainfall runoff	0.10	0.31	0.28	0.14	0.14	0.22	0.32	.69	.80
Rivers	-0.04	-0.09	-0.04	0.29	0.23	0.20	-	-0.18	.72
Drainage	-	-.21	-.04	-.11	-	-.13	-	-.25	-.24
Salt addition (million tons) from:									
Surface water	.44	2.15	0.95	2.15	3.05	2.73	3.38	3.53	11.31
Rain runoff	-.05	-.17	-.15	-.10	-.09	-.15	-.22	-.37	-.54
River seepage	-.11	-.21	-.16	-.19	-.20	-.06	.00	-.48	-.46
Drainage	-	-.86	-.14	-.60	-	-.73	-	-1.00	-1.33
Net per zone	.28	.91	.50	1.26	2.76	1.79	3.16	1.68	8.98
Net per acre	.38	.33	.33	.65	.85	.69	1.06	.33	.83

Source : IBMR run WSISRG1A

calculations; these do not represent a net import or export of salt to the zone, only a possible redistribution between groundwater and the topsoil.

The total salt addition for the Basin thus calculated amounts to 24.27 million tons (excluding Kabul-Swat canals), or .69 tons per acre of CCA per year. The only other attempt to measure this total was under the RAP, using a quite different method of comparing salt inflows at the rim stations and salt export below Kotri barrage. That estimate totaled 17 million tons for 1976-77.[1] The difference between the two estimates is due to the considerable increase in the canal diversions after Tarbela. It could also be due to the method used, or to the fact that the system is getting progressively more saline. As soil salinity increases, the runoff to rivers and hence canals within the basin becomes more saline, and more is retained.

From the salt addition sections, it can be seen that all salts are brought by surface water - namely canal water for irrigation. Rainfall runoff carries off a small amount. River seepage on balance exports salt from Punjab zones where rivers are deep and cut through the groundwater aquifer and imports salt in Sind zones where river reaches are shallow and sometimes even higher than the surrounding areas. The amount of salt moved by the river losses and gains are small relative to the total. The table also reveals the still small effect that subsurface drainage has on exporting salt: 28% of that brought by surface water in the saline areas of the Punjab, and 12% in Sind. Apart from drainage, there is little difference in total salinity buildup between the fresh and saline areas, as can be seen from the closeness of the per-acre figures at the bottom of each section. There is probably a very large difference, however, in where the salts end up after evaporation and tubewell operations are taken into effect. Tubewell operations, if large compared to surface flows, serve to redistribute some of the salts in the groundwater to the topsoil. Evaporation, far more intense in areas of high watertable, is the most important factor in soil salinity buildup as it bring groundwater and deep soil salts directly to the surface. In the following section, we attempt to quantity these effects in an effort to predict how changes to the irrigation system might affect both soil and groundwater salinity changes.

As discussed above, the addition of salts is a necessary evil associated with irrigation. Surface water will always bring substantial quantities of salts, and even with adequate drainage, some will be left behind. *Where*, is the question, the answer to which determines the sustainability of an irrigation system. Continuing deposits in the topsoil will lead to soil salinity and eventual cessation of crop growth. Continuing deposits in the groundwater will lead to eventual contamination. If the groundwater is already unusable as an irrigation source, then additional salts seeped or leached will have no immediate consequence for agricultural productivity. If the groundwater is currently usable, then additional salt seepage is a cause for concern. As Bresler put it in a review of irrigation and soil salinity:

> Water soluble salts entering the soil profile through the process of irrigation
> may accumulate in the root zone or may be leached out of this zone,

---

[1] Revised Action Program for Irrigated Agriculture, Volume II, <u>Irrigation, Drainage, and Flood Management</u>, pp. III-161.

may accumulate in the root zone or may be leached out of this zone, depending on the convective diffusive transport process and solute interactions within the soil. Understanding these processes is important in establishing management practices directed toward preventing the hazardous effect of salt on the plant and the soil.[1]

He goes on to describe several methods of estimating salt transport in soils, all requiring detailed information about the physical properties of the salt solute and the soils. Numerous methods can be found in the soil-science literature, none of which have been tested in the context of the Indus Basin. If they were, they would be not necessarily be helpful in meeting our broadbrush objectives of understanding which factors play the dominant role in leading to soil and groundwater salinity, and how they may be alleviated. Instead, we have developed a rather simplistic approach which may hold promise for quantifying the effects of surface irrigation, tubewell operations, and evaporation on the distribution of salts between the root zone and the groundwater. The method is derived more from a rudimentary understanding of the direction of movement of water and salts than from any theory of how and what rate they move. The method is suited for spreadsheet application, and can be modified to the extent better information about salt and water movements becomes available in the Indus Basin case. It deals with an entire agricultural year and makes heroic assumptions about how infiltration of surface water and other flows mix with available soil-water moisture. Most of the assumptions, however, can be modified if more information about the soil-physics becomes available. In the meantime, the method may be useful in identifying the sources of soil and groundwater salinity, and the measures which may be taken to reduce them.

As examples, calculations for the PSW fresh zone are given in Table 5.3. The section (1) of the table contains basic parameters and assumptions pertaining to the zone. GCA is gross commanded area of the zone with fresh groundwater. Depth of soil is set to six feet in all zones, since this is approximately equal to the maximum root depth of the most common crops in Pakistan. We loosely refer to this top six feet as "topsoil". Soil moisture storage coefficient is an assumption, and refers to the average over the crop year. The volume of water in the topsoil is computed by multiplying the GCA with the depth of soil and the soil moisture storage coefficient. The groundwater storage coefficient is the moisture content of the underlying aquifer, assumed to be .25 for Punjab zones, and .3 for Sindh. Aquifer depth has been estimated from various WAPDA reports, as has Beginning WTD, Salt Concentration of the surface water, Soil Water and Groundwater. Salt contents of the soil water are computed from the volume of water in the soil and salt concentration of soil water. Salts in the groundwater were derived from the volume of water contained in the aquifer and its concentration. Net recharge was described for Table 5.1.

The key calculations appear in the middle part of the spreadsheet (section 2). The flows affecting salt movement have been divided into six categories for fresh zones: rain seepage, rain runoff, canal water, evaporation, tubewell withdrawals, and river seepage. The

---

[1]Eshel Bresler, "Irrigation and Soil Salinity", in Dan Yaron, ed., <u>Salinity in Irrigation and Water Resources</u> New York: Dekker, 1981, pp. 65.

first line is input, taken from Table 5.1. The method differs somewhat for each, so each is described in turn.

Of total rainfall (4.293 MAF) 3.65 infiltrates the topsoil (the remaining 15% runs off). Salt concentration of rainfall is taken to be zero. This "clean" infiltration is assumed to mix with the 4.23 MAF of soil moisture with a PPM of 1280, to yield a volume of 7.88 MAF. This mixture contains the same amount of salt, 7.35 million tons (Mtons), as before, but the volume has increased temporarily. The seepage to groundwater from this component is estimated using the field efficiencies and groundwater contribution factors. It turns out to be 0.3 MAF for this zone. The salt seeped due to this flow is .28 Mtons (.28 = 0.3*7.35/7.88). In "Salt additions" rows, the results are summarized: rainfall seepage has a pure leaching effect, taking .28 Mtons of salt from the soil and depositing it in the groundwater. There is no net addition of salt from rainfall, although there could be if rainfall has a positive PPM.

Rainfall runoff, however, has an exportation effect. We assume that runoff remains on the soil long enough to acquire a PPM taken as twice that of surface water, in this case, 400 PPM. This assumption can be easily replaced by relating runoff PPM to soil moisture PPM, or by measurement of surface drainage effluent PPM. The assumption is not critical given the small quantities of runoff involved (particularly in Sind with its very low rainfall) relative to the total flows. The .64 MAF of runoff thus exports .35 Mtons of salt, all from the soil.

The column titled "Surface water" refers to all surface water entering the zone whether by irrigation canal or losses from the link canals transversing the zone. The total of 7.93 MAF canal diversions and link canal losses infiltrates the topsoil, of which 3.94 will eventually seep to groundwater. The new soil-water volume is 12.15 MAF, containing 9.51 Mtons of salt. The 3.94 MAF of seepage leaches a total of 3.08 Mtons of salt to the groundwater. This component is the main reason for the existence of saline groundwater in areas of poor natural sub-surface drainage and decades of continued irrigation. Note the effect on the soil: 2.15 Mtons of salt were brought by the surface water, 3.08 Mtons were leached, resulting in a net leaching of .93 Mtons from the topsoil. Other things equal, then, surface irrigation has a leaching effect on soil.

Evaporation from groundwater has been discussed earlier, and the numbers are taken straightforwardly from Tables 5.1. In this zone, evaporation is .40 MAF at a concentration equal to that of groundwater (900 PPM). The effect is negative leaching: salts are removed from groundwater through the capillary action involved in evaporation, and left in the soil. In other zones with more shallow watertables, evaporation is much higher, and the soil salinity buildup due to it are much more severe.

Tubewell withdrawals total 7.43 MAF (private plus government) of which 2.12 is returned as seepage. Withdrawals bring all salts to the surface at a PPM equal to that of groundwater (total salt withdrawn is 9.09 Mtons). The 2.12 MAF of seepage contains 2.99 Mtons of salt. Taking into account the salt initially withdrawn by tubewells, the net effect is to increase soil salts by 6.10 (9.09-2.99) Mtons and reduce groundwater salts by 6.10 (-9.09+2.99). The result of this component agrees with that observed in practice: heavy use of groundwater for irrigation will tend to increase soil salinity.

Finally, river seepage exports a small amount of salt from this zone's groundwater over the year. The total is taken from Table 5.2 and is the sum of the monthly flows (not reported). When seepage is from the groundwater to the river, the PPM is equal to that of the groundwater. When the direction is reversed, the PPM is equal to that of the surface water.

Table 5.3: Simulated Salt Distribution for PSW Fresh Zone.

Section 1 : Basic Parameters

GCA	3.52 MA	Groundwater storage Coefficient	0.25
Depth of Soil	6.00 Feet	Aquifer depth	400.00 Feet
Soil Moisture storage coefficient	0.20	Beginning WTD	11.70 Feet
Volume of water in Soil	4.23 MAF	Volume of GW	341.90 MAF
Salt concentration of soil water	1280 PPM	Salt Concentration	900.00 PPM
Salts in soil water	7.36 Mill. tons	Salts in Groundwater	418.48 MAF
Soil Weight	85.00 Lbs	Net Recharges (excluding evaporation)	-1.17 MAF

Section 2 : Salt Transport

(all flows are in MAF and salts in Million tons)

	Rain[1]	Rain[1] Runoff	Surface[2] Water	Evaporation	Tubewell Pumpage	River[7] seepage	Row Sum
Total flow	3.65	.64	7.92	0.40	7.43	-0.10	
Seepage to Groundwater	0.30	-	3.94	-	2.12	-0.10	
Salt Concentration (PPM)	0	400	200	900	900	900	
Salt addition soil-water	-	-0.35	2.15	0.49	9.09	-	
Volume of soil-Water	7.88	-	12.15	-	11.66	-	
New salt contents of soil water[3]	7.36	-	9.51	-	16.45	-	
Salts seeped to GW[4]	.28	-	3.08	-	2.99	-0.09	
Salt additions							
To soil[5]	-0.28	-0.35	-0.93	0.49	6.10	-	5.03
To Groundwater[6]	0.28	-	3.08	-0.49	-6.10	-0.09	-3.32
To soil & GW system	-	-0.35	2.15	-	-	-0.09	1.71
Salts per acre	-	-0.10	0.61	-	-	-0.03	0.49

Section 3: Salt Balance

	Soil			Groundwater		
	Mtons	PPM	EC[8]	Volume MAF	Mtons	PPM
Beginning of the Year	50.08	1280	2.0	341.9	418.48	900
End of the year	55.11	1408	2.2	340.7	415.16	896

Notes:
1. Total rainfall is 4.293 MAF 15% of which (.64) goes to runoff and the rest (3.65) infiltrates into the soil.
2. Includes canal diversions at the canal head (6.85 MAF) and the losses in the link canals(1.076).
3. Original salt contents from section 1 plus the salts brought by this component.
4. Salts moved by the seepage to GW with concentration of the soil-water mix e.g. in case of surface water 9.51/12.15*3.94.
5. Salts addition to soil-water minus the salts seeped to GW.
6. Net salt addition, salts seeped minus salt withdrawl in case of evaporation and tubewell pumpage.
7. Rivers move salts directly in and out of groundwater reservoir.
8. $Ec = PPM/640$

The Salt Additions rows reflect the total impact of salt movement by category of flow on the zone. Note that the effects cancel out for rainfall seepage, evaporation, and tubewell withdrawals for the total system of soil and groundwater. Rain runoff always exports salt, and canal water always imports salts. River seepage may be in either direction, or non-existent for zones not bordering a river. The row sums reflect the total impact on soil and groundwater. In PSW fresh, the net effect on soil salt is positive, while the effect on groundwater is negative. Note that this method clearly identifies which factors are most important in salt movement: tubewell operations are far and away the most important contributor to soil salinity in this zone; in their absence, the other factors would leach about one million tons from the topsoil annually.

In section (3), the results are computed for the annual effect on soil and groundwater. With the small net addition, total salts in the topsoil (not just those in the soil-water mixture) increased to 55.11 million tons, raising the PPM of the soil-water mixture to 1408. The EC of the soil increases from 2.0 to 2.2. Groundwater PPM falls by a small amount, given the net reduction groundwater salts.

Table 5.4 reports the same exercise for the SRWS saline zone, one of the worst affected by salinity. There is no tubewell activity, and no drainage. The main factors in salt movement are canal water and evaporation. Soil salinity is high to begin with at an EC of 8.8 and a PPM of 5620 for the soil-water mixture. Water table depth average 6 feet (or less). The relatively high canal diversions bring 3.38 Mtons of salt at a PPM of 250. The seepage of 3.89 MAF contains 8.84 Mtons of salt, resulting in a net reduction of soil salt of 5.46 Mtons. Others things equal, then, the high canal water deliveries have a strong "cleansing" effect on the soil.

Other things are not equal. Evaporation from groundwater is very high (3.54 MAF, more than one foot per acre), and with a high initial groundwater PPM of 4000, leaves 19.24 Mtons of salt near the surface. This effect more than cancels the leaching from rain and canal water, resulting in a total soil salinity addition of 12.44 Mtons. This result is due to a combination of high watertable and high potential evaporation of this zone. Soil salinity rises markedly to EC = 9.4. Since potential evaporation is not a variable influenced by policy, the only apparent prescription to reduce evaporation and hence soil salinity is drainage.

Table 5.4: Simulated Salt Distribution for SRWS Saline Zone.

Section 1 : Basic Parameters

GCA	2.98 MA	Groundwater storage	
Depth of Soil	6.00 Feet	Coefficient	0.30
Soil Moisture storage coefficient	0.20	Aquifer depth	185.00 Feet
		Beginning WTD	6.00 Feet
Volume of water in Soil	3.58 MAF	Volume of GW	160.03 MAF
Salt concentration of soil water	5620 PPM	Salt Concentration	4000.00 PPM
		Salts in Groundwater	870.54 MAF
Salts in soil water	27.33 Mill. tons	Net Recharges (excluding evaporation)	0.57 MAF
Soil Weight	85.00 Lbs		

Section 2 : Salt Transport
(all flows are in MAF and salts in million tons)

	Rain[1]	Rain[1] Runoff	Surface Water	Evaporation	River seepage	Sub-surface drainage	Row Sum
Total flow	1.79	.32	9.95	3.54	-	-	
Seepage to Groudwater	0.22	-	3.89				
Salt Concentration (PPM)	-	500	250	4000			
Salt addition soil-water	-	-0.22	3.38	19.24	-		
Volume of soil-Water	5.37	-	13.53				
New salt contents of soil water	27.33	-	30.71				
Salts seeped to GW[2]	1.12	-	8.84		-		
Salt additions							
To soil	-1.12	-0.22	-5.46	19.24	-	-	12.44
To Groundwater	1.12	-	8.84	-19.24	-	-	-9.28
To soil & GW system	-	-0.22	3.38	-	-	-	3.16
Salts per acre	-	-0.07	1.14	-	-	-	1.06

Section 3 : Salt Balance

	Soil			Groundwater		
	Mtons	PPM	EC	Volume	Mtons	PPM
Beginning of the Year	186.03	5620	8.8	160.0	870.56	4000
End of the year	198.47	5994	9.4	160.6	861.36	3943

Notes:
1. Total rainfall is 2.116 MAF 15% of which (.32) goes to runoff and the rest (3.65) infiltrates into the soil.
2. Salts moved by the seepage to GW with concentration of the soil-water mix e.g. in case of surface water 30.71/13.53*3.89.

Dynamics of Salt Accumulation

The method described above is useful for approximating the distribution of salts between topsoil and groundwater as a function of initial conditions and water flows. For fresh areas with relatively low watertables, it emphasizes the need for sufficient surface water supplies, which flush salts from the topsoil to the groundwater, to offset the effects of tubewell withdrawals, which effectively redistribute salts from the groundwater to the topsoil. In saline areas, the method quantifies the onerous effects of high evaporation from groundwater, which, like tubewell withdrawals, results in groundwater salts being deposited at or near the surface through capillary action.

The results of the model, while highlighting the importance of the various flows, are highly sensitive to the initial conditions assumed, in particular, the watertable depth (which, given potential evaporation, determines actual evaporation), and the initial salinity levels of the topsoil and groundwater. The reader may have noted that the method is quite amenable to dynamic simulation of salt distribution: one needs only assume values for next year's surface inflows and tubewell withdrawals (if applicable), and update the WTD, evaporation, and salinity levels from the previous solution.

In the two examples presented in the last section, PSW Fresh and SWRS Saline areas, it is clear that quite large changes can be expected when the model is solved again in iterative fashion. Neither zone was in equilibrium for the conditions assumed: PSW showed a net discharge large enough to lower the watertable by about a foot over the course of the year, and SRWS showed the opposite tendency: positive recharge in an area where the watertable was already virtually at the surface. Hence evaporation will change (lower in PSW; even higher in SRWS). Moreover, the very large accumulations of salt in the topsoil in both areas imply that those flows which "flush" will be more effective in the following year. This is because the seepage from the soil-water mixture will have a higher salt concentration, and hence more salts will be leached.

Two conditions must be met for the model to produce stable, year-to-year results. One, the area must be in equilibrium with respect to watertable level, which will occur only if net recharge is zero. Two, the initial estimates of soil and groundwater salinity must be exact. Neither condition is met. Net recharge is not zero in either of the two example areas, and probably no where in the Indus Basin. The data from which the soil and groundwater salinities were derived are more than ten years old, and were only crude approximations when they were collected. In fact, the results strongly suggest that soil salinities (and possibly groundwater salinities as well) are sharply underestimated.

Table 5.5 reports the salient results from running the salt distribution model over several years with all flow data (except evaporation) held constant. Consider first zone PSW Fresh. The negative recharge (1.17 MAF in the first year) declines slowly as the watertable falls and evaporation reduces. By 1995,[1] evaporation from groundwater is near zero (.08 MAF). Soil salinity takes a large jump in the first year, but the rate of increase is slowed, partly because of the lower evaporation, but mostly because of the greater effectiveness of canal water in flushing. Had the initial estimates of soil salinity been higher, the rate of change would be much slower. This effect corresponds to the rapid drop in salts deposited in the topsoil. From five Mtons, they also approach zero by 1995. The change in groundwater salinity is much less drastic, leading us to believe that these estimates have more validity. After a small drop in the second year, they slowly rise through 1995, as can be expected.

Table 5.5
Dynamics of the Salt Transport Model

PSW Fresh:

			salt addition to:		salinity of (PPM)	
Initial	WTD	Evaporation	soil	GW	soil	GW
Condition	11.70				1,280	900
1988	12.68	0.40	5.03	-3.32	2,156	897
1989	14.36	0.29	2.13	-0.42	2,528	900
1990	15.95	0.22	0.91	0.80	2,687	905
1991	17.48	0.18	0.41	1.31	2,757	912
1992	18.98	0.14	0.20	1.51	2,792	919
1993	20.44	0.12	0.12	1.59	2,813	926
1994	21.89	0.10	0.09	1.62	2,828	933
1995	23.31	0.08	0.08	1.63	2,843	940

SRWS Saline:

			salt addition to:		salinity of (PPM)	
Initial	WTD	Evaporation	soil	GW	soil	GW
Condition	6.00				5,620	4,000
1988	5.16	3.54	12.47	-9.31	8,183	3,943
1989	5.00	4.03	10.74	-7.58	10,391	3,907
1990	4.98	4.10	7.40	-4.24	11,912	3,887
1991	4.98	4.11	4.91	-1.75	12,920	3,879
1992	4.98	4.11	3.26	-0.10	13,590	3,878
1993	4.98	4.11	2.19	0.97	14,040	3,883
1994	4.98	4.11	1.49	1.67	14,347	3,891
1995	4.98	4.11	1.05	2.11	14,562	3,900

In SRWS, the rate of change in the watertable is much slower. We assumed six feet as an initial condition, and the model quickly stabilizes at just under five. At this level, all of the recharge evaporates - the epitomal definition of waterlogged soils. Salts continue to build up in the topsoil despite the more effective leaching by canal water. In two cycles, soil salinity nearly doubles, and nearly triples in seven. Groundwater, however, progressively improves in quality because salts are being exported upward to the topsoil.

---

[1] Although we have keyed the initial conditions to 1988 data to the extent the information permits, and solved the model sequentially for seven years, the results should not be taken as a prediction for the actual year 1995. Given the uncertainties about the inital salinities, the results may be more descriptive of an earlier year, say 1990.

By the year termed 1995, both examples have nearly stabilized, although for different reasons. Results for this year, for all zones and fresh/saline distinction, are presented in Table 5.6. In the fresh areas, soil salinity continues to build up everywhere at moderate rates except PMW and PCW, which are approximately in balance. However, the effect of the prior years' buildup was to substantially increase soil salinity in all zones. In the Sind zones, the numbers indicate EC's sufficiently high to adversely affect crop yields. For the reference year 1995, the model shows that most salts in the fresh areas end up in the groundwater. But given the size of the aquifer, the total impact, at least in the short run, is not serious. Of course, the model does not include the lateral intrusion of saline groundwater, which is by far the more serious threat to fresh groundwater.

In most saline zones, additional salts are about evenly distributed between soil and groundwater, with the Sind zones, on average, exhibiting much higher buildups in both. Soil EC's in all zones, but in the Sind in particular, are probably high enough to choke off plant growth. Groundwater PPM's in the saline areas show a slight drop everywhere. This is due to two factors: one, the volume of the saline aquifers is ever increasing as the watertable continues to rise, and two, the salinity of the recharge is, on average, lower than that of the groundwater.

For the Basin as a whole, the model shows that, of the 24.4 million net tons of salt brought into the system,[1] 24.6% is deposited in the soil, and 75.4% in the groundwater. Slightly more than half of the total ends up in the fresh groundwater aquifers, which explains their declining quality. The effects of soil salinity on crop yields is already severe in many parts of the Basin, and in all likelihood will get worse. Although increasing salinity of saline groundwater is of no consequence, that of fresh groundwater, which supplies about one-third of total crop requirements in the fresh areas, certainly is.

The sheer quantity of salts (.7 tons per acre) to be dealt with is sobering. If this irrigation system is to survive, salinity must either be exported via drainage or controlled through water management to control watertables, and system rehabilitation to reduce losses. In the following chapter, we look at the dimensions of the water management problem, and then employ the IBMR to investigate the potential for controlling waterlogging and salinity.

---

[1] Net of salts drained away and flushed by rainfall runoff.

Table 5.6
Distribution of Salinity Change

Fresh Areas

	PMW	PCW	PSW	PRW	SCWN	SRWN	SCWS	Punjab	Sind
Gross area (M acres)	2.35	9.87	3.52	3.14	2.07	1.81	0.45	18.88	4.33
Annual change in Soil salinity (Million Tons) due to:									
Rain runoff	-0.17	-0.61	-0.35	-0.39	-0.09	-0.05	-0.03	-1.52	-0.17
Rain seepage	-0.45	-1.80	-0.61	-0.81	-0.41	-0.30	-0.12	-3.68	-0.83
Surface Water	-1.65	-11.03	-3.82	-4.62	-3.70	-2.68	-0.95	-21.12	-7.33
Tubewell pumpage	1.89	13.52	4.76	6.10	0.74	0.29	1.14	26.27	2.17
Evaporation	0.36	0.29	0.10	0.01	3.59	3.06	0.01	0.77	6.65
Net change	-0.01	0.36	0.08	0.29	0.13	0.31	0.06	0.72	0.49
Annual change in Groundwater salinity (Million Tons) due to:									
Rain seepage	0.45	1.80	0.61	0.81	0.41	0.30	0.12	3.68	0.83
Surface water	2.77	17.90	5.97	6.49	5.79	4.19	1.42	33.13	11.40
Tubewell pumpage	-1.89	-13.52	-4.76	-6.10	-0.74	-0.29	-1.14	-26.27	-2.17
Evaporation	-0.36	0.29	-0.11	-0.01	-3.59	-3.06	-0.01	-0.77	-6.65
River seepage	-0.10	-0.22	-0.09	-0.43	-0.04	-	0.01	-0.83	0.05
Net change	0.86	5.67	1.63	0.77	1.91	1.14	0.40	8.93	3.46
Net annual change per acre (Tons):									
Soil salt	-.005	0.04	.023	.092	.061	0.17	0.13	0.04	0.12
GW salt	.367	0.57	.464	.244	.923	.632	0.89	0.47	0.79
Soil EC (Millimohs per cm):									
"1988"	2.0	2.0	2.0	2.0	2.0	2.0	2.0	2.0	2.0
"1995"	3.2	5.0	4.4	6.1	6.7	8.0	7.7	4.9	7.4
Groundwater PPM:									
"1988"	900	900	900	900	1200	1500	1000	900	1250
"1995"	920	941	940	984	1219	1497	1131	945	1328

Table 5.6 (continued)

Saline Areas

	PMW	PCW	PSW	SCWN	SRWN	SCWS	SRWS	Punjab	Sind
Gross area (M acres)	0.73	2.75	1.53	1.96	3.24	2.61	2.98	5.01	10.79
Annual change in soil salinity (Million Tons) due to:									
Rain runoff	-0.05	-0.17	-0.15	-0.09	-0.09	-0.15	-0.22	-0.37	-0.55
Rain seepage	-0.53	-1.42	-0.74	-1.34	-1.43	-1.98	-2.80	-2.69	-7.56
Surface water	-2.82	-11.95	-5.76	-15.69	-16.12	-18.81	-17.68	-20.53	-68.29
Evaporation	3.49	14.14	7.06	17.63	19.16	21.58	21.74	24.69	80.11
Net change	0.09	0.59	4.12	0.51	1.52	0.64	1.05	1.09	3.71
Annual change in Groundwater Salinity (Million Tons) Due to:									
Rain seepage	0.53	1.42	0.74	1.34	1.43	1.98	2.80	2.69	7.56
Surface water	3.26	14.10	6.70	17.84	19.17	21.54	21.06	24.06	79.60
Evaporation	-3.49	-14.14	-7.06	-17.63	-19.16	-21.58	-21.74	-24.69	-80.11
River seepage	-0.11	-0.21	-0.16	-0.19	-0.20	-0.06	-	-0.48	-0.46
Drainage	-	-0.83	-0.14	-0.54	-	-0.68	-	-0.96	-1.22
Net Change	0.18	0.35	0.09	0.82	1.23	1.19	2.12	0.62	5.37
Net annual change per acre (Tons):									
Soil salt	0.13	0.21	0.27	0.26	0.47	0.24	0.35	0.22	0.34
GW salt	0.25	0.13	0.06	0.42	0.38	0.46	0.71	0.12	0.50

	PMW	PCW	PSW	SCWN	SRWN	SCWS	SRWS	Punjab	Sind
Soil EC (Millimohs per cm):									
"1988"	4.7	4.7	4.7	6.3	6.3	6.3	6.3	4.7	6.3
"1995"	12.2	14.4	12.5	23.3	21.6	21.7	22.7	13.5	22.2
Groundwater PPM									
"1988"	3000	3000	3000	4000	4000	4000	4000	3000	4000
"1995"	2951	2914	2911	3814	3888	3846	3900	2918	3868

Seepage of surface water includes seepage from link canals, irrigation canals, watercourses and fields.

CHAPTER 6
OPTIONS FOR GROUNDWATER CONTROL

Introduction

From the analysis of the last chapter, several conclusions emerge:

a) waterlogging is a serious problem in much of the saline areas of the Indus Basin, and it undoubtedly is worsening rapidly.

b) The main cause of waterlogging in saline areas is the application of surface irrigation water combined with the lack of natural subsurface drainage.

c) High watertables induce soil salinity in saline groundwater areas, largely through the influence of groundwater evaporation.

d) In fresh groundwater areas, soil salinity is probably building up rapidly as a result of intensive tubewell operations. However, given the low watertables, it can be rectified by reducing groundwater withdrawals while leaching with surface water.

e) Tubewell operations are inducing rapid lowering of the watertables which may soon reach the point where it is no longer technically feasible to continue withdrawing with existing pumps, and/or it is no longer economic.

f) The growing differential in watertable depths between neighboring fresh and saline groundwater areas implies that saline intrusion will occur sooner or later and contaminate fresh groundwater supplies.

Project planners have long been aware of these processes, and several projects have been designed to alleviate them. Beginning in the late 1960's, Salinity Control and Reclamation Projects (SCARP's) sought to lower watertables through the introduction of government-operated tubewells. Most were located in FGW areas, and the water was returned to the system for re-use. Most water-system oriented projects such as ISRP and CWM, as well as the SCARPs have surface drainage components designed to reduce erosion and seepage from excessive rainfall, and involve some canal lining to reduce losses. The Fourth Drainage Project was one of the first to recognize that serious waterlogging problems in saline areas must be addressed through sub-surface drains. The LBOD includes most of the above measures plus the construction of an outfall drain to dispose of the effluent to the sea.

Apart from the completed SCARP's, the WSIP includes various phases of all the above-mentioned projects. However, the WSIP program, if achieved, will still fall far short of that required to alleviate the groundwater problems. Only 15% of the saline area will be served by sub-surface drainage, system losses will be reduced only marginally, and waterlogging and salinity buildup will continue virtually unchecked. Free of the financial and political constraints under which the WSIP operated, we turn to a simulation of projects which may have the desired impact.

Benefits from such projects are assumed, in project documents and by the WSIP, to accrue from two sources: the gain in yield from lowered watertables and soil salinity, and the gain in production from the water saved by the project. Unfortunately, the IBMR is not currently set up to simulate endogenous yield improvement as a result of drainage. Conceptually, it can be done, although at a severe cost in model size and solution cost. Practically, we simply do not have sufficient confidence in the yield improvement projections available to implement them. Therefore we cannot at present simulate benefits from yield improvements with any degree of confidence. We can, however, look at the second category of benefits - that from water loss reductions.

In this chapter, we describe several types of projects and policy interventions, and simulate, to the best of the model's ability, the impacts of:

1- subsurface drainage
2- water allocation policies
3- canal lining
4- watercourse improvements

But first, we will recap the dimensions of the problem, and the scope of the various means available to deal with.

Dimensions of the Policy Problem

The single most important variable leading to waterlogging and salinity problems is annual net recharge to groundwater. Where this variable has been positive over an extended period of time, waterlogging has occurred, combined with soil salinity induced by high evaporation from groundwater. Where recharge has been continuously negative, the aquifer has been mined, and soil salinity has been induced through tubewell operations. Annual net recharge varies widely across the Indus Basin, from sharply positive to sharply negative, as shown in the first line in each fresh-saline section of Table 6.1.

First, let us focus on canal diversions, which bring the vast bulk of irrigation water to the system. Theoretically, there exists a pattern of diversions which could bring recharge into balance in each fresh/saline zone category. Table 6.1 reports the steps involved in determining such a pattern. When canal water is diverted to a given zone, a given percentage is lost in the main canals, branches and distributaries (e.g., 31.6% for PMW), of which 70% infiltrates to groundwater as "seepage".[1] Of the remaining water reaching the watercourses, additional losses occur based on the watercourse command efficiency, of which 80% results in seepage.

---

[1] Seepage and loss factors were estimated under the RAP and tested extensively for consistency with earlier versions of the IBM. See RAP, <u>Irrigation, Drainage, and Flood Management</u>, op. cit., and Duloy and O'Mara, op. cit.

Table 6.1: Changes in Canal Diversions to Achieve Groundwater Balance

Fresh Areas

	PMW	PCW	PSW	PRW	SCWN	SRWN	SCWS
Net Recharge	-0.41	-3.94	-1.57	-3.83	0.69	0.23	-0.56
Canal Efficiency	0.684	0.725	0.712	0.765	0.783	0.818	0.809
Watercourse Eff.	0.485	0.555	0.534	0.547	0.453	0.55	0.452
Canal Diversions	3.85	24.84	6.85	6.06	6.14	4.44	1.40
Losses from "	2.58	14.84	4.24	3.52	3.99	2.45	0.89
Total seepage	1.94	11.19	3.20	2.68	3.03	1.87	0.68
Desired seepage	2.35	15.13	4.77	6.51	2.35	1.64	1.24
Required diversions	4.67	33.58	10.21	14.74	4.75	3.90	2.54
pct. change	21.3%	35.2%	49.2%	143.2%	-22.7%	-12.3%	81.4%
change in MAF	0.82	8.74	3.37	8.68	-1.39	-0.55	1.14

Saline Areas

	PMW	PCW	PSW	SCWN	SRWN	SCWS	SRWS
Net Recharge	0.41	2.06	1.34	1.24	0.72	1.40	0.57
Canal Efficiency	0.684	0.725	0.712	0.783	0.818	0.809	0.805
Watercourse Eff.	0.485	0.555	0.534	0.453	0.550	0.452	0.550
Canal Diversions	1.52	7.78	3.02	6.32	8.97	8.03	9.95
Losses from "	1.02	4.65	1.87	4.08	4.93	5.09	5.54
Total seepage	0.77	3.51	1.41	3.12	3.78	3.92	4.24
Desired seepage	0.36	1.45	0.07	1.89	3.06	2.51	3.67
Required diversions	0.71	3.21	0.14	3.82	7.26	5.15	8.60
pct. change	-53.6%	-58.7%	-95.2%	-39.6%	-19.1%	-35.8%	-13.5%
change in MAF	-0.82	-4.57	-2.87	-2.50	-1.71	-2.88	-1.34

Summary of required changes in canal diversions:

Total Punjab Fresh	21.61
Total Sind Fresh	-2.53
Total Fresh	19.08
Total Punjab Saline	-8.26
Total Sind Saline	-8.43
Total Saline	-16.69
Punjab net	13.35
Sind net	-10.96
Basin net	2.38

Combined, these factors reveal the very low overall efficiency of the irrigation system: only 35-40% of diverted canal water is available for crop consumption, nearly half is lost to groundwater, and the rest either evaporates or is drained off by the surface drainage system. "Total seepage" in the table is that component of canal diversions which reaches the groundwater. "Desired seepage" is that which would permit zero net recharge, and "Required diversions" are those necessary to reduce or increase seepage to achieve the required recharge. Finally, the percentage changes in diversions and their absolute magnitudes are reported.

In all fresh areas except the northern Sind, substantial increases in diversions would be required; in all other areas, reductions would be necessary - up to 95% in PSW saline. In total, about 19 MAF in additional diversions would be required in the fresh areas to offset the ongoing aquifer mining, and about 16.7 MAF in reduced diversions to saline areas would be required to arrest waterlogging. For the Basin as a whole (excluding NWFP) about 2.4 MAF in additional supplies would be required to bring all areas into balance.

These calculations assume, unrealistically, that other things would remain constant. As the IBMR has repeatedly demonstrated, they do not. When canal diversions change, virtually all other variables change as well, notably cropping patterns and intensities, and tubewell operations. Clearly, with additional surface supplies, tubewell withdrawals would decline. These results should only be considered as indicative of the directions and crude magnitudes which would be required. Given the large magnitudes, and the associated disruption costs, they are clearly out of the question. Politically, they are also out of the question given the sharply divergent effect they would have on the two provinces concerned (a reduction in diversions of 11 MAF to Sind, and an increase of 13.4 to Punjab). Although clearly an important variable to which we will later return, canal diversion policies alone cannot provide a solution.

In the fresh areas, which would require 19 MAF of increased diversions to offset the negative recharges, an alternative measure is to reduce tubewell withdrawals. Table 6.2 reports calculations, similar to those above, of how much tubewell withdrawals would need to be reduced. Total tubewell withdrawals include private and government. "Seepage" is the combined losses to groundwater, and "net withdrawals" is the difference. It can be seen that it is possible to achieve zero recharge in the fresh areas by reducing tubewell operations - by between 57 and 91%. (The northern Sindh areas, having a small positive recharge, would need to increase withdrawals to some extent). The required reductions for all fresh zones amount to 27.5 MAF.

Again, this option appears to be out of the question: severe output effects would be incurred, and farmers would have no motive for cooperating with such measures in the absence of compensation from increased surface supplies. In fact, their motivations for investing in tubewells in the first place was the unreliable, untimely, and inadequate canal water supplies. However, some combination of increased canal diversions coupled with reduced tubewell withdrawals which would halt aquifer mining is obviously feasible, and ought to be achievable.

The saline areas present different problems and require different solutions. As noted above, it is theoretically possible to reduce canal diversions to eliminate the upward pressure on watertables. But, as with tubewell reductions, such a policy would entail unacceptable output and income losses. The obvious alternative, alluded to above, and simulated with the IBMR below, is to increase the efficiency of delivery. As revealed in Table 6.1, the overall efficiencies (the product of canal efficiencies, watercourse and field efficiencies, and seepage factors) range from about 33 - 45%, resulting in 42 - 50% of diversions at the canal head ending up as seepage to groundwater.

Table 6.3 reports a simplified analysis of the potential effects of improving the efficiencies on seepage and net recharge. The top portion, labeled "present situation" reproduces the relevant results from Table 6.1 for the saline areas, plus the existing canal and watercourse efficiencies. In the next two sections, seepage and net recharge are recalculated assuming that canals and watercourses, respectively, are remodeled to achieve the reported efficiencies. The combined seepage rate, computed as total seepage to groundwater divided by total canal diversions, reflects the overall impact of each intervention. Note that both canal and watercourse improvements reduce this rate, as well as net recharge (all other items contributing to net recharge held constant), but neither can totally eliminate the net recharge.

The last section of Table 6.3 assumes that both canals and watercourses are improved. In most zones, total seepage declines by about one-third, and recharge is cut in half. But only in the northern Sindh zones can these measures eliminate positive recharge and thus the upward pressure on the watertable. The last row reveals that probably no realistic project options exist which alone can eliminate the problem in most saline areas. However, this exercise assumes that canal diversions are held constant, implying that the improvements result in substantially greater water supplies at the root zone. They thus provide a means of reducing canal supplies to saline areas without incurring output or income losses, while making water available for increased diversions to the fresh areas.

These exercises have shown that no panacea exists for correcting the imbalances leading to, or perpetuating, waterlogging and salinity. But they also show that the scope exists, through some carefully conceived combination of policies and projects, to redress the imbalances. We now use the IBMR to attempt to quantify the ability of some of those interventions to achieve an environmentally sound, yet productive, irrigation system.

Sub-Surface Drainage

As noted in chapter 4, at least 7.7 million acres (12.2 million if NCA's figures are taken) are severely waterlogged. In the last section we noted that measures to reduce recharge to zero (but not necessarily to eliminate waterlogging) would require drastic measures which would probably be both unproductive and politically unacceptable. For many areas, drainage is probably the only recource if full productivity is to be restored.

Recent World Bank cost estimates for sub-surface drainage are US$ 1,200 per acre, although it is believed that this could be reduced to $1,000 if drainage were undertaken within a well-managed program approach rather the peicemeal efforts attempted to date. Even at the lower figure, the total cost of waterlogged area drainage is US$ 7.7 billion based on the 7.7 million acre figure - nearly twice the cost of the WSIP expected plan. For this reason, the WSIP recommended sub-surface drainage only in severely affected, non-rice areas. It is not apparent from the plan documentation how large these areas refer to, but it is noteworthy that LBOD, the largest drainage project in Pakistan to date, would be cut off after Stage I (about one million acres). Non-rice areas amount to about half the saline area, and it seems reasonable that these should be singled out. From Table 4.2 it was evident that cotton (the primary source of export earnings) and sugarcane (a major item in the import bill) are strongly affected by high water tables. Wheat and rabi fodder are not particularly sensitive to high water tables. Thus maximum gains from sub-surface drainage can probably be obtained from 3-5 million acres of non-rice, saline land, costing about $US 3-6 billion.

Table 6.2:
Reductions in Tubewell Withdrawals Required to Achieve Groundwater Balance

	PMW	PCW	PSW	PRW	SCWN	SRWN	SCWS
Net recharge	-1.43	-8.39	-3.69	-5.95	0.14	0.11	-1.01
TW withdrawals	3.13	19.21	7.43	8.62	1.80	0.48	1.60
Seepage from "	1.01	4.45	2.12	2.11	0.55	0.12	0.45
Net withdraw.	2.11	14.77	5.31	6.51	1.25	0.37	1.15
Required change	2.11	10.92	5.17	7.87	-0.20	-0.15	1.41
% change	-67.5%	-56.8%	-69.5%	-91.3%	11.0%	31.1%	-88.2%

Table 6.3:
Potential of Improved Delivery Efficiencies for Achieving Groundwater Balance

	PMW	PCW	PSW	SCWN	SRWN	SCWS	SRWS
Present situation:							
Canal efficiencies	0.684	0.725	0.712	0.783	0.818	0.809	0.805
Watercourse efficiencies	0.485	0.555	0.534	0.453	0.550	0.452	0.550
Combined seepage rate	50.3%	45.1%	46.7%	49.5%	42.2%	48.8%	42.6%
Total seepage	0.77	3.51	1.41	3.12	3.78	3.92	4.24
Net recharge	0.41	2.06	1.34	1.24	0.72	1.40	0.57
With Canal lining:							
Canal efficiencies	0.90	0.93	0.92	0.93	0.94	0.94	0.91
Watercourse efficiencies	0.48	0.56	0.53	0.45	0.55	0.45	0.55
Combined seepage rate	44.1%	38.0%	39.9%	45.7%	38.0%	45.3%	39.2%
Total seepage	0.67	2.96	1.20	2.89	3.41	3.64	3.90
Net recharge	0.32	1.51	1.14	1.00	0.35	1.12	0.23
With Watercourse Improvements:							
Canal efficiencies	0.68	0.72	0.71	0.78	0.82	0.81	0.81
Watercourse efficiencies	0.58	0.64	0.62	0.55	0.63	0.55	0.63
Combined seepage rate	45.1%	40.4%	41.8%	43.4%	36.9%	42.5%	37.3%
Total seepage	0.69	3.14	1.26	2.74	3.31	3.41	3.71
Net recharge	0.33	1.70	1.20	0.85	0.24	0.90	0.04
With Both Canal and Watercourse Improvements:							
Canal efficiencies	0.90	0.93	0.92	0.93	0.94	0.94	0.91
Watercourse efficiencies	0.58	0.64	0.62	0.55	0.63	0.55	0.63
Combined seepage rate	37.3%	32.0%	33.6%	38.5%	31.9%	38.0%	33.2%
Total seepage	0.57	2.49	1.01	2.43	2.86	3.05	3.30
Net recharge	0.21	1.05	0.95	0.55	-0.20	0.53	-0.37
Required seepage rate	23.3%	18.6%	2.2%	29.9%	34.1%	31.3%	36.9%
(for zero net recharge)							

We say probably, because the IBMR is not suited to the estimation of benefits from changes in either the water table depth or soil salinity.[1] From Tables 4.2 and 4.4 we can speculate that sub-surface drainage would increase yields by about 50% on average for the most severely affected areas. Since these areas are currently producing about 20% of Pakistan's agricultural output, value added might increase by about US$ 700 million per year, based on the value added projections from run #43. If so, then sub-surface drainage would be economically viable, as well as "solving" the waterlogging and salinity problems.[2]

But this level of commitment to drainage is not likely to be forthcoming, at least in the short run. Drainage projects are not as "exciting" as new reservoirs which generate power and flood control benefits as well as supplying more water in the periods of shortage, or as canal extension projects which open up new settlement areas. And they have the associated problems of effluent disposal - an environmental (not to mention political) problem in its own right. Furthermore, drainage projects must be location-specific, which always raises inter-provincial investment allocation issues. It is not difficult to equitably spread the benefits of additional power and water from a reservoir, nor to conceive of canal extension projects for each province. But given that the waterlogging and salinity problems are far more severe and widespread in Sindh than elsewhere, drainage projects will inevitably take lower precedence as a national priority. Perhaps, though unfortunately, WSIP was realistic in recommending that a relatively small component of its program be devoted to drainage. Given past experience, even that extent may be optimistic.

Water Allocations

Water allocation policies - or the lack thereof - have been a source of contention and an impediment to progress since the founding of the nation.[3] In previous studies

---

[1] The analysis of the benefits from LBOD was undertaken by components of the IBM appropriately modified. There are no insurmountable difficulties involved in endogenizing yield with respect to watertable depth and soil salinity, although experts are not in agreement on how to handle the synergistic effects of reducing watertables and soil salinity simultaneously . Solution costs would be high, however, and available data on yield response due not warrant its undertaking at this time.

[2] LBOD I had an estimated ERR of 13.6% for the overall project, and 21.9% for the most severely affected region. "Staff Appraisal Report Pakistan: Left Bank Outfall Drain Stage I Project", World Bank Report No. 5185-PAK (November 5, 1984), p. 42. (restricted circulation).

[3] In March, 1991, after most of the analytical work for this paper had been completed, the provinces reached agreement on the apportionment of Indus waters between them. The implications of this agreement are discussed in the next chapter.

employing the Indus Basin Model Revised[1] the authors have examined this issue in some detail, and under a variety of assumptions as to the parameters of a more rational policy. We found tremendous variation in the marginal value of water throughout the Basin, implying that substantial national output and income gains could be obtained through water allocations more closely approximating the optimal. We also discovered that many areas have been receiving canal diversions in excess of their needs, while in other areas land is idle for lack of water. Those studies concentrated on the efficiency and equity aspects of water allocation policies, but not on the environmental aspects. Given that the crude calculations in Section 6.2 above indicated that the *potential* exists for achieving balanced recharge through water allocations alone, we revisit this topic again briefly.

In the solutions to the IBMR reported in Chapter 3, water allocations, by month and canal, were forced to be proportional to those recorded in the post-Tarbela period. Any deviation from proportionality would have violated the so-called ad hoc arrangements whereby water was allocated according to historical precedent, unless agreed to otherwise. This restriction was carried forward to the 2000 solutions with (run # 43) and without (run # 41) the WSIP "Expected" Plan. We now relax this requirement to examine the environmental impacts.

There are several options, based on the above-mentioned studies:

1) **Pareto-optimal** allocation of any additional supplies available to the system. In this case, the canal diversions from the proportional case are used as lower bounds, and any excess water is allocated to the most productive canals.

2) **Global-optimal** allocations with no lower bounds. I.e., water is allocated to the most productive canals, regardless of equity considerations or provincial rights. This version was never reported, for obvious reasons.

3) **Provincial-optimal** allocations. As in (2), but maintaining fixed shares among the provinces.

4) **Pareto-optimal (80%)**. As in (1), but the lower bounds were set at 80% of the diversions from the proportional case.

We have selected the last case for the experiment reported here. It minimizes the equity problems of a globally-optimum solution because no canal will suffer more than a 20% reduction in supplies, yet it permits some flexibility in allocations between the provinces. Recall from Table 6.1 that balancing recharge through diversions alone would require a substantial inter-provincial reallocation.

---

[1] For Kalabagh Dam and for the WSIP. World Bank Report No. 6884-PAK, op.cit., and Guide, op.cit.

The results, labeled run # 45, compared to run # 43, are reported in Table 6.4. Run # 45 includes the WSIP projects, and thus differs only in water allocations. Definitions of variables are identical to those reported in Chapter 3 for Table 3.8.

First, note that the efficiency gains (production and value added) are relatively small (1.0 and 1.3%), but in absolute terms, significant at about 100 million US$ of output. The previous studies reported gains of 2-3% for these variables. Apparently, the WSIP investments make more efficient use of the available water, thus reducing the gains from water allocation changes. Production in Punjab rises by 1.6%, and value added and farm income rise by about the same magnitude. Output in the Sind falls by a small amount (0.7%), but in spite of this, value added and farm incomes both rise. Overall cropping intensity improves marginally.

On the second page of Table 6.4, it can be seen that the model employs the limited flexibility given it to increase canal diversions to the Punjab (by 14.6%) and reduce them to Sind (by 7.8%). This is precisely the direction, and about half the magnitude, as suggested in Table 6.1. Even though the model's objective function contains no element pertaining to environmental considerations, the direction it desires to go to maximize net social product is the same as that required to improve water balances. Clearly, environmental considerations need not run counter to efficiency objectives, although both may conflict with equity concerns.

The model also finds that about five MAF in additional diversions are possible when the proportionality restrictions are relaxed; Table 6.1 required 2.4 MAF additional diversions to balance recharge everywhere. Tubewell withdrawals in the Punjab fall by four MAF in response to the additional canal water. Again, this is the direction (but not nearly the necessary magnitude) indicated by the spreadsheet calculations of Table 6.2. Withdrawals rise by about one MAF in Sind, again in line with the desired direction, given that recharge is positive even in some fresh areas of Sind. Total water supply is virtually unchanged between the two simulations, implying the reduced tubewell withdrawals can be entirely replaced with more efficient canal diversions. Not only would this reduce the rate of aquifer mining, but the rate of soil salinity buildup as well. In fact, as we shall see later in this chapter, the combined effect of replacing tubewell water with canal water will serve to leach salts from the soils in the fresh areas.

"Slack water at the root zone" is the excess of water available over crop requirements. It is a measure of wasted water induced by restrictive water allocations imposed on the model, and contributes to recharge along with losses from canals, etc. In run #43, it amounts to 4.64 MAF, virtually all of which is in the saline areas. This is a clear indication that canal supplies to saline areas are too high, regardless of waterlogging concerns. In run # 45, slack water drops 34% to 3.06 MAF. This indicates that there is further scope, beyond the maximum 20% reduction allowed by the simulated allocation policy, to reduce diversions to saline areas. And because the slack water is truly that, its elimination would not result in any production or income losses, but only environmental improvement.

Net discharge declines by a very large amount (six MAF) in the Punjab fresh zones. This result is most encouraging given our concerns with fresh water aquifer mining. Even though mining would continue, its rate would be cut in half through these allocations. Of course, farmers would be far better off since they would save tubewell operating and maintenance costs. Recharge in Punjab saline areas rises slightly because of the increased supplies they receive. Recharge in the Sind (positive everywhere) falls slightly - a move in the right direction, but obviously far short of what is required. Finally, the nemesis of evaporation from groundwater, which brings salts to the surface, is not much affected by these allocations. Because the IBMR deals only with one year, any changes in WTD brought about by changes in water allocations are not large enough to significantly effect it at the provincial level.

Canal Lining

Virtually all of the 100 plus MAF of surface water feeding the CCA of Pakistan passes through the canal system as described in chapter 2. Losses, most (70%) of which seep to groundwater as recharge, range from 20 to 40% of flow, depending on length of the canal, soil type of the earthen bed, and condition. Thus roughly 29 MAF is lost by the canal system, of which about 20 MAF becomes groundwater recharge.

Canal losses through fresh groundwater areas is not of concern: they serve to offset tubewell pumping and hence dampen the process of aquifer mining. Of course, there is a delay in achieving the ultimate benefits from use of the water as it goes into temporary storage instead of immediate use. Canal losses through saline areas are much more destructive. The water is not only lost forever for purposes of irrigation as it mixes with highly saline groundwater, it is environmentally detrimental as it contributes to waterlogging. Although the WSIP contains several projects which include canal lining in fresh areas, we will confine ourselves to the saline areas.

Table 6.4:
Effects of More Efficient Water Allocations

	Run #	43	45	
	Year	2000	2000	
	Allocation	prop.	P.O. 80%	

Performance (millions of 1988 Rupees valued at economic prices):
and annualized growth rates from 1988 base

All irrigated agriculture:				
	Value of production	204,264	206,381	1.0%
	Value added	146,774	148,680	1.3%
	Hired labor wagebill	5,955	5,937	-0.3%
	Non-wage costs	51,534	51,763	0.4%
	Farm real income	124,957	126,351	1.1%
	Hydel output (BKWH)	16.793	16.342	-2.7%
Punjab:				
	Value of production	149,637	152,085	1.6%
	Value added	107,882	109,673	1.7%
	Hired labor wagebill	3,974	4,238	6.6%
	Non-wage costs	37,781	38,175	1.0%
	Farm real income	92,335	93,680	1.5%
Sindh:				
	Value of production	50,382	50,051	-0.7%
	Value added	36,143	36,258	0.3%
	Hired labor wagebill	1,981	1,699	-14.2%
	Non-wage costs	12,258	12,093	-1.3%
	Farm real income	30,550	30,598	0.2%

Production (thousands of tons, all canal-irrigated areas)

	Wheat	12,188.8	11,889.4	-2.5%
	Basmati	1,103.8	1,103.8	0.0%
	Irri	2,725.5	2,702.9	-0.8%
	Maize	280.0	333.3	19.0%
	total grains	16,298.1	16,029.4	-1.6%
	Cotton	11,763.4	11,672.5	-0.8%
	Gram	308.1	316.3	2.7%
	Mustard & Rape	214.9	209.6	-2.5%
	Sugarcane (raw)	67,634.1	71,304.8	5.4%
	Orchards	6,416.0	6,416.0	0.0%
	Potatoes	750.0	764.9	2.0%
	Onions	818.5	818.5	0.0%
	Chillis	144.3	147.6	2.3%
	Cow milk	1,491.2	1,511.5	1.4%
	Buffalo milk	8,118.7	8,577.5	5.7%
	Meat	366.0	376.7	2.9%

Use of selected inputs:

Labor (million man-days)

	Family	1,499	1,537	2.6%
	Hired	171	171	0.0%

Land (cropped area, thousands of acres)

	Rabi	26,073	25,691	-1.5%
	Kharif	23,378	24,148	3.3%
	Total	49,451	49,840	0.8%
	Intensity	132.9	134.0	0.8%

Fertilizer (thousand tons)

	Nitrogen	2357.7	2354	-0.2%
	Phosphate	714.3	717.4	0.4%

Table 6.4 (continued)

	Run #	43	45	
		Water Inputs (MAF):		
Canal diversions (measured at canal head)				
	NWFP	3.31	3.31	0.0%
	Punjab	60.03	68.82	14.6%
	Sind	50.60	46.65	-7.8%
	Pakistan	113.95	118.79	4.2%
Tubewell extractions (measured at watercourse head)				
	Punjab	57.13	52.59	-8.0%
	Sind	5.58	6.51	16.8%
	Pakistan	62.71	59.10	-5.8%
Total water supply (measured at the root zone)				
	NWFP	2.26	2.26	0.0%
	Punjab	70.31	71.99	2.4%
	Sind	32.02	30.38	-5.1%
	Pakistan	104.59	104.63	0.0%
Slackwater at root zone				
	Punjab Fresh	0.00	0.00	0.0%
	Punjab Saline	1.51	1.18	-21.9%
	Sind Fresh	0.29	0.08	-71.4%
	Sind Saline	2.84	1.80	-36.7%
	Basin sum	4.64	3.06	-34.1%
Net recharge (excluding evaportion from groundwater)				
	Punjab Fresh	-11.78	-5.83	-49.5%
	Punjab Saline	7.05	7.10	11.4%
	Sind Fresh	3.19	2.69	-15.8%
	Sind Saline	16.33	14.70	-10.0%
Evaporation from Groundwater				
	Punjab Fresh	0.45	0.45	0.0%
	Punjab Saline	5.96	6.25	4.9%
	Sind Fresh	3.08	3.28	6.5%
	Sind Saline	13.98	13.58	-2.9%

Canal lining has not been a major focus of sector planners. It is often argued that losses of the order of 25% for canals of 100 miles or more are not significant, and canal lining is highly disruptive to the production process as canals must be shut down for lengthy periods.[1] The ISRP series of projects and CWM do contain canal remodeling components, but it is not their major focus. Given the increasingly tight water supply situation, and the groundwater issues raised in the previous chapter, we feel that a close look at this option is warranted.

---

[1] Warren Fairchild has pointed out that the installation of interceptor drains is a viable alternative to canal lining, and probably costs less. They serve to retrieve the lost water before it becomes contaminated, and return it to the system. Their installation would not require canal closure, and thus would not suffer the same disruptive effects of canal lining.

The "saline" canals (those passing through saline groundwater areas) are listed in Table 6.5 along with their current efficiencies and those expected to prevail if remodeled with concrete linings.[1] These, of course, are main canals, and each has a number of branch, minors, and distributaries associated with it.

Table 6.5:
"Saline" Canals in Canal Lining Simulation

Canal	Length[1] (000 feet)	Efficiency Without	with	Canal	Length[1] (000 feet)	Efficiency Without	With
06 Sadiqia	5237	.733	.920	11 Jhang	991	.713	.914
12 Gugera	2023	.751	.925	14 Lower Jehlum	2767	.644	.893
15 Bahawal	915	.783	.935	16 Mialsi	1809	.783	.935
18 Haveli	930	.776	.933	20 Panjnad	2413	.730	.919
26 Thal	3986	.671	.901	28 Muzuffargarh	3787	.725	.917
32 Begari	965	.820	.946	33 Ghotki	2464	.765	.929
34 Northwest	3765	.804	.941	35 Rice	2532	.854	.956
36 Dadu	3218	.804	.941	38 Khairpur East	2384	.738	.921
39 Rohri	4996	.811	.943	41 Nara	8557	.816	.945
42 Kalari	3310	.800	.940	43 Lined Channel	2351	.800	.800
44 Fuleli	3537	.800	.940	45 Pinyari	4004	.821	.946

[1]Estimated using the total length and CCA in saline areas.

The total length of all canals in the system is 37,200 miles, of which approximately 14,880 miles pass through saline areas. Cost estimates of lining range from Rs. 1.25 million (US$58,140) to Rs. 1.6 million (US$74,420) per mile, depending on capacity.[2] The total cost estimate for all saline canals amounts to US$926 million. This figure does not include the cost of disruption of agricultural production. We assume that annual maintenance costs are 5% of the project costs, and commence four years after completion. The solution to the IBMR with the saline canals (run #46) lined is given in the middle columns of Table 6.6 along with the percentage changes from the 2000 "with WSIP" run #43.

Canal lining offers substantial production gains, amounting to 3.6% for the country, with Sind, where most of the saline canals are located, rising by 5.0%. Production of all commodities rises, with particularly strong gains in maize and sugarcane. Use of all types of inputs grows, and the overall cropping intensity rises from 132.9 to 138.4 Although canal diversions are held constant, total water supply at the root zone rises by 3.6% in Punjab, 10.2% in Sind, and 5.5% overall. In all, saline canal lining increases root-zone water supply by nearly 6 MAF.

---

[1]Derived from various WAPDA and World Bank internal reports. "Efficiency" refers to the proportion of water available at the tail from a one unit discharge at the head.

[2]Source: WAPDA.

A simplified cost-benefit analysis of this "project" is reported in Table 6.7. We assume that the construction time is six years, and the costs are equally distributed over that period. Benefits are defined as the increment in value added of run #46 over run #43. They are phased in over a six year period with a one-year delay from project start. As mentioned above, maintenance costs begin four years after completion, and cost 5% of the total project annually. The economic rate of return (ERR) comes out to an attractive 25.6%, and the N/K ratio, as used for WSIP project analysis, is 1.75, based on a cost of capital rate of 12%.[1]

This analysis does not include environmental benefits, which we do not attempt to value. However, on the basis of the analysis in the last chapter, we look for

a) a reduction in net recharge in saline areas,
b) a reduction in evaporation in saline areas,
c) a reduction in tubewell withdrawals in fresh areas.

The first will tend to lower the water table in saline areas, and hence reduce waterlogging and evaporation-induced soil salinity buildup. The latter will slow groundwater mining and delay intrusion from saline areas.

From Table 6.6, it can be seen that total tubewell withdrawals are little changed. They rise slightly in Punjab and fall modestly in Sind. Net recharge increases slightly in fresh areas of Sind and remains about the same in Punjab. Net recharge in saline areas drops sharply (about 11% in Sind and about 13% in Punjab). The latter induces a small (2%) fall in evaporation, but evaporation in Sind saline areas is left virtually unchanged. This is because water tables are already at saturation levels, and net recharge remains positive. Although canal lining in saline areas appears to be a highly viable investment activity, alone it will have little impact on the environmental problem.

Watercourse Improvements

The largest percentage losses in the irrigation system occur below the level of canals - on watercourses and fields. It has long been recognized that so-called on-farm water management (OFWM) projects have great promise for increasing the delivery efficiency and reducing losses to groundwater.[2] Under these projects watercourses are aligned, lined with masonry, and provided with improved control measures. A USAID pilot project renovated

---

[1] "N/K" is essentially a benefit-cost ratio sometimes adjusted for non-economic criteria in the WSIP rating procedures. Discount (cost-of-capital) rates used by WSIP are 12% throughout.

[2] A series of USAID-financed experiments in the early 1970's found that 30 - 50% of diversions were "lost" in the communal watercourses, with additional losses in the farmer-maintained branches and field ditches. "Staff Appraisal Report Pakistan: Third On-Farm Water Management Project", World Bank Report No. 9142-PAK (December 17, 1990), p. 3. Restricted circulation.

1,330 watercourses in the late 1970's, followed by the World Bank-supported OFWM I (3,230) and OFWM II (4,431). Many other water sector projects include OFWM components, so that the total of watercourses renovated to date is about 15,500 - about 14.5% of the 107,000 in the country (about 100,000 lie in the canal-irrigated area). Plans exist for the renovation of another 17,700 under ongoing projects.[1] A proposed OFWM III (a component of the WSIP "expected" plan) will renovate 5,350 watercourses in canal-commanded areas at an average cost of about US$16,000.[2]

We estimate that 35,600 watercourses remain to be renovated in saline areas. At the average cost given for OFWM III, this amounts to a total project costs of US$570 million. As in the case of canal lining, we assume the "project" would take six years, and require 5% maintenance beginning four years after completion. Benefits derive from increased watercourse efficiencies as given in the box at right, and amount to US$ 235 million per year upon completion (from Table 6.6).

Watercourse Efficiencies (saline areas only)		
Zone	base case	2000 Efficiency with WC Renovation
PMW	.509	.607
PCW	.577	.661
PRW	.568	.654
PSW	.555	.644
SCWN	.482	.586
SRWN	.576	.661
SCWS	.483	.587
SCWN	.570	.656

As reported in Table 6.6, the economic impacts are substantial, and closely parallel those of the canal-lining experiment. This is because both types of projects reach the same ends - the increase in water supply at the root zone. For watercourse renovation, the total increase is about 4.6 MAF, compared with about 5.8 MAF for canal lining. But the watercourse renovations cost only about two-thirds as much, and produce an even more impressive return: more than 40%, with an N/K ratio of 2.67.

The environmental benefits are generally positive: tubewell extractions decline marginally in all areas, as does net recharge in saline areas. Evaporation remains about the same in all areas except Sind-saline, where it declines by about 1%. But again, these improvements are too small to bring about any measurable improvement in the waterlogging and salinity problems. It appears that the combination of poor natural drainage and water tables that are already "disastrously" high will require sub-surface drainage in the saline areas.

---

[1] Ibid., p. 4.

[2] The total project cost works out to about US$20,000 per watercourse, but includes several components in addition to watercourse renovation.

Table 6.6: Effects of Loss-Reducing Projects

	Run #	WSIP EXP 43	+ CANAL LINING 46		+ OFWM 47	

Performance (millions of 1988 Rupees valued at economic prices)
and percentage change from WSIP base:

All irrigated agriculture:						
Value of production		204,264	211,588	3.6%	210,907	3.3%
Value added		146,774	152,145	3.7%	151,831	3.4%
Hired labor wagebill		5,955	6,076	2.0%	5,987	0.5%
Non-wage costs		51,534	53,367	3.6%	52,326	1.5%
Farm real income		124,957	129,455	3.6%	130,046	4.1%
Hydel output (BKWH)		16.793	16.728	-0.4%	16.751	-0.3%
Punjab:						
Value of production		149,637	154,443	3.2%	153,196	2.4%
Value added		107,882	111,453	3.3%	110,695	2.6%
Hired labor wagebill		3,974	3,994	0.5%	3,915	-1.5%
Non-wage costs		37,781	38,996	3.2%	38,587	2.1%
Farm real income		92,335	95,281	3.2%	94,734	2.6%
Sind:						
Value of production		50,382	52,901	5.0%	53,466	6.1%
Value added		36,143	37,943	5.0%	38,387	6.2%
Hired labor wagebill		1,981	2,082	5.1%	2,072	4.6%
Non-wage costs		12,258	12,871	5.0%	13,007	6.1%
Farm real income		30,550	32,109	5.1%	32,477	6.3%

Production (thousands of tons, all canal-irrigated areas)

Wheat		12,188.8	12,604.7	3.4%	12,531.0	2.8%
Basmati		1,103.8	1,120.5	1.5%	1,103.8	0.0%
Irri		2,725.5	2,733.0	0.3%	2,728.2	0.1%
Maize		280.0	368.0	31.4%	355.5	27.0%
total grains		16,298.1	16,826.2	3.2%	16,718.5	2.6%
Cotton		11,763.4	12,150.8	3.3%	12,176.3	3.5%
Gram		308.1	317.7	3.1%	309.5	0.5%
Mustard & Rape		214.9	228.1	6.1%	225.8	5.1%
Sugarcane (raw)		67,634.1	74,368.7	10.0%	74,158.6	9.6%
Orchards		6,416.0	6,416.0	0.0%	6,416.0	0.0%
Potatoes		750.0	756.9	0.9%	758.7	1.2%
Onions		818.5	819.8	0.2%	819.9	0.2%
Chilies		144.3	146.4	1.5%	147.2	2.0%
Cow milk		1,491.2	1,508.5	1.2%	1,508.9	1.2%
Buffalo milk		8,118.7	8,347.7	2.8%	8,233.0	1.4%
Meat		366.0	377.2	3.1%	373.6	2.1%

Use of selected inputs:

Labor (million man-days)

Family		1,499	1,560	4.1%	1,549	3.3%
Hired		171	174	1.9%	171	0.3%

Land (cropped area, thousands of acres)

Rabi		26,073	27,064	3.8%	26,873	3.1%
Kharif		23,378	24,445	4.6%	24,270	3.8%
Total		49,451	51,509	4.2%	51,143	3.4%
Intensity		132.9	138.4	4.2%	137.5	3.4%

Fertilizer use (thousand tons)

Nitrogen		2357.7	2447.4	3.8%	2440.5	3.5%
Phosphate		714.3	744.2	4.2%	740.4	3.7%

Table 6.6 (continued)

Water Inputs (MAF)

Canal diversions (measured at canal head)

NWFP	3.31	3.31	0.0%	3.31	0.0%
Punjab	60.03	60.03	0.0%	60.03	0.0%
Sind	50.60	50.60	0.0%	50.60	0.0%
Pakistan	113.95	113.95	0.0%	113.95	0.0%

Tubewell extractions (measured at watercourse head)

Punjab	57.13	57.24	0.2%	56.98	-0.3%
Sind	5.58	5.50	-1.4%	5.50	-1.4%
Pakistan	62.71	62.74	0.0%	62.49	-0.4%

Total water supply (measured at the root zone)

NWFP	2.26	2.26	0.0%	2.26	0.0%
Punjab	70.31	72.85	3.6%	71.80	2.1%
Sind	32.02	35.29	10.2%	35.13	9.7%
Pakistan	104.64	110.45	5.5%	109.25	4.4%

Net recharge (excluding evaporation from groundwater)

Punjab Fresh	-11.78	-11.84	0.49%	-11.67	-0.89%
Punjab Saline	7.10	6.21	-12.65%	6.39	-10.19%
Sind Fresh	3.19	3.23	1.10%	3.23	1.10%
Sind Saline	16.33	14.57	-10.76%	14.26	-12.67%

Evaporation from Groundwater

Punjab Fresh	0.45	0.45	0.00%	0.45	0.00%
Punjab Saline	5.96	5.85	-1.93%	5.87	-1.49%
Sind Fresh	3.08	3.08	-0.03%	3.08	-0.03%
Sind Saline	13.98	13.98	0.00%	13.83	-1.09%

Table 6.7: Cost-Benefit Analysis of Water Conservation Projects

	-- WSIP + Canal Lining --			---- WSIP + OFWM ----		
		ERR =	25.6%		ERR =	40.6%
		N/K =	1.75		N/K =	2.67
	cost	benefit	B - C	cost	benefit	B - C
1992	-154		-154	-95		-95
1993	-154	42	-113	-95	39	-56
1994	-154	83	-71	-95	78	-17
1995	-154	125	-29	-95	118	23
1996	-154	167	12	-95	157	62
1997	-154	208	54	-95	196	101
1998		250	250		235	235
1999		250	250		235	235
2000		250	250		235	235
2001	-46	250	204	-29	235	207
2002	-46	250	204	-29	235	207
2003	-46	250	204	-29	235	207
.	.	.	.	.	.	.
2017	-46	250	204	-29	235	207

Summary of Environmental Impacts

In chapter 3 we used the IBMR to examine the efficiency implications of the WSIP Expected Plan (Run #43) compared to the case (Run #41) without additional investment in the Indus Basin. In this chapter we have used the model to examine the performance of three additional interventions: more flexible water allocations (Run #45), canal lining in saline areas (Run #46), and watercourse renovations in saline areas (Run #47) both with respect to efficiency and environmental impact. All three were shown to have strong merit on performance grounds (agricultural output, farm incomes, etc.), and each were shown to make some degree of inroad on the waterlogging and salinity problems beyond that which might be expected from the WSIP investments.

We conclude this chapter by comparing some of the environmental variables from each of the solutions which the analysis of chapter 5 found to be significant. These are reported in Table 6.9.

Fresh Areas In the Punjab fresh areas, the major problems are rapidly declining watertables due to an excess of tubewell withdrawals over recharge from rain and surface water losses, and a buildup in soil salinities brought about by excessive tubewell activity. By comparing #43 with #41 in Table 6.9, it appears that the WSIP plan will hasten the decline in watertables. This result is probably due to the fact that the package of investments makes water more valuable, and induces farmers to use more from any source available - in this case, groundwater. That soil salinities improve ever so slightly in spite of the increased tubewell activity is probably due to the additional surface water available (with its leaching characteristics) through improvements in the distribution system. Runs #46 and #47, which deal only with saline areas, make no impact on the Punjab fresh problems. But Run #45 makes a strong impact, as it reduces net discharge by 50%. The result is a watertable about ten feet higher, and a further marginal improvement in soil salinity after the effects have had a chance to work themselves out. The underlying aquifer also becomes slightly "cleaner" but only because its volume would be greater (the salt content would be in fact be higher given that more salts were flushed from the topsoil). In the Sind fresh areas, mining is less of a problem except in specific areas, but soil salinity and groundwater quality are more severe. Current levels of tubewell activity are not high enough for the interventions to make much of a difference at all in addition to those benefits brought about by the WSIP. It is questionable how long the Sind fresh areas can remain in the "fresh" category in any event: soil salinities are already dangerously high, and adjacent saline area watertables so high that lateral intrusion is likely to contaminate most pockets of fresh water.

Saline Areas The WSIP makes some progress in lowering watertable depth in the Punjab from 5.3 feet on average to 5.7 feet (primarily through its drainage components), and in lowering the rate of increase in soil salinity. The improved water allocations would cancel out these gains, but the loss-reducing projects of #46 and #47 would herald further improvement, particularly through reducing recharge. In the Sind, the WSIP has virtually no impact on recharge, WTD, or salinity levels. Each of our suggested interventions make some improvement, but not nearly enough. If our projected soil salinities are anywhere close to being correct, the Sind saline areas will likely be abandoned for productive purposes.

Table 6.9: Summary of Environmental Effects

Fresh Areas

Run number (see text)	41	43	45	46	47	41	43	45	46	47
			Punjab					Sind		
Net recharge MAF	-9.81	-11.78	-5.83	-11.84	-11.67	3.09	3.19	2.69	3.23	3.23
WTD (feet)*	53.2	56.1	46.1	56.1	56.1	30.3	28.6	32.1	24.2	30.6
Change in soil salinity (Million Tons) due to:										
Surface water	-24.19	-25.37	-24.74	-25.43	-25.30	-7.59	-7.54	-7.66	-7.68	-7.68
Tubewell pumpage	30.66	32.61	29.65	32.68	32.52	2.66	2.58	3.19	2.55	2.55
Evaporation	0.32	0.18	2.51	0.18	0.18	6.47	6.48	6.09	6.69	6.69
Net change	1.19	1.28	1.39	1.29	1.27	0.47	0.45	0.53	0.48	0.48
E-O-Y EC**	5.4	5.3	5.2	5.3	5.3	7.9	7.7	7.9	7.8	7.8
Change in Groundwater salinity (Million Tons) due to:										
Surface water	37.62	38.83	40.13	38.88	38.76	12.20	12.15	12.39	12.29	12.29
Tubewell pumpage	-30.66	-32.61	-29.65	-32.68	-32.52	-2.66	-2.58	-3.19	-2.55	-2.55
Evaporation	-0.32	-0.18	-2.51	-0.18	-0.18	-6.47	-6.48	-6.09	-6.69	-6.69
Net change	10.13	9.95	11.78	9.95	9.97	3.96	3.98	4.02	3.96	3.95
E-O-Y PPM***	954	955	943	956	955	1347	1350	1355	1351	1352

Saline Areas

Run number (see text)	41	43	45	46	47	41	43	45	46	47
			Punjab					Sind		
Net recharge MAF	7.05	7.10	7.90	6.21	6.39	16.34	16.33	14.70	14.57	14.26
WTD 7 years feet*	5.3	5.7	5.3	5.9	5.8	4.8	4.7	5.2	5.0	5.1
Change in soil salinity (Million Tons) due to:										
Surface water	-22.29	-22.33	-25.17	-19.28	-19.89	-70.04	-69.97	-62.59	-60.78	-59.59
Evaporation	26.11	26.41	29.70	23.60	24.17	81.85	81.80	73.16	73.52	72.49
Net change	0.54	0.49	0.61	0.68	0.65	3.37	3.39	2.82	4.27	4.39
E-O-Y EC**	14.4	13.3	14.3	13.6	13.5	23.2	23.2	21.1	23.3	23.4
Change in Groundwater Salinity (Million Tons) Due to:										
Surface water	22.29	26.27	29.55	23.22	23.83	82.61	82.56	73.72	73.38	72.18
Evaporation	-26.11	-26.41	-29.70	-23.60	-24.17	-81.85	-81.80	-73.16	-73.52	-72.49
Net change	0.37	0.43	0.94	0.96	0.90	4.01	3.94	3.06	3.71	4.24
E-O-Y PPM***	2943	2955	2949	2965	2963	3867	3867	3885	3877	3882

* average watertable depth, in feet, in reference year 2007
** soil salinity in EC, at end of reference year
*** groundwater salinity in PPM, at end of reference year

CHAPTER 7
A SUGGESTED PROGRAM

Developmental Objectives

Pakistan looks to its agricultural sector for typical contributions to national well-being and economic growth. It desires a varied, adequate, and stable food supply. It desires more foreign exchange earnings from its export products cotton and rice. It hopes to reduce the dependency on imports for sugar and oilseeds. And it desires adequate and growing incomes for its farming population.

Thus far in its short history, agriculture has met most of these objectives. Somehow the sector has managed to grow in excess of a population growing at about 3.2% and showing little sign of slowing. In general, the population is adequately fed and clothed. Agriculture has always exhibited a positive trade balance, at least when cotton-derivative manufactured goods are included. Severe poverty still exists among non-tenured agricultural workers, but many farmers, particularly those who own tractors and tubewells, are relatively affluent.

But most of the growth over the past 40 years can be traced to two dramatic events. First was the extension and rehabilitation of the irrigation system under the Indus Basin Plan which culminated in the commissioning of Tarbela. These investments markedly increased the water supply, area under cultivation, and area cultivated year-round. Second was the green revolution in grains, through which yields of wheat and rice tripled. Today, there are few means left to increase water supply, and water availability has become the single most significant constraint on agriculture. Within the past few years, yields of most crops, wheat included, have stagnated or even declined. For the first time, agricultural growth has slowed to less than population growth. There may be positive surprises in store for Pakistani farmers, but they have yet to evidence themselves.

A variety of explanations for the disappointing performance are possible. The land tenure system is a legal nightmare. Land ownership is undoubtedly highly skewed, but the extent is largely unknown because of fragmentation, phantom ownership, and widespread litigation. Institutions charged with supporting agriculture, such as credit agencies, research centers, and extension services, can show little success. They are widely ridiculed by farmers, and their staffs are demoralized. Marketing services remain primitive, as does most rural infrastructure. That the sector has performed as well as it has is testimony to its extremely fertile soils, its highly favorable growing climate, and its industrious and motivated farming population.

The Need for Environmental Concern

But this population can do little without the two basic inputs to agriculture: water and land. That water is the single most constraining factor to Pakistani agriculture is well documented.[1] That poor water management has compounded the problem is also well known.[2] We have shown that, in all probability, the quality of groundwater, on which Pakistan depends for about one-third of its irrigation supplies, is deteriorating. Within a few years, most of it may be unusable, or out of reach of existing pumps. Given that tubewell farmers have been the singular bright spot in the sector for the last two decades, this prospect is most discomforting.

We have also quantified what others have long observed: that waterlogging and salinity have rended, and continue to rend vast areas unsuitable for crop production. Additional areas, possibly 20-25% of the Basin's cultivable land, are suffering yield declines of 25-50% due to these problems. The trends that brought about these conditions are still in place. Drainage and system rehabilitation efforts have made but a small dent in the problem, hardly noticeable in the overall picture. Water allocation policies continue to pour excess water into areas where it is not needed which only serves to raise the watertable further (except where it is already saturated) and bring more salts to the once fertile topsoil. These policies also seem to ignore the _existence_ of the quarter-million tubewells and the fact that they are rapidly mining the limited groundwater supply.[3] Inter-provincial squabbles, only partly over water rights and investment allocations, have thwarted efforts to improve system management and delayed some much-needed investments.

But the breezes which precede the winds of change may have already been felt. The WSIP will not be remembered for its analytical foundations (although these contained significant advances) or the quality of its recommended package of investments (although we have shown that it will likely exhibit high returns), but because of the inter-governmental cooperation it spawned. Under the WSIP, planning cells were established with each province as well as within the federal government, and all worked closely together and with the Ministry of Water and Power, WAPDA, the UNDP, and the World Bank to achieve the best plan possible under the circumstances. If these cells survive and are given adequate support, sector planning in Pakistan could become a model for the developing world.

Even more noteworthy, and not coincidental,[4] is the recent agreement on water

---

[1] In a recent survey, 64% of farmers cited water shortage as their most critical constraint. Second was waterlogging at 15%. "Report on Farm Survey - 1988", WAPDA/WSIP, Planning Division, Lahore, 1991.

[2] For an execllent review of water management issues, see S. S. Kirmani, "Comprehensive Water Resources Management: A Prerequisite for Progress in Pakistan's Irrigated Agriculture", paper presented to the Consultative Meeting on Water Sector Investment Planning, March 12-14, 1991, Islamabad.

[3] Kirmani, op. cit., p. 12.

[4] The agreement on inter-provincial water allocations was reached within a month of the final consultative meetings on the WSIP.

apportionment. After decades of negotiations, and the input of numerous studies, Sindh and Punjab agreed to share the Indus waters in given proportions, trade water rights when supply-demand conditions warrant, and cooperate on water projects of potential mutual benefit. Furthermore, they set up a joint commission to arbitrate any possible future disputes. This development may be as important as any in the water sector since Tarbela, and certainly provides the potential for great progress in water management and development.

This agreement now makes possible the introduction of intra-provincial optimization of water allocations. In a previous study, we estimated the gains from such a change to be equal to that obtainable from a project costing about 750 million US$.[1] Excess water can now be released to the sea, used to offset tubewell mining, or sold to another province. Water shortages can be reduced by re-allocation within a province or by purchases from another province, all without fear that any change in demand or allocation might set an irrevocable precedent of future detriment. Major projects such as Kalabagh or Basha Dams, sorely needed both for power and rabi irrigation supplies by all provinces, are now possible. Projects which <u>require</u> additional water in one province can now be balanced against projects which <u>save</u> water in another province. Drainage projects in Sindh or Baluchistan, for example, might be financed through water sales to Punjab. In short, this agreement removes the greatest obstacles to efficient water use and rational investment planning. Within this context, the environmental considerations must also be addressed, if the sector is to survive.

Elements of a Suggested Program

We have taken the WSIP Expected Plan as the basis for the interventions investigated in this study. Without the investments contained in that Plan, agricultural growth might well become negative by the end of the decade. The Plan is balanced, both among provinces, and among needs. It provides for limited efforts in drainage, system rehabilitation, irrigated area extension, and water supply. Each of its components is defensible on its own, and the package exhibits considerable synergism.

But the Plan does little to alleviate the environmental problems we have elaborated. The drainage components will relieve watertable pressure in selected areas, and if canal water is available for leaching, these may be brought back to full productivity. System rehabilitation and continuing OFWM efforts will reduce recharge by some degree, but nowhere near to zero, meaning that waterlogging and salinity buildup will continue virtually unabated. New water-using projects will need to be allocated already scarce water, probably from fresh areas which will necessitate accentuated aquifer mining by profit-maximizing farmers, resulting in yet more soil salinity. Far more is needed.

---

[1]<u>Guide</u>, <u>op</u>. <u>cit</u>., pp. 104-106.

Our suggested program contains the following elements:

Drainage. For those areas already severely waterlogged and highly saline, drainage is probably the only recourse other than abandonment. The roughly 3 - 6 billion US$ cost to provide sub-surface drainage to all severely affected areas is undoubtedly out of the question, but priority areas can be identified. WSIP's recommendation of draining only the non-rice areas deserves further study. Some areas near alluvial channels may have sufficient natural drainage that, if irrigation water is carefully controlled, they may survive with less expensive measures such as scavenger wells. These options presumably will be studied under a recently launched National Drainage Program.

Canal Lining in Saline Areas. We have shown that the lining of canals passing through saline areas will have a substantial effect on recharge to groundwater, and that the water saved has a high opportunity value for use elsewhere. Just on efficiency grounds, we have estimated the economic return to such projects at about 25% - twice that of Tarbela. Indeed, the water which could be saved annually would nearly equal the live storage capacity of the proposed Kalabagh reservoir. Although such a program might cost one billion US$, it could clearly pay for itself. The less costly alternative of interceptor drains might accomplish the same environmental ends, and avoid the disruption caused by canal closures.

On Farm Water Management. The benefits from water saved from OFWM have long been recognized, and projects designed to extend it are ongoing, albeit at a painfully slow pace. We have shown that OFWM, if extended to all saline areas, would produce nearly the environmental benefits of canal lining, and have an estimated economic return of about 40% based on a cost of about 600 million US$. Problems associated with OFWM are complicated because it involves considerable farmer participation and cooperation. But they should not be insurmountable if financial incentives and increased awareness by farmers of the potential disaster are made widely available.

Efficient, Environment-Sensitive Water Allocations. A vast potential exists for increasing output and incomes, reducing waterlogging, and replacing mined aquifer water through more efficient, environment-sensitive water allocation policies. We have demonstrated the broadbrush directions such allocations should take: reduced diversions to waterlogged areas, increased diversions to fresh areas, and an overall reallocation from Sind to Punjab. The water apportionment agreement moves the issue from the theoretical to the possible. Sind can now contemplate selling water, for example, to pay for drainage. Punjabi farmers may soon be able to purchase additional canal water in lieu of operating existing tubewells and/or investing in new ones capable of reaching the falling watertable.[1]

---

[1]Herve Plusquellec has pointed out that we have ignored another important aspect of water management for groundwater control: the antiquated waribundi rotation system at the watercourse level. This system, which has been in effect for over a century, undoubtedly results in localized waterlogging as well as inequitable distribution of water below the moghas. See Ch. M. H. Jahania, Canal and Drainage Act of 1873, Lahore, 1973, p. 89.

None of these measures alone can "solve" the environmental problems. But together, and combined with the WSIP investments, in all likelihood they can. The program is a costly one, probably amounting to 150% of the WSIP projected cost. But it is clearly economically viable. It would be financially viable as well if farmers were required to pay realistic charges for the water they use, and an active water market could be established, both among farmers and among provinces.[1]

Need for Future Work

<u>Soil and Groundwater Surveys</u>. This study had to rely in part on a soil and water survey that was taken thirteen years ago, and updated only in piecemeal fashion since. The modeling work reported in chapter 5 indicates that our estimates of soil and water salinity, extrapolated from that survey work, are probably far too low. This is substantiated by recent watercourse-level field work in the Punjab.[2] A comprehensive re-survey, based on sound methodology, is long overdue.

<u>Research on Yield Effects</u>. We have estimated, in very crude fashion, that waterlogging and salinity may be costing Pakistan one-fourth of its productive potential in agriculture. Yet there has been very little reliable local research which quantifies the yield/WTD and yield/salinity relationships.[3] Given the potentially enormous returns to projects which alleviate waterlogging and salinity, such research ought to be given the highest priority.

<u>Research on Salt-Tolerant Crops</u>. Some successful research has been undertaken and applied in Pakistan on salt-tolerant crops, in particular, kallar grass for fodder. But scientists in other countries have found that a wide variety of grains, oilseeds, legumes, and fuelwood crops can grow in highly saline soils and with highly saline water.[4] If it turns out that large areas of the Basin cannot be economically reclaimed, research into such alternatives also deserve a high priority.

<u>Computerized System Management</u>. If integrated, comprehensive management (ICM) of water resources is to become a reality, both physical controls within the system will need

---

[1] The water apportionment agreement now makes a water market among <u>canal commands</u> within a province conceivable which could exploit the widely differing returns to water within provinces.

[2] By Jacob W. Kijne, Edward J. Vander Velde, and others with International Irrigation Management Pakistan. Draft report not authorized for citation.

[3] The failures of agricultural research in Pakistan are perplexing given that the federal government operates twenty one research facilities while sixteen provincial agencies operate another 161! (NCA, p. 254).

[4] <u>Saline Agriculture: Salt-Tolerant Plants for Developing Countries</u>, National Academy Press, 1990.

to be upgraded or constructed, and computerized operation of the day-to-day decisions will need to be implemented. Although the IBMR is a planning tool, the techniques it employs can be adapted for operational purposes, both at the Basin and provincial level.[1]

---

[1]The first steps toward the reality of an operational IBMR were undertaken under the WSIP as staff from each provincial and the federal cell were trained in its use, and the model installed on a WAPDA computer.

APPENDICES:
GAMS LISTING OF DATA AND MODELS

Introduction

<u>A.1</u>. The IBMR has been implemented using the General Algebraic Modelling System (GAMS). This is a computer language that allows mathematical models to be concisely represented and easily transferred between different types of computers. GAMS is available on most of the widely used machines, including DOS based personal computers, IBM mainframes, and the DEC line of mini- and super-mini computers. The experiments documented in this paper were executed on an IBM-3090 mainframe under the VM/CMS operating system.

<u>A.2</u>. The program defining the IBMR has been broken into four parts, containing respectively the data (file WSISD*), the model setup and solve (file WSISM*), the report generator (file WSISR*) and the groundwater simulation program (file WSISRG*). A separate model file is usually created for each experiment. It contains the data and model changes needed to modify the original model. The GAMS "restart" option (see the GAMS User's Guide for details) is then used to execute the experiment, and the results are saved using the "save" option. The report program is then executed using these saved results to tabulate the results. Although the report programs are substantially the same for all experiments, for the sake of proper documentation a separate report file was created for each. Table A.1 lists the input, output and work files for each experiment.

Description of Experiments:

For this paper, several IBMR runs were made for model validation, to analyze the agricultural performance in 1978-88 and year 2000 and to analyze alternative waterlogging and salinization scenarios. The results of these runs were discussed in detail in Chapters 3 through 6. A brief description of each run is given below for quick reference.

Run 1. This run simulates 1987-88 conditions and is used to validate the model. The linearized model for all agroclimatic zones (model WSISZ) was solved for this run. Canal water diversions were fixed to the historic average for each canal in every month. Because surface water availability for each canal is fixed, the network equations were excluded and all nine agroclimatic zones were solved simultaneously. New investments in tractors, and tubewells were not allowed. This was a restart from file WSISD1 in which data setup was for 1987-88.

Run 1a. The surface network system was added to the specifications of Run 1. Model was configured to use the 50 percent probability inflows.

Run 33. This run simulates 1987-88 scenario with proportional water allocation. This means that each canal receives at least the historical average diversion in each month. Any extra water in the system is allocated proportional to the historical share determined on a seasonal basis. Relative shares are strictly maintained in this case, which means that if one canal is at capacity, extra water is allowed to flow to the sea. The model is configured to allow investment in new tubewells and tractors, and export limits for wheat are relaxed. Model WSISN was solved using a restart from run

Run 41.   This is first of the year 2000 runs and provides the without Water Sector Investment Plan (WSIP) case. Data for year 2000 are established by executing file WSISD41. Proportional water allocations were made in this scenario. Water shares were determined from average historic allocation on the seasonal basis.

Run 43.   This is as run 41 with Water Sector Investment Plan (WSIP).

Run 45A. This run is a Pareto-Optimal allocation case with the WSIP. Each canal is guaranteed only 80% of the historic average diversions in each season and extra water is allocated optimally over the whole basin.

Run 46.   Same as Run 43 with canal losses reduced by 70% in the saline areas(canal lining).

Run 47.   Same as run 43 with Watercourse command losses reduced by 20% in the saline areas (OFWM).

Table A.1: List of Input and Output Files

Run Number	Input file Name	Output file Name		Work File Input	Work File Output
1A.	WSISD1	WSISD1	Listing		WSISD1
	WSISM1A	WSISM1A	Listing	WSISD1	WSISM1A
	WSISR1A	WSISR1A	Listing	WSISM1A	WSISR1A
	WSISRG1	WSISRG1A	Listing	WSISR1A	
33.	WSISD31	WSISD31	Listing		WSISD31
	WSISM33	WSISM33	Listing	WSISD31	WSISM33
	WSISR33	WSISR33	Listing	WSISM33	WSISR33
	WSISRG33	WSISRG33	Listing	WSISR33	
41.	WSISD41	WSISD41	Listing		WSISD41
	WSISM41	WSISM41	Listing	WSISD41	WSISM41
	WSISR41	WSISR41	Listing	WSISM41	WSISR41
	WSISRG41	WSISRG41	Listing	WSISR41	
43.	WSISD43	WSISD43	Listing		WSISD43
	WSISM43	WSISM43	Listing	WSISD43	WSISM43
	WSISR43	WSISR43	Listing	WSISM43	WSISR43
	WSISRG43	WSISRG43	Listing	WSISR43	
45A	WSISD43	WSISD43	Listing		WSISD43
	WSISM43	WSISM43	Listing	WSISD43	WSISM43
	WSISR43	WSISR43	Listing	WSISM43	WSISR43
	WSISRG43	WSISRG43	Listing	WSISR43	
46.	WSISM46	WSISM46	Listing	WSISM43	WSISM46
	WSISR46	WSISR46	Listing	WSISM46	WSISR46
	WSISRG46	WSISRG46	Listing	WSISR46	
47.	WSISM47	WSISM47	Listing	WSISM43	WSISM47
	WSISR47	WSISR47	Listing	WSISM47	WSISR47
	WSISRG47	WSISRG47	Listing	WSISR47	

Table A.2: listing of Files Listed in this Appendix

Section A.1 Data and Scenario selection.
Section A.2 Model specification.
Section A.3 Report Program.
Section A.4 Groundwater Simulation Program.

Appendix A.1:

Data and Scenario Selection

January 29, 1991
COMDEF table for year 2000 cahnged to separate the Water Sector Investment Plan. Parameters Plan1 and Plan2 were created for coefficients changed due to the plan. Chotiari reservoir is part of the plan (LBOD project).

November 29, 1990
WSPD3 file from WSIPS Lahore was updated to include the Basic Plan, Pehur High Level (22A-PHL) canal was separated from the Upper Swat Canal. Shares of PHL were computed using monthly diversion equal to the capacity of the canal(.0589 maf/month).
Power Characteristics were included.

Following updates were made by the WSIPS Lahore.
Canal diversions Updated, Prices of Maize and Sugercane to Mill update Kalabagh the Maximum level 915 feet (6.1 MAF) and Rule curves adjusted for sluicing.

```
INDUS BASIN MODEL REVISED (IBMR) FILENAME=WSISD41                    07/25/91 17:17:54
SET DEFINITION                                                       GAMS 2.21 IBM CMS


25   SETS
26   Z       AGROCLIMATIC ZONES /
27               NWFP    NORTHWEST FROINTIER MIXED CROPPING
28               PMW     PUNJAB WHEAT-MIXED CROPPING
29               PCW     PUNJAB COTTON-WHEAT
30               PSW     PUNJAB SUGARCANE-WHEAT
31               PRW     PUNJAB RICE-WHEAT
32               SCWN    SIND COTTON-WHEAT NORTH
33               SRWN    SIND RICE-WHEAT NORTH
34               SCWS    SIND COTTON WHEAT SOUTH
35               SRWS    SIND RICE-WHEAT SOUTH  /
36   PV      PROVINCES AND COUNTRY /NWFP, PUNJAB, SIND, PAKISTAN /
37   PV1(PV)  PROVINCES             /NWFP, PUNJAB, SIND/
38   PV2(PV)  PUNJAB AND SIND       /PUNJAB, SIND/
39   PVZ(PV,Z) PROVINCE TO ZONE MAP /NWFP.NWFP
40                                   PUNJAB.(PMW,PCW,PSW,PRW)
41                                   SIND.  (SCWN,SCWS,SRWN,SRWS) /
42   CQ  CROP AND LIVESTOCK PRODUCTS
43           /BASMATI,   IRRI,       COTTON,     RAB-FOD,
44            GRAM,      MAIZE,      MUS+RAP,    KHA-FOD,
45            SC-MILL,   SC-GUR,     WHEAT,      ORCHARD,
46            POTATOES,  ONIONS,     CHILLI
47            COW-MILK,  BUFF-MILK,  MEAT /
48   CC(CQ)  CONSUMABLE COMODITIES /
49            BASMATI,   IRRI,       GRAM ,      MAIZE ,
50            MUS+RAP,   SC-GUR,     WHEAT ,     POTATOES,
51            ONIONS,    CHILLI     /
52   C(CQ)   CROPS /
53               BASMATI   RICE CROP
54               IRRI      RICE CROP
55               COTTON
56               RAB-FOD   FODDER CROP
57               GRAM
58               MAIZE
59               MUS+RAP
60               KHA-FOD   FODDER CROP
61               SC-GUR    SUGARCANE PROCESSED AT THE FARM
62               SC-MILL   SUGARCANE FOR MILL
63               WHEAT
64               ORCHARD
65               POTATOES
66               ONIONS
67               CHILLI    /
68   CF(C)   FODDER CROPS      /RAB-FOD, KHA-FOD /
69   CNF(C)  NON-FODDER CROPS
70
71   T       TECHNOLOGY       /BULLOCK,   SEMI-MECH /
72   S       SEQUENCE         /STANDARD   STANDARD SEQUENCE
73                             LA-PLANT   LATE PLANTING
74                             EL-PLANT   EARLY PLANTING
75                             QK-HARV    QUICK HARVESTING /
76   W    WATER STRESS LEVEL /
77                             STANDARD   NO STRESS
78                             LIGHT      LIGHT STRESS
79                             HEAVY      HEAVY STRESS
80                             JANUARY    WATER STRESS IN JANUARY/
81   G      GROUND WATER QUALITY TYPES/ FRESH, SALINE/
82   GF(G) FRESH GROUND WATER SUB-ZONE/FRESH /
83   GS(G) SALINE GROUND WATER SUB-ZONE/SALINE/
84   T1    SUB ZONES BY GW QUALITY/ FRESH, SALINE, TOTAL /
85   R1    RESOURCES /
86          CCA      CULTURABLE COMMANDED AREA OF THE CANAL
87          CCAP     CANAL CAPACITY AT THE CANAL HEAD
```

```
 88              CEFF       CANAL EFFICIENCY FROM BARRAGE TO THE WATER COURSE HEAD
 89              WCE-R      WATER COURSE COMMAND EFFICINECY IN RABI SEASON
 90              WCE-K      WATER COURSE COMMAND EFFICINECY IN KHARIF SEASON
 91              FLDE       FIELD EFFICIENCY
 92              FARMPOP    FARM POPULATION IN THE IRRIGATED
 93              FARMHH     NUMBER OF AGRICULTURAL HOUSEHOLDS
 94              TRACTORS   TRACTOR POPULATION IN THE IRRIGATED AREA
 95              TUBEWELLS  NUMBER OF TUBEWELLS
 96              TWC        EXISTING PRIVATE TUBEWELL CAPACITY
 97              BULLOCKS
 98              COWS
 99              BUFFALOS                                                /
100     DC(R1)   CHARACTERISTICS OF CANAL COMMAND  /
101              CCA        CULTURABLE COMMANDED AREA OF THE CANAL
102              CCAP       CANAL CAPACITY AT THE CANAL HEAD
103              CEFF       CANAL EFFICIENCY FROM BARRAGE TO THE WATER COURSE HEAD
104              WCE-R      WATER COURSE COMMAND EFFICINECY IN RABI SEASON
105              WCE-K      WATER COURSE COMMAND EFFICINECY IN KHARIF SEASON
106              FLDE       FIELD EFFICIENCY /
107     SA       SUBAREAS   /S1*S4/
108     WCE(DC)             WATERCOURSE EFFECIENCIES /WCE-R, WCE-K /
109     M1       MONTHS AND SEASONS /JAN,FEB,MAR,APR,MAY,JUN,
110                             JUL,AUG,SEP,OCT,NOV,DEC,RABI,KHARIF,ANNUAL/
111     M(M1)    MONTHS    /JAN,FEB,MAR,APR,MAY,JUN
112                         JUL,AUG,SEP,OCT,NOV,DEC/
113     WCEM(WCE,M)   MAPPING FROM SEASON TO MONTHS FOR WATERCOURSE EFFICIENCES  /
114                        WCE-R. (OCT,NOV,DEC,JAN,FEB,MAR)
115                        WCE-K. (APR,MAY,JUN,JUL,AUG,SEP)/
116     SEA(M1) SEASONS    /RABI, KHARIF /
117     SEAM(SEA,M)        MAPPING FROM SEASONS TO MONTHS/
118                        RABI.  (OCT,NOV,DEC,JAN,FEB,MAR)
119                        KHARIF.(APR,MAY,JUN,JUL,AUG,SEP)/
120        SEA1 /RABI, KHARIF, ANNUAL/
121        SEA1M(SEA1,M) /RABI.  (OCT,NOV,DEC,JAN,FEB,MAR)
122                       KHARIF.(APR,MAY,JUN,JUL,AUG,SEP)/
123     CI       CROP INPUT OUTPUTS   / STRAW-YLD, NITROGEN, PHOSPHATE , SEED /
124     P2(CI)                        / NITROGEN, PHOSPHATE /
125     A        ANIMAL TYPES         / COW, BULLOCK, BUFFALO /
126     AI       ANIMALS INPUT OUTPUT /TDN, DP, LABOR, COW-MILK, BUFF-MILK, MEAT /
127     Q(CQ)    LIVESTOCK COMODITIES /COW-MILK   MILK FROM CATTLE COW
128                                    BUFF-MILK  MILK FROM BUFFALE COW
129                                    MEAT       FROM COWS BUFFALOES AND BULLOCKS/
130     NT       NUTRIENTS FOR ANIMALS /TDN  TOTAL DIGESTABLE NUTRIENTS
131                                     DP   DIGESTABLE PROTEIN   /
132     IS       IRRIGATION SYSTEM SCENARIOS /1980*2000 /
133     PS       PRICE SECENARIOS / 87-88 /
134  *--
135  * CHANGE THE SET ISR TO SETUP DATA FOR DESIRED YEAR.
136  *
137     ISR(IS) IRRIGATION SYSTEM SCENARIO FOR THIS RUN /2000/
138     PLAN    WATER SECTOR PLAN INCLUDED IF INCP IS WITH  /WITH, WITHOUT/
139     INCP(PLAN) SET TO INCLUDE OR EXCLUDE PLAN FOR 2000 /WITHOUT/
140          ;
141     SCALAR BASEYEAR  BASE YEAR FOR CROP YIELDS /1988/;
142
143             CNF(C) = YES;  CNF(CF) = NO ;
144             PVZ("PAKISTAN",Z) = YES ;
145             SEA1M("ANNUAL",M) = YES;
146
147  * PARAMETERS TO EXPORT DATA FOR ZONE MODEL
148  *&&Z SET SET1 /TDN,DP,LABOR,COW-MILK,BUFF-MILK,MEAT,FIX-COST/
149  *&&Z       ;
150  * FOLLOWING PARAMETERS ARE TO STORE DATA FOR EXPORT TO THE ZONE MODELS.
151  *&&Z PARAMETER
152  *&&Z    ZONE1XXXXX(Z,C,T,S,W,M)   BULLOCK REQURENENTS (BULLOCK PAIR HRS PER MONTH)
```

INDUS BASIN MODEL REVISED (IBMR) FILENAME=WSISD41
SET DEFINITION

```
153  *&&Z   ZONE2XXXXX(Z,A,SET1)    INPUT OUTPUT COEFFICIENTS FOR LIVESTOCK
154  *&&Z   ZONE3XXXXX(Z,CQ, * )    DEMAND DATA
155  *&&Z   ZONE4XXXXX(Z,C,CI)      CROP INPUT OUTPUT
156  *&&Z   ;
```

INDUS BASIN MODEL REVISED (IBMR) FILENAME=WSISD41
CROP DATA

```
158
159      TABLE LAND(C,Z,T,S,W,M) LAND OCCUPATION BY MONTH
160
161                                                          JAN FEB MAR APR MAY JUN JUL AUG SEP OCT NOV DEC
162    (BASMATI,IRRI).(PMW,PCW,PSW,PRW).(BULLOCK,
163                     SEMI-MECH).STANDARD.STANDARD                              1   1   1   1   1  .5
164    IRRI.  SCWN.(BULLOCK,SEMI-MECH).STANDARD.STANDARD                      1   1   1   1   1   1
165    IRRI.  SRWN.(BULLOCK,SEMI-MECH).STANDARD.STANDARD                     .5   1   1   1   1   1  .5
166    IRRI.  SCWS.(BULLOCK,SEMI-MECH).STANDARD.STANDARD                 .5   1   1   1   1   1   1
167    IRRI.  SRWS.(BULLOCK,SEMI-MECH).STANDARD.STANDARD                      1   1   1   1   1   1
168
169    MAIZE. NWFP.BULLOCK.              STANDARD.STANDARD                         1   1   1   1   1
170    MAIZE. (PCW,PSW,PRW).BULLOCK.     STANDARD.STANDARD                        .5   1   1   1   1   1  .5
171    MAIZE. (PCW,PSW,PRW).SEMI-MECH.   STANDARD.STANDARD                             1   1   1   1   1  .5
172    MAIZE. SCWN.     (BULLOCK,SEMI-MECH).STANDARD.STANDARD                     .5   1   1   1   1  .5
173    MAIZE. SCWS.     (BULLOCK,SEMI-MECH).STANDARD.STANDARD                 .5   1   1   1   1   1  .5
174    MAIZE. NWFP.SEMI-MECH.STANDARD.STANDARD                                    .5   1   1   1   1
175
176    MUS+RAP.NWFP.(BULLOCK,SEMI-MECH).         STANDARD.STANDARD                                1   1   1   1   1   1
177    MUS+RAP.(PMW,PCW).(BULLOCK,SEMI-MECH).    STANDARD.STANDARD   1   1   1                        1   1   1
178    MUS+RAP. PSW.    (BULLOCK,SEMI-MECH).     STANDARD.STANDARD   1   1  .5                       .5   1   1   1
179    MUS+RAP. PRW.    (BULLOCK,SEMI-MECH).     STANDARD.STANDARD   1  .5                                1   1   1
180    MUS+RAP.(SCWN,SRWN,SRWS).(BULLOCK,SEMI-MECH).STANDARD.STANDARD 1   1  .5                            1   1   1
181    MUS+RAP.SCWS.(BULLOCK,SEMI-MECH).         STANDARD.STANDARD   1   1   1                       .5   1   1   1
182
183    (SC-GUR,SC-MILL).(NWFP,PMW,PCW,PSW,PRW, SCWN,SRWN,SCWS,SRWS).
184               (BULLOCK,SEMI-MECH).STANDARD.STANDARD            1   1   1   1   1   1   1   1   1   1   1   1
185    KHA-FOD.(NWFP,PMW,PCW).(BULLOCK,SEMI-MECH).STANDARD.STANDARD         .5  .5  .5   1   1   1  .5  .5
186    KHA-FOD.PSW.(BULLOCK,SEMI-MECH).     STANDARD.STANDARD              .5  .5  .5   1   1  .5  .5
187    KHA-FOD.PRW.(BULLOCK,SEMI-MECH).     STANDARD.STANDARD              .5  .5  .5   1   1   1  .5  .5
188    KHA-FOD.(SCWN).   (BULLOCK,SEMI-MECH).STANDARD.STANDARD              .5  .5   1   1  .5   1   1 .75
189    KHA-FOD.SCWS.     (BULLOCK,SEMI-MECH).STANDARD.STANDARD              .5  .5   1   1  .5   1   1   1
190    KHA-FOD.(SRWN,SRWS).(BULLOCK,SEMI-MECH).STANDARD.STANDARD            .5  .5   1   1   1   1   1
191    KHA-FOD.SRWS.(BULLOCK,SEMI-MECH).    LA-PLANT. STANDARD                       1   1   1   1   1   1
192
193    RAB-FOD.NWFP.BULLOCK.               STANDARD.STANDARD         1   1   1   1   1                .5   1   1   1
194    RAB-FOD.(NWFP,PCW).SEMI-MECH.       STANDARD.STANDARD         1   1   1   1   1               .25   1   1   1
195    RAB-FOD.PMW. BULLOCK.               STANDARD.STANDARD         1   1   1   1   1                     1   1   1
196    RAB-FOD.PCW. BULLOCK.               STANDARD.STANDARD         1   1   1   1   1                .5   1   1   1
197    RAB-FOD.PSW. BULLOCK.               STANDARD.STANDARD         1   1   1   1  .5                     1   1   1
198    RAB-FOD.PRW. BULLOCK.               STANDARD.STANDARD         1   1   1   1   1                     1   1   1
199    RAB-FOD.SCWN.(BULLOCK,SEMI-MECH).   STANDARD.STANDARD         1   1   1   1   1                .5   1   1   1
200    RAB-FOD.SRWN.(BULLOCK,SEMI-MECH).   STANDARD.STANDARD         1   1 1.75                      .5   1   1   1
201    RAB-FOD.SCWS.(BULLOCK,SEMI-MECH).   STANDARD.STANDARD         1   1   1   1 .25                .5   1   1   1
202    RAB-FOD.SRWS.(BULLOCK,SEMI-MECH).   STANDARD.STANDARD         1   1   1   1  .5                .5   1   1   1
203    RAB-FOD.PMW. SEMI-MECH.             STANDARD.STANDARD         1   1   1   1   1                     1   1   1
204    RAB-FOD.PSW. SEMI-MECH.             STANDARD.STANDARD         1   1   1   1  .5               .25   1   1   1
205    RAB-FOD.PRW. SEMI-MECH.             STANDARD.STANDARD         1   1   1   1   1                .5   1   1   1
206    RAB-FOD.(SRWN,SRWS).(BULLOCK,SEMI-MECH).STANDARD.HEAVY         1   1                           .5   1   1   1
207    RAB-FOD.(SRWN,SRWS).(BULLOCK,SEMI-MECH).STANDARD. LIGHT        1   1   1                       .5   1   1   1
208
209    COTTON. PMW.(BULLOCK,SEMI-MECH).        STANDARD.STANDARD                         1   1   1   1   1   1  .5
210    COTTON. (PCW,PSW,PRW).(BULLOCK,SEMI-MECH).STANDARD.STANDARD                       1   1   1   1   1   1  .5
211    COTTON. SCWN.BULLOCK.                   STANDARD.STANDARD                    .5   1   1   1   1   1   1  .5
212    COTTON. SCWS.BULLOCK.                   STANDARD.STANDARD                     1   1   1   1   1   1   1
213    COTTON. SRWS.BULLOCK.                   STANDARD.STANDARD                    .5   1   1   1   1   1   1
214    COTTON. SCWN.SEMI-MECH.                 STANDARD.STANDARD                         1   1   1   1   1   1  .5
215    COTTON. SCWS.SEMI-MECH.                 STANDARD.STANDARD                    .5   1   1   1   1   1   1
216    COTTON. SRWS.SEMI-MECH.                 STANDARD.STANDARD                         1   1   1   1   1   1
217    COTTON. PCW. BULLOCK.   EL-PLANT. STANDARD                                   .5   1   1   1   1   1   1  .5
218    COTTON. PCW. SEMI-MECH.LA-PLANT. STANDARD                                             1   1   1   1   1   1  .5
219
220    GRAM.(NWFP,PMW,PCW).BULLOCK.            STANDARD.STANDARD     1   1   1                        .5   1   1   1
221    GRAM.(NWFP,PMW,PCW).SEMI-MECH.          STANDARD.STANDARD     1   1   1                       .25   1   1   1
```

INDUS BASIN MODEL REVISED (IBMR) FILENAME=WSISD41
CROP DATA
07/25/91 17:17:54
GAMS 2.21 IBM CMS

```
222   GRAM.(PSW,PRW).BULLOCK.              STANDARD.STANDARD          1  1  1                    .5  1  1
223   GRAM.(PSW,PRW).SEMI-MECH.            STANDARD.STANDARD          1  1  1                   .25  1  1
224   GRAM.(SCWN,SRWN,SCWS,SRWS).(BULLOCK,SEMI-MECH).STANDARD.STANDARD  1  1  1           .25    1  1  1
225
226   WHEAT. (NWFP,PMW,PCW,PSW).BULLOCK.LA-PLANT.(STANDARD,LIGHT,
227                                         HEAVY,JANUARY)            1  1  1  1                    1  1
228   WHEAT.  PRW. BULLOCK. LA-PLANT.(STANDARD,LIGHT,HEAVY,JANUARY)    1  1  1  1                      1
229   WHEAT. (SCWN,SRWN).BULLOCK.LA-PLANT.(STANDARD,LIGHT,HEAVY,JANUARY) 1 1 1 .5                    1  1
230   WHEAT. (SCWS,SRWS).BULLOCK.LA-PLANT.(STANDARD,LIGHT,HEAVY,JANUARY) 1 1 1 .25                   1  1
231   WHEAT. (NWFP,PMW,PCW,PSW).BULLOCK.QK-HARV.
232         (STANDARD,LIGHT,HEAVY,JANUARY)                             1  1  1 .5          .5       1  1
233
234   WHEAT.  PRW. BULLOCK. QK-HARV.(STANDARD,LIGHT,HEAVY,JANUARY)     1  1  1 .5          .5          1
235   WHEAT. (SCWN,SRWN,SCWS,SRWS).BULLOCK.QK-HARV.(STANDARD,LIGHT,
236                                         HEAVY,JANUARY)            1  1  1              .5       1  1
237
238   WHEAT. (NWFP,PMW,PCW,PSW).BULLOCK.STANDARD.(STANDARD,LIGHT,HEAVY,
239                                         JANUARY)                  1  1  1  1                   .5  1
240   WHEAT.  PRW. BULLOCK.  STANDARD.(STANDARD,LIGHT,HEAVY,JANUARY)   1  1  1  1                   .5
241   WHEAT. (SCWN,SRWN).BULLOCK.STANDARD.(STANDARD,LIGHT,HEAVY,JANUARY) 1 1 1 .5           .5       1  1
242   WHEAT. (SCWS,SRWS).BULLOCK.STANDARD.(STANDARD,LIGHT,HEAVY,JANUARY) 1 1 1 .25          .5       1  1
243
244   WHEAT. (NWFP,PMW,PCW,PSW).SEMI-MECH.LA-PLANT.(STANDARD,LIGHT,HEAVY) 1 1 1 1                       1
245   WHEAT. (NWFP,PMW,PSW).SEMI-MECH.LA-PLANT.JANUARY                 1  1  1  1                      1
246   WHEAT.  PCW.SEMI-MECH.LA-PLANT.JANUARY                           1  1  1  1                      1
247   WHEAT.  PRW. SEMI-MECH.LA-PLANT.(STANDARD,LIGHT,HEAVY,JANUARY)   1  1  1  1                      1
248   WHEAT. (SCWN,SRWN).SEMI-MECH.LA-PLANT.
249         (STANDARD,LIGHT,HEAVY,JANUARY)                             1  1  1 .5                   .5  1
250   WHEAT. (SCWS,SRWS).SEMI-MECH.LA-PLANT.
251         (STANDARD,LIGHT,HEAVY,JANUARY)                             1  1  1 .25                  .5  1
252
253   WHEAT. (NWFP,PMW,PSW).SEMI-MECH.QK-HARV.
254         (STANDARD,LIGHT,HEAVY,JANUARY)                             1  1  1 .5         .25       1  1
255   WHEAT. (PCW,PRW).SEMI-MECH.QK-HARV.
256         (STANDARD,LIGHT,HEAVY,JANUARY)                             1  1  1 .5          .5       1  1
257   WHEAT. (SCWN,SRWN,SCWS,SRWS).SEMI-MECH.
258         QK-HARV.(STANDARD,LIGHT,HEAVY,JANUARY)                     1  1  1                       1  1
259
260   WHEAT.(NWFP,PMW,PSW).SEMI-MECH.STANDARD.
261         (STANDARD,LIGHT,HEAVY,JANUARY)                             1  1  1  1                   .5  1
262
263   WHEAT. (PCW,PRW).SEMI-MECH.STANDARD.(STANDARD,LIGHT,HEAVY,JANUARY) 1 1 1 1                    .5  1
264   WHEAT.(SCWN,SRWN).SEMI-MECH.STANDARD.(STANDARD,LIGHT,HEAVY,JANUARY) 1 1 1 .5                   1  1
265   WHEAT.(SCWS,SRWS).SEMI-MECH.STANDARD.(STANDARD,LIGHT,HEAVY,JANUARY) 1 1 1 .25                  1  1
266   (ORCHARD). (NWFP,PMW,PCW,PSW,PRW, SCWN,SRWN,SCWS,SRWS).
267              (BULLOCK,SEMI-MECH).STANDARD.STANDARD                 1  1  1  1  1  1  1  1  1  1  1  1
268   POTATOES.(SCWN,SCWS,SRWN,SRWS) .SEMI-MECH.STANDARD.STANDARD      1  1                   .5    1  1  1
269   POTATOES.(NWFP,PMW,PCW,PSW,PRW).SEMI-MECH.STANDARD.STANDARD      1  1  1  1
270   ONIONS. (NWFP,PMW,PCW,PSW,PRW).SEMI-MECH.STANDARD.STANDARD       1  1  1  1  1
271   ONIONS. (SCWN,SCWS,SRWN,SRWS). SEMI-MECH.STANDARD.STANDARD       1                           1  1  1  1
272   CHILLI. (NWFP,PMW,PCW,PSW,PRW).SEMI-MECH.STANDARD.STANDARD             .5 1  1  1  1  1  1  1 .5
273   CHILLI. (SCWN,SCWS,SRWN,SRWS). SEMI-MECH.STANDARD.STANDARD       1  1  1 .5                      .5  1
274         ;
275   SET TECH(Z,C,T,S,W)  TECHNOLOGY AVAILABILITY INDICATOR ;
276       TECH(Z,C,T,S,W)$SUM(M, LAND(C,Z,T,S,W,M))  = YES;
277
278
279       TABLE BULLOCK(C,Z,T,S,W,M) BULLOCK POWER REQUIREMENTS(BULLOCK PAIR HOURS PER MONTH )
280
281                                         JAN FEB MAR APR MAY JUN  JUL AUG SEP OCT NOV  DEC
282   BASMATI.PMW. BULLOCK.  STANDARD.STANDARD                   22.0 17.2         2.0 15.6
283   BASMATI.PCW. BULLOCK.  STANDARD.STANDARD                   22.0 17.2         2.0 15.6
284   BASMATI.PSW. BULLOCK.  STANDARD.STANDARD                   22.6 17.3         2.0 17.4
285   BASMATI.PRW. BULLOCK.  STANDARD.STANDARD                   32.5 24.6         2.0 23.0
286   BASMATI.PMW. SEMI-MECH.STANDARD.STANDARD                                         13.6
```

INDUS BASIN MODEL REVISED (IBMR) FILENAME=WSISD41
CROP DATA

```
287  BASMATI.PCW. SEMI-MECH.STANDARD.STANDARD                                                                 13.6
288  BASMATI.PSW. SEMI-MECH.STANDARD.STANDARD                                                                 15.4
289  BASMATI.PRW. SEMI-MECH.STANDARD.STANDARD                                                                 20.0
290
291  RAB-FOD.NWFP.BULLOCK.    STANDARD.STANDARD   3.0  3.0  3.0  3.0  2.0              8.0 17.9  2.0  2.0
292  RAB-FOD.PMW. BULLOCK.    STANDARD.STANDARD   3.0  3.0  3.0  3.0  2.0                  17.9  7.6  2.0
293  RAB-FOD.PCW. BULLOCK.    STANDARD.STANDARD   2.0  4.0  4.0  2.0  2.0             10.1 13.0  7.6  2.0
294  RAB-FOD.PSW. BULLOCK.    STANDARD.STANDARD   4.0  4.0  4.0  2.0  2.0                  16.5  9.7  2.0
295  RAB-FOD.PRW. BULLOCK.    STANDARD.STANDARD   4.0  4.0  4.0  2.0  2.0                  12.7  9.4  2.0
296  RAB-FOD.SCWN.BULLOCK.    STANDARD.STANDARD   2.0  2.0  2.0  1.0                   9.6 16.4  2.0  2.0
297  RAB-FOD.SRWN.BULLOCK.    STANDARD.STANDARD   2.0  2.0  2.0  1.0                   8.9 14.8  2.0  2.0
298  RAB-FOD.SCWS.BULLOCK.    STANDARD.STANDARD   2.0  2.0  2.0  2.0  1.0             10.3 18.4  2.0  2.0
299  RAB-FOD.SRWS.BULLOCK.    STANDARD.STANDARD   2.0  2.0  2.0  2.0  1.0             10.7 18.3  2.0  2.0
300  RAB-FOD.SCWN.SEMI-MECH.STANDARD.STANDARD     2.0  2.0  2.0  1.0                                  2.0  2.0
301  RAB-FOD.SRWN.SEMI-MECH.STANDARD.STANDARD     2.0  2.0  2.0  1.0                                  2.0  2.0
302  RAB-FOD.SCWS.SEMI-MECH.STANDARD.STANDARD     2.0  2.0  2.0  2.0  1.0                              2.0  2.0
303  RAB-FOD.SRWS.SEMI-MECH.STANDARD.STANDARD     2.0  2.0  2.0  2.0  1.0                              2.0  2.0
304  COTTON. PMW. BULLOCK.    STANDARD.STANDARD                  14.1  6.0              1.0  1.0  1.0  1.0
305  COTTON. PCW. BULLOCK.    STANDARD.STANDARD                  17.1 15.2              1.0  1.0  1.0  1.0
306  COTTON. PSW. BULLOCK.    STANDARD.STANDARD                   9.0 13.0              1.0  1.0  1.0  1.0
307  COTTON. PRW. BULLOCK.    STANDARD.STANDARD                  10.3 17.1              1.0  1.0  1.0  1.0
308  COTTON. SCWN.BULLOCK.    STANDARD.STANDARD             7.4 19.0  5.0               2.0  2.0  4.0
309  COTTON. SCWS.BULLOCK.    STANDARD.STANDARD            26.0       5.0               2.0  4.0
310  COTTON. SRWS.BULLOCK.    STANDARD.STANDARD             9.3 21.4  5.0               2.0  4.0
311  COTTON. (SCWN,SCWS,SRWS).SEMI-MECH.
312                           STANDARD.STANDARD                       5.0
313  COTTON. PCW. BULLOCK.    EL-PLANT. STANDARD           16.0  4.0 12.3               8.9 14.5    1.0  1.0  1.0  1.0
314  GRAM.   NWFP.BULLOCK.    STANDARD.STANDARD             7.0                        10.7  5.6
315  GRAM.   PMW. BULLOCK.    STANDARD.STANDARD        7.0                              10.7  5.6
316  GRAM.   PCW. BULLOCK.    STANDARD.STANDARD        7.0                              10.8  5.6
317  GRAM.   PSW. BULLOCK.    STANDARD.STANDARD        7.0                                   10.2  5.2
318  GRAM.   PRW. BULLOCK.    STANDARD.STANDARD        7.0                                   10.3  5.3
319  GRAM.   SCWN.BULLOCK.    STANDARD.STANDARD        4.0  5.7                          6.4  8.1
320  GRAM.   SCWS.BULLOCK.    STANDARD.STANDARD        4.0  5.7                          6.4  8.1
321  GRAM.   SRWN.BULLOCK.    STANDARD.STANDARD        4.9  6.4                          7.1  9.3
322  GRAM.   SRWS.BULLOCK.    STANDARD.STANDARD        4.9  6.4                          7.1  9.3
323
324  IRRI.   PMW. BULLOCK.    STANDARD.STANDARD                  18.9 19.2                    1.5 18.4
325  IRRI.   PCW. BULLOCK.    STANDARD.STANDARD                  18.9 19.2                    1.5 18.4
326  IRRI.   PSW. BULLOCK.    STANDARD.STANDARD                  16.3 18.4                    1.5 20.2
327  IRRI.   PRW. BULLOCK.    STANDARD.STANDARD                  32.6 24.0                    1.5 22.0
328  IRRI.   SCWN.BULLOCK.    STANDARD.STANDARD            10.5 18.3                          3.5 18.4
329  IRRI.   SRWN.BULLOCK.    STANDARD.STANDARD            10.7 18.6                         13.4 10.0
330  IRRI.   SCWS.BULLOCK.    STANDARD.STANDARD        9.1 10.3 17.1                          3.6 14.7
331  IRRI.   SRWS.BULLOCK.    STANDARD.STANDARD            19.0 17.5                          3.7 10.0 10.0
332  IRRI.   PMW. SEMI-MECH.STANDARD.STANDARD                                                    16.4
333  IRRI.   PCW. SEMI-MECH.STANDARD.STANDARD                                                    16.4
334  IRRI.   PSW. SEMI-MECH.STANDARD.STANDARD                                                    18.2
335  IRRI.   PRW. SEMI-MECH.STANDARD.STANDARD                                                    20.0
336  IRRI.   SCWN.SEMI-MECH.STANDARD.STANDARD                                                    16.4
337  IRRI.   SRWN.SEMI-MECH.STANDARD.STANDARD                                                    10.0 10.0
338  IRRI.   SCWS.SEMI-MECH.STANDARD.STANDARD                                                    12.7
339  IRRI.   SRWS.SEMI-MECH.STANDARD.STANDARD                                                    10.0 10.0
340
341  MAIZE.  NWFP.BULLOCK.    STANDARD.STANDARD                  41.4                     5.0
342  MAIZE.  PCW. BULLOCK.    STANDARD.STANDARD                  10.8  4.5 14.2                    5.0
343  MAIZE.  PSW. BULLOCK.    STANDARD.STANDARD                  10.2  4.1 13.4                    5.0
344  MAIZE.  PRW. BULLOCK.    STANDARD.STANDARD                  10.2  4.1 13.4                    5.0
345  MAIZE.  SCWN.BULLOCK.    STANDARD.STANDARD                  15.1 23.9                    3.0
346  MAIZE.  SCWS.BULLOCK.    STANDARD.STANDARD             14.3 21.7                          3.5
347  MAIZE.  NWFP.SEMI-MECH.STANDARD.STANDARD                    3.5
348  MAIZE.  PCW. SEMI-MECH.STANDARD.STANDARD                                  3.0
349  MAIZE.  PSW. SEMI-MECH.STANDARD.STANDARD                                  3.0
350  MAIZE.  PRW. SEMI-MECH.STANDARD.STANDARD                                  3.0
351  MAIZE.  SCWN.SEMI-MECH.STANDARD.STANDARD                                                   3.0
```

INDUS BASIN MODEL REVISED (IBMR) FILENAME=WSISD41
CROP DATA

#	Description	C1	C2	C3	C4	C5	C6	C7	C8	C9	C10	C11	C12	C13
352	MAIZE. SCWS.SEMI-MECH.STANDARD.STANDARD											3.5		
354	MUS+RAP.NWFP.BULLOCK. STANDARD.STANDARD							7.4	8.0	14.5	1.0	1.0	1.0	
355	MUS+RAP.PMW. BULLOCK. STANDARD.STANDARD	1.0	1.0	1.5							15.6	6.2	1.0	
356	MUS+RAP.PCW. BULLOCK. STANDARD.STANDARD	1.0	1.0	1.0							10.8	10.1	1.0	
357	MUS+RAP.PSW. BULLOCK. STANDARD.STANDARD	1.0	1.0	1.0						10.2	9.3	1.0	1.0	
358	MUS+RAP.PRW. BULLOCK. STANDARD.STANDARD	1.0	1.0								15.6	1.0	1.0	1.0
359	MUS+RAP.SCWN.BULLOCK. STANDARD.STANDARD	1.0	1.0	1.0							12.9	8.1	1.0	1.0
360	MUS+RAP.SRWN.BULLOCK. STANDARD.STANDARD	1.0	1.0	1.0							13.7	7.5	1.0	1.0
361	MUS+RAP.SRWS.BULLOCK. STANDARD.STANDARD	1.0	1.0	1.0							13.7	7.5	1.0	1.0
362	MUS+RAP.SCWS.BULLOCK. STANDARD.STANDARD	1.0	1.0	1.0							12.8	6.9	1.0	1.0
364	SC-GUR. NWFP.BULLOCK. STANDARD.STANDARD	14.6	11.3	11.0		2.0	2.0				8.0	11.0	14.0	18.0
365	SC-GUR. (PMW,PCW,PSW,PRW).BULLOCK.													
366	STANDARD.STANDARD	17.8	11.6	8.2		1.8	1.8					11.5	13.5	
367	SC-GUR. SCWN.BULLOCK. STANDARD.STANDARD	8.2	6.1	10.9			4.9							
368	SC-GUR. SRWN.BULLOCK. STANDARD.STANDARD	8.5	6.3	11.2			5.6							
369	SC-GUR. SCWS.BULLOCK. STANDARD.STANDARD	9.1	7.4	12.1			5.8							
370	SC-GUR. SRWS.BULLOCK. STANDARD.STANDARD	9.3	7.2	12.5			5.7							
371	SC-GUR. NWFP.SEMI-MECH.STANDARD.STANDARD	7.0	8.0	3.0		2.0	2.0				10.5	6.0	14.0	11.0
372	SC-GUR. (PMW,PCW,PSW,PRW).SEMI-MECH.													
373	STANDARD.STANDARD	11.0	6.5	2.5		1.8	1.8					11.5	11.0	
374	SC-GUR. SCWN.SEMI-MECH.STANDARD.STANDARD						4.9							
375	SC-GUR. SRWN.SEMI-MECH.STANDARD.STANDARD						5.6							
376	SC-GUR. SCWS.SEMI-MECH.STANDARD.STANDARD						5.8							
377	SC-GUR. SRWS.SEMI-MECH.STANDARD.STANDARD						5.7							
379	SC-MILL.NWFP.BULLOCK. STANDARD.STANDARD	18.0	18.0	14.0	12.0						15.9	20.0	15.0	16.0
380	SC-MILL.(PMW,PCW,PSW,PRW).BULLOCK.													
381	STANDARD.STANDARD	16.8	15.1	15.2	12.0	8.0	1.8					10.0	12.5	
382	SC-MILL.SCWN.BULLOCK. STANDARD.STANDARD	8.2	6.0	10.9			4.9							
383	SC-MILL.SRWN.BULLOCK. STANDARD.STANDARD	8.5	6.3	11.2			5.6							
384	SC-MILL.SCWS.BULLOCK. STANDARD.STANDARD	9.1	7.4	12.1			5.8							
385	SC-MILL.SRWS.BULLOCK. STANDARD.STANDARD	9.3	7.2	12.5			5.7							
386	SC-MILL.NWFP.SEMI-MECH.STANDARD.STANDARD	2.0	2.0								3.0			
387	SC-MILL.(PMW,PCW,PSW,PRW).SEMI-MECH.													
388	STANDARD.STANDARD		2.5			1.8	1.8							
389	SC-MILL.SCWN.SEMI-MECH.STANDARD.STANDARD						4.9							
390	SC-MILL.SRWN.SEMI-MECH.STANDARD.STANDARD						5.6							
391	SC-MILL.SCWS.SEMI-MECH.STANDARD.STANDARD						5.8							
392	SC-MILL.SRWS.SEMI-MECH.STANDARD.STANDARD						5.7							
394	KHA-FOD.NWFP.BULLOCK. STANDARD.STANDARD				15.4	1.0	1.0	16.0	1.0	15.0	1.0	0.5		
395	KHA-FOD.PMW. BULLOCK. STANDARD.STANDARD				15.0	1.0	1.0	15.5	1.0	16.0	1.0	0.5		
396	KHA-FOD.PCW. BULLOCK. STANDARD.STANDARD				16.0	1.0	1.0	16.0	1.5	15.0	1.0	0.5		
397	KHA-FOD.PSW. BULLOCK. STANDARD.STANDARD				17.0	1.0	1.0	16.5	1.0	16.0	1.0	0.5		
398	KHA-FOD.PRW. BULLOCK. STANDARD.STANDARD				16.0	1.0	1.2	15.5	1.5	17.0	1.2	0.5		
399	KHA-FOD.SCWN.BULLOCK. STANDARD.STANDARD				10.7	2.0	6.3	2.0	2.0	6.3	2.0	2.0		
400	KHA-FOD.SCWS.BULLOCK. STANDARD.STANDARD				10.3	2.0	6.4	2.0	2.0	6.4	2.0	2.0		
401	KHA-FOD.SRWN.BULLOCK. STANDARD.STANDARD				11.3	2.0	8.7	2.0	2.0	5.8	2.0	2.0		
402	KHA-FOD.SRWS.BULLOCK. STANDARD.STANDARD				11.3	2.0	8.7	2.0	2.0	5.8	2.0	2.0		
403	KHA-FOD.SRWS.BULLOCK. LA-PLANT. STANDARD					10.3	2.0	2.0	6.4	2.0				
405	RAB-FOD.SRWN.BULLOCK. STANDARD. HEAVY	2.0	2.0								8.9	14.8	2.0	2.0
406	RAB-FOD.SRWS.BULLOCK. STANDARD. HEAVY	2.0	2.0								10.7	18.3	2.0	2.0
407	RAB-FOD.SRWN.BULLOCK. STANDARD. LIGHT	2.0	2.0	2.0							8.9	14.8	2.0	2.0
408	RAB-FOD.SRWS.BULLOCK. STANDARD. LIGHT	2.0	2.0	2.0							10.7	18.3	2.0	2.0
409	RAB-FOD.SRWN.SEMI-MECH.STANDARD. HEAVY	2.0	2.0										2.0	2.0
410	RAB-FOD.SRWS.SEMI-MECH.STANDARD. HEAVY	2.0	2.0										2.0	2.0
411	RAB-FOD.SRWN.SEMI-MECH.STANDARD. LIGHT	2.0	2.0	2.0									2.0	2.0
412	RAB-FOD.SRWS.SEMI-MECH.STANDARD. LIGHT	2.0	2.0	2.0									2.0	2.0
414	WHEAT. NWFP.BULLOCK. LA-PLANT. HEAVY					4.6	4.6					38.4		
415	WHEAT. PMW. BULLOCK. LA-PLANT. HEAVY					4.0	4.0					27.1		
416	WHEAT. PCW. BULLOCK. LA-PLANT. HEAVY					4.8	4.8					39.2		

```
INDUS BASIN MODEL REVISED (IBMR) FILENAME=WSISD41                                    07/25/91 17:17:54
  CROP DATA                                                                            GAMS 2.21 IBM CMS

  417    WHEAT.  PSW. BULLOCK.   LA-PLANT. HEAVY               4.7  4.7            37.9
  418    WHEAT.  PRW. BULLOCK.   LA-PLANT. HEAVY               4.5  4.5            17.2  12.9
  419    WHEAT.  SCWN.BULLOCK.   LA-PLANT. HEAVY               5.4                 31.2
  420    WHEAT.  SRWN.BULLOCK.   LA-PLANT. HEAVY               5.5                 33.4
  421    WHEAT.  SCWS.BULLOCK.   LA-PLANT. HEAVY     2.9  3.0                      32.3
  422    WHEAT.  SRWS.BULLOCK.   LA-PLANT. HEAVY     3.1  3.1                      31.6
  423    WHEAT.  NWFP.BULLOCK.   LA-PLANT. JANUARY             4.6  4.6            38.4
  424    WHEAT.  PMW. BULLOCK.   LA-PLANT. JANUARY             4.0  4.0            27.1
  425    WHEAT.  PCW. BULLOCK.   LA-PLANT. JANUARY             4.8  4.8            39.2
  426    WHEAT.  PSW. BULLOCK.   LA-PLANT. JANUARY             4.7  4.7            37.9
  427    WHEAT.  PRW. BULLOCK.   LA-PLANT. JANUARY             4.5  4.5            17.2  12.9
  428    WHEAT.  SCWN.BULLOCK.   LA-PLANT. JANUARY             5.4                 31.2
  429    WHEAT.  SRWN.BULLOCK.   LA-PLANT. JANUARY             5.5                 33.4
  430    WHEAT.  SCWS.BULLOCK.   LA-PLANT. JANUARY   2.9  3.0                      32.3
  431    WHEAT.  SRWS.BULLOCK.   LA-PLANT. JANUARY   3.1  3.1                      31.6
  432    WHEAT.  NWFP.BULLOCK.   LA-PLANT. LIGHT               6.0  6.0            38.4
  433    WHEAT.  PMW. BULLOCK.   LA-PLANT. LIGHT               5.2  5.2            27.1
  434    WHEAT.  PCW. BULLOCK.   LA-PLANT. LIGHT               6.3  6.3            39.2
  435    WHEAT.  PSW. BULLOCK.   LA-PLANT. LIGHT               6.1  6.1            37.9
  436    WHEAT.  PRW. BULLOCK.   LA-PLANT. LIGHT               5.9  5.9            17.2  12.9
  437    WHEAT.  SCWN.BULLOCK.   LA-PLANT. LIGHT               7.2                 31.2
  438    WHEAT.  SRWN.BULLOCK.   LA-PLANT. LIGHT               7.2                 33.4
  439    WHEAT.  SCWS.BULLOCK.   LA-PLANT. LIGHT     3.8  3.9                      32.3
  440    WHEAT.  SRWS.BULLOCK.   LA-PLANT. LIGHT     4.0  4.1                      31.6
  441    WHEAT.  NWFP.BULLOCK.   LA-PLANT. STANDARD            7.1  7.0            38.4
  442    WHEAT.  PMW. BULLOCK.   LA-PLANT. STANDARD            6.2  6.1            27.1
  443    WHEAT.  PCW. BULLOCK.   LA-PLANT. STANDARD            7.4  7.4            39.2
  444    WHEAT.  PSW. BULLOCK.   LA-PLANT. STANDARD            7.2  7.2            37.9
  445    WHEAT.  PRW. BULLOCK.   LA-PLANT. STANDARD            6.9  6.9            17.2  12.9
  446    WHEAT.  SCWN.BULLOCK.   LA-PLANT. STANDARD            8.4                 31.2
  447    WHEAT.  SRWN.BULLOCK.   LA-PLANT. STANDARD            8.5                 33.4
  448    WHEAT.  SCWS.BULLOCK.   LA-PLANT. STANDARD  4.5  4.6                      32.3
  449    WHEAT.  SRWS.BULLOCK.   LA-PLANT. STANDARD  4.7  4.8                      31.6
  450    WHEAT.  NWFP.BULLOCK.   QK-HARV.  HEAVY               10.4                22.3  16.0
  451    WHEAT.  PMW. BULLOCK.   QK-HARV.  HEAVY               9.1                 15.9  11.2
  452    WHEAT.  PCW. BULLOCK.   QK-HARV.  HEAVY               10.9                18.6  20.6
  453    WHEAT.  PSW. BULLOCK.   QK-HARV.  HEAVY               10.7                18.3  19.6
  454    WHEAT.  PRW. BULLOCK.   QK-HARV.  HEAVY               10.3                30.1
  455    WHEAT.  SCWN.BULLOCK.   QK-HARV.  HEAVY     3.1  3.1                      17.5  13.7
  456    WHEAT.  SRWN.BULLOCK.   QK-HARV.  HEAVY     3.2  3.2                      18.3  15.1
  457    WHEAT.  SCWS.BULLOCK.   QK-HARV.  HEAVY     6.7                           13.4  18.9
  458    WHEAT.  SRWS.BULLOCK.   QK-HARV.  HEAVY     7.0                           13.7  17.9
  459    WHEAT.  NWFP.BULLOCK.   QK-HARV.  JANUARY             10.4                22.3  16.0
  460    WHEAT.  PMW. BULLOCK.   QK-HARV.  JANUARY             9.1                 15.9  11.2
  461    WHEAT.  PCW. BULLOCK.   QK-HARV.  JANUARY             10.9                18.6  20.6
  462    WHEAT.  PSW. BULLOCK.   QK-HARV.  JANUARY             10.7                18.3  19.6
  463    WHEAT.  PRW. BULLOCK.   QK-HARV.  JANUARY             10.3                30.1
  464    WHEAT.  SCWN.BULLOCK.   QK-HARV.  JANUARY   3.1  3.1                      17.5  13.7
  465    WHEAT.  SRWN.BULLOCK.   QK-HARV.  JANUARY   3.2  3.2                      18.3  15.1
  466    WHEAT.  SCWS.BULLOCK.   QK-HARV.  JANUARY   6.7                           13.4  18.9
  467    WHEAT.  SRWS.BULLOCK.   QK-HARV.  JANUARY   7.0                           13.7  17.9
  468    WHEAT.  NWFP.BULLOCK.   QK-HARV.  LIGHT               13.6                22.3  16.0
  469    WHEAT.  PMW. BULLOCK.   QK-HARV.  LIGHT               11.9                15.9  11.2
  470    WHEAT.  PCW. BULLOCK.   QK-HARV.  LIGHT               14.3                18.6  20.6
  471    WHEAT.  PSW. BULLOCK.   QK-HARV.  LIGHT               13.9                18.3  19.6
  472    WHEAT.  PRW. BULLOCK.   QK-HARV.  LIGHT               13.4                30.1
  473    WHEAT.  SCWN.BULLOCK.   QK-HARV.  LIGHT     4.1  4.0                      17.5  13.7
  474    WHEAT.  SRWN.BULLOCK.   QK-HARV.  LIGHT     4.1  4.1                      18.3  15.1
  475    WHEAT.  SCWS.BULLOCK.   QK-HARV.  LIGHT     8.6                           13.4  18.9
  476    WHEAT.  SRWS.BULLOCK.   QK-HARV.  LIGHT     9.1                           13.7  17.9
  477    WHEAT.  NWFP.BULLOCK.   QK-HARV.  STANDARD            16.0                22.3  16.0
  478    WHEAT.  PMW. BULLOCK.   QK-HARV.  STANDARD            14.0                15.9  11.2
  479    WHEAT.  PCW. BULLOCK.   QK-HARV.  STANDARD            16.8                18.6  20.6
  480    WHEAT.  PSW. BULLOCK.   QK-HARV.  STANDARD            16.4                18.3  19.6
  481    WHEAT.  PRW. BULLOCK.   QK-HARV.  STANDARD            15.8                30.1
```

INDUS BASIN MODEL REVISED (IBMR) FILENAME=WSISD41   07/25/91 17:17:54
CROP DATA   GAMS 2.21 IBM CMS

482	WHEAT.	SCWN.BULLOCK.	QK-HARV.	STANDARD	4.7	4.8		17.5	13.7
483	WHEAT.	SRWN.BULLOCK.	QK-HARV.	STANDARD	4.7	4.8		18.3	15.1
484	WHEAT.	SCWS.BULLOCK.	QK-HARV.	STANDARD	10.3			13.4	18.9
485	WHEAT.	SRWS.BULLOCK.	QK-HARV.	STANDARD	10.8			13.7	17.9
486	WHEAT.	NWFP.BULLOCK.	STANDARD.	HEAVY		5.2	5.2	22.3	16.0
487	WHEAT.	PMW. BULLOCK.	STANDARD.	HEAVY		4.5	4.6	15.9	11.2
488	WHEAT.	PCW. BULLOCK.	STANDARD.	HEAVY		5.5	5.5	18.6	20.6
489	WHEAT.	PSW. BULLOCK.	STANDARD.	HEAVY		5.3	5.4	18.3	19.6
490	WHEAT.	PRW. BULLOCK.	STANDARD.	HEAVY		5.1	5.2		30.1
491	WHEAT.	SCWN.BULLOCK.	STANDARD.	HEAVY		6.2		17.5	13.7
492	WHEAT.	SRWN.BULLOCK.	STANDARD.	HEAVY		6.3		18.3	15.1
493	WHEAT.	SCWS.BULLOCK.	STANDARD.	HEAVY	3.3	3.4		13.4	18.9
494	WHEAT.	SRWS.BULLOCK.	STANDARD.	HEAVY	3.5	3.5		13.7	17.9
495	WHEAT.	NWFP.BULLOCK.	STANDARD.	JANUARY		5.2	5.2	22.3	16.0
496	WHEAT.	PMW. BULLOCK.	STANDARD.	JANUARY		4.5	4.6	15.9	11.2
497	WHEAT.	PCW. BULLOCK.	STANDARD.	JANUARY		5.5	5.5	18.6	20.6
498	WHEAT.	PSW. BULLOCK.	STANDARD.	JANUARY		5.3	5.4	18.3	19.6
499	WHEAT.	PRW. BULLOCK.	STANDARD.	JANUARY		5.1	5.2		30.1
500	WHEAT.	SCWN.BULLOCK.	STANDARD.	JANUARY		6.2		17.5	13.7
501	WHEAT.	SRWN.BULLOCK.	STANDARD.	JANUARY		6.3		18.3	15.1
502	WHEAT.	SCWS.BULLOCK.	STANDARD.	JANUARY	3.3	3.4		13.4	18.9
503	WHEAT.	SRWS.BULLOCK.	STANDARD.	JANUARY	3.5	3.5		13.7	17.9
504	WHEAT.	NWFP.BULLOCK.	STANDARD.	LIGHT		6.8	6.8	22.3	16.0
505	WHEAT.	PMW. BULLOCK.	STANDARD.	LIGHT		6.0	6.0	15.9	11.2
506	WHEAT.	PCW. BULLOCK.	STANDARD.	LIGHT		7.1	7.1	18.6	20.6
507	WHEAT.	PSW. BULLOCK.	STANDARD.	LIGHT		6.9	7.0	18.3	19.6
508	WHEAT.	PRW. BULLOCK.	STANDARD.	LIGHT		6.7	6.7		30.1
509	WHEAT.	SCWN.BULLOCK.	STANDARD.	LIGHT		8.1		17.5	13.7
510	WHEAT.	SRWN.BULLOCK.	STANDARD.	LIGHT		8.2		18.3	15.1
511	WHEAT.	SCWS.BULLOCK.	STANDARD.	LIGHT	4.3	4.3		13.4	18.9
512	WHEAT.	SRWS.BULLOCK.	STANDARD.	LIGHT	4.5	4.6		13.7	17.9
513	WHEAT.	NWFP.BULLOCK.	STANDARD.	STANDARD		8.0	8.0	22.3	16.0
514	WHEAT.	PMW. BULLOCK.	STANDARD.	STANDARD		7.0	7.0	15.9	11.2
515	WHEAT.	PCW. BULLOCK.	STANDARD.	STANDARD		8.4	8.4	18.6	20.6
516	WHEAT.	PSW. BULLOCK.	STANDARD.	STANDARD		8.2	8.2	18.3	19.6
517	WHEAT.	PRW. BULLOCK.	STANDARD.	STANDARD		7.9	7.9		30.1
518	WHEAT.	SCWN.BULLOCK.	STANDARD.	STANDARD		9.5		17.5	13.7
519	WHEAT.	SRWN.BULLOCK.	STANDARD.	STANDARD		9.7		18.3	15.1
520	WHEAT.	SCWS.BULLOCK.	STANDARD.	STANDARD	5.1	5.2		13.4	18.9
521	WHEAT.	SRWS.BULLOCK.	STANDARD.	STANDARD	5.4	5.4		13.7	17.9
522	WHEAT.	NWFP.SEMI-MECH.LA-PLANT.		HEAVY		3.2	3.2		
523	WHEAT.	PMW. SEMI-MECH.LA-PLANT.		HEAVY		3.6	3.6		
524	WHEAT.	PCW. SEMI-MECH.LA-PLANT.		HEAVY		3.4	3.4		
525	WHEAT.	PSW. SEMI-MECH.LA-PLANT.		HEAVY		3.2	3.2		
526	WHEAT.	PRW. SEMI-MECH.LA-PLANT.		HEAVY		6.3	6.3		
527	WHEAT.	SCWN.SEMI-MECH.LA-PLANT.		HEAVY		5.4			
528	WHEAT.	SRWN.SEMI-MECH.LA-PLANT.		HEAVY		5.5			
529	WHEAT.	SCWS.SEMI-MECH.LA-PLANT.		HEAVY	2.9	3.0			
530	WHEAT.	SRWS.SEMI-MECH.LA-PLANT.		HEAVY	3.1	3.1			
531	WHEAT.	NWFP.SEMI-MECH.LA-PLANT.		JANUARY		3.2	3.2		
532	WHEAT.	PMW. SEMI-MECH.LA-PLANT.		JANUARY		3.6	3.6		
533	WHEAT.	PCW. SEMI-MECH.LA-PLANT.		JANUARY		3.4	3.4		
534	WHEAT.	PSW. SEMI-MECH.LA-PLANT.		JANUARY		3.2	3.2		
535	WHEAT.	PRW. SEMI-MECH.LA-PLANT.		JANUARY		6.3	6.3		
536	WHEAT.	SCWN.SEMI-MECH.LA-PLANT.		JANUARY		5.4			
537	WHEAT.	SRWN.SEMI-MECH.LA-PLANT.		JANUARY		5.5			
538	WHEAT.	SCWS.SEMI-MECH.LA-PLANT.		JANUARY	2.9	3.0			
539	WHEAT.	SRWS.SEMI-MECH.LA-PLANT.		JANUARY	3.1	3.1			
540	WHEAT.	NWFP.SEMI-MECH.LA-PLANT.		LIGHT		4.1	4.2		
541	WHEAT.	PMW. SEMI-MECH.LA-PLANT.		LIGHT		4.7	4.7		
542	WHEAT.	PCW. SEMI-MECH.LA-PLANT.		LIGHT		4.4	4.4		
543	WHEAT.	PSW. SEMI-MECH.LA-PLANT.		LIGHT		4.2	4.2		
544	WHEAT.	PRW. SEMI-MECH.LA-PLANT.		LIGHT		8.2	8.3		
545	WHEAT.	SCWN.SEMI-MECH.LA-PLANT.		LIGHT		7.2			
546	WHEAT.	SRWN.SEMI-MECH.LA-PLANT.		LIGHT		7.2			

INDUS BASIN MODEL REVISED (IBMR) FILENAME=WSISD41
CROP DATA

547	WHEAT.	SCWS.SEMI-MECH.LA-PLANT.	LIGHT	3.8	3.9
548	WHEAT.	SRWS.SEMI-MECH.LA-PLANT.	LIGHT	4.0	4.1
549	WHEAT.	NWFP.SEMI-MECH.LA-PLANT.	STANDARD		4.9 4.9
550	WHEAT.	PMW. SEMI-MECH.LA-PLANT.	STANDARD		5.5 5.5
551	WHEAT.	PCW. SEMI-MECH.LA-PLANT.	STANDARD		5.2 5.2
552	WHEAT.	PSW. SEMI-MECH.LA-PLANT.	STANDARD		4.9 4.9
553	WHEAT.	PRW. SEMI-MECH.LA-PLANT.	STANDARD		9.7 9.7
554	WHEAT.	SCWN.SEMI-MECH.LA-PLANT.	STANDARD		8.4
555	WHEAT.	SRWN.SEMI-MECH.LA-PLANT.	STANDARD		8.5
556	WHEAT.	SCWS.SEMI-MECH.LA-PLANT.	STANDARD	4.5	4.6
557	WHEAT.	SRWS.SEMI-MECH.LA-PLANT.	STANDARD	4.7	4.8
558	WHEAT.	NWFP.SEMI-MECH.QK-HARV.	HEAVY		7.3
559	WHEAT.	PMW. SEMI-MECH.QK-HARV.	HEAVY		8.2
560	WHEAT.	PCW. SEMI-MECH.QK-HARV.	HEAVY		7.7
561	WHEAT.	PSW. SEMI-MECH.QK-HARV.	HEAVY		7.3
562	WHEAT.	PRW. SEMI-MECH.QK-HARV.	HEAVY		7.2
563	WHEAT.	SCWN.SEMI-MECH.QK-HARV.	HEAVY	3.1	3.1
564	WHEAT.	SRWN.SEMI-MECH.QK-HARV.	HEAVY	3.2	3.2
565	WHEAT.	SCWS.SEMI-MECH.QK-HARV.	HEAVY	6.7	
566	WHEAT.	SRWS.SEMI-MECH.QK-HARV.	HEAVY	7.0	
567	WHEAT.	NWFP.SEMI-MECH.QK-HARV.	JANUARY		7.3
568	WHEAT.	PMW. SEMI-MECH.QK-HARV.	JANUARY		8.2
569	WHEAT.	PCW. SEMI-MECH.QK-HARV.	JANUARY		7.7
570	WHEAT.	PSW. SEMI-MECH.QK-HARV.	JANUARY		7.3
571	WHEAT.	PRW. SEMI-MECH.QK-HARV.	JANUARY		7.2
572	WHEAT.	SCWN.SEMI-MECH.QK-HARV.	JANUARY	3.1	3.1
573	WHEAT.	SRWN.SEMI-MECH.QK-HARV.	JANUARY	3.2	3.2
574	WHEAT.	SCWS.SEMI-MECH.QK-HARV.	JANUARY	6.7	
575	WHEAT.	SRWS.SEMI-MECH.QK-HARV.	JANUARY	7.0	
576	WHEAT.	NWFP.SEMI-MECH.QK-HARV.	LIGHT		9.5
577	WHEAT.	PMW. SEMI-MECH.QK-HARV.	LIGHT		10.7
578	WHEAT.	PCW. SEMI-MECH.QK-HARV.	LIGHT		10.0
579	WHEAT.	PSW. SEMI-MECH.QK-HARV.	LIGHT		9.5
580	WHEAT.	PRW. SEMI-MECH.QK-HARV.	LIGHT		9.4
581	WHEAT.	SCWN.SEMI-MECH.QK-HARV.	LIGHT	4.1	4.0
582	WHEAT.	SRWN.SEMI-MECH.QK-HARV.	LIGHT	4.1	4.1
583	WHEAT.	SCWS.SEMI-MECH.QK-HARV.	LIGHT	8.6	
584	WHEAT.	SRWS.SEMI-MECH.QK-HARV.	LIGHT	9.1	
585	WHEAT.	NWFP.SEMI-MECH.QK-HARV.	STANDARD		11.2
586	WHEAT.	PMW. SEMI-MECH.QK-HARV.	STANDARD		12.6
587	WHEAT.	PCW. SEMI-MECH.QK-HARV.	STANDARD		11.8
588	WHEAT.	PSW. SEMI-MECH.QK-HARV.	STANDARD		11.2
589	WHEAT.	PRW. SEMI-MECH.QK-HARV.	STANDARD		11.1
590	WHEAT.	SCWN.SEMI-MECH.QK-HARV.	STANDARD	4.7	4.8
591	WHEAT.	SRWN.SEMI-MECH.QK-HARV.	STANDARD	4.7	4.8
592	WHEAT.	SCWS.SEMI-MECH.QK-HARV.	STANDARD	10.3	
593	WHEAT.	SRWS.SEMI-MECH.QK-HARV.	STANDARD	10.8	
594	WHEAT.	NWFP.SEMI-MECH.STANDARD.	HEAVY		3.6 3.7
595	WHEAT.	PMW. SEMI-MECH.STANDARD.	HEAVY		4.1 4.1
596	WHEAT.	PCW. SEMI-MECH.STANDARD.	HEAVY		3.8 3.8
597	WHEAT.	PSW. SEMI-MECH.STANDARD.	HEAVY		3.6 3.7
598	WHEAT.	PRW. SEMI-MECH.STANDARD.	HEAVY		3.6 3.6
599	WHEAT.	SCWN.SEMI-MECH.STANDARD.	HEAVY		6.2
600	WHEAT.	SRWN.SEMI-MECH.STANDARD.	HEAVY		6.3
601	WHEAT.	SCWS.SEMI-MECH.STANDARD.	HEAVY	3.3	3.4
602	WHEAT.	SRWS.SEMI-MECH.STANDARD.	HEAVY	3.5	3.5
603	WHEAT.	NWFP.SEMI-MECH.STANDARD.	JANUARY		3.6 3.7
604	WHEAT.	PMW. SEMI-MECH.STANDARD.	JANUARY		4.1 4.1
605	WHEAT.	PCW. SEMI-MECH.STANDARD.	JANUARY		3.8 3.8
606	WHEAT.	PSW. SEMI-MECH.STANDARD.	JANUARY		3.6 3.7
607	WHEAT.	PRW. SEMI-MECH.STANDARD.	JANUARY		3.6 3.6
608	WHEAT.	SCWN.SEMI-MECH.STANDARD.	JANUARY		6.2
609	WHEAT.	SRWN.SEMI-MECH.STANDARD.	JANUARY		6.3
610	WHEAT.	SCWS.SEMI-MECH.STANDARD.	JANUARY	3.3	3.4
611	WHEAT.	SRWS.SEMI-MECH.STANDARD.	JANUARY	3.5	3.5

INDUS BASIN MODEL REVISED (IBMR) FILENAME=WSISD41
CROP DATA

```
612   WHEAT.  NWFP.SEMI-MECH.STANDARD. LIGHT                         4.7  4.8
613   WHEAT.  PMW. SEMI-MECH.STANDARD. LIGHT                         5.4  5.3
614   WHEAT.  PCW. SEMI-MECH.STANDARD. LIGHT                         5.0  5.0
615   WHEAT.  PSW. SEMI-MECH.STANDARD. LIGHT                         4.7  4.8
616   WHEAT.  PRW. SEMI-MECH.STANDARD. LIGHT                         4.7  4.7
617   WHEAT.  SCWN.SEMI-MECH.STANDARD. LIGHT                         8.1
618   WHEAT.  SRWN.SEMI-MECH.STANDARD. LIGHT                         8.2
619   WHEAT.  SCWS.SEMI-MECH.STANDARD. LIGHT                    4.3  4.3
620   WHEAT.  SRWS.SEMI-MECH.STANDARD. LIGHT                    4.5  4.6
621   WHEAT.  NWFP.SEMI-MECH.STANDARD. STANDARD                      5.6  5.6
622   WHEAT.  PMW. SEMI-MECH.STANDARD. STANDARD                      6.3  6.3
623   WHEAT.  PCW. SEMI-MECH.STANDARD. STANDARD                      5.9  5.9
624   WHEAT.  PSW. SEMI-MECH.STANDARD. STANDARD                      5.6  5.6
625   WHEAT.  PRW. SEMI-MECH.STANDARD. STANDARD                      5.5  5.6
626   WHEAT.  SCWN.SEMI-MECH.STANDARD. STANDARD                      9.5
627   WHEAT.  SRWN.SEMI-MECH.STANDARD. STANDARD                      9.7
628   WHEAT.  SCWS.SEMI-MECH.STANDARD. STANDARD                 5.1  5.2
629   WHEAT.  SRWS.SEMI-MECH.STANDARD. STANDARD                 5.4  5.4
630   ORCHARD.(NWFP,PMW,PCW,PSW,PRW,SCWN,SCWS,
631       SRWN,SRWS).BULLOCK.STANDARD.STANDARD   5.   5.   3.5  3.5  3.5  3.5  3.5  3.5  3.5  3.5  3.5  3.5
632   ORCHARD.(NWFP,PMW,PCW,PSW,PRW,SCWN,SCWS,
633       SRWN,SRWS).SEMI-MECH.STANDARD.STANDARD 3.   3.   3.   2.   1.                 2.   2.   2.   2.
634
635   CHILLI.(NWFP,PMW,PSW,PCW,PRW).SEMI-MECH.
636                       STANDARD.STANDARD      16.
637   CHILLI.(SCWN,SCWS,SRWN,SRWS).SEMI-MECH.
638                       STANDARD.STANDARD                                                             8.
639   ;
640   *- CONVERT BULLOCK PAIR HOURS TO BULLOCK HOURS
641   *&&Z  ZONE1XXXXX(Z,C,T,S,W,M) = BULLOCK(C,Z,T,S,W,M) ;
642     BULLOCK(C,Z,T,S,W,M) = BULLOCK(C,Z,T,S,W,M)*2;
644     TABLE LABOR(C,Z,T,S,W,M) LABOR REQUIREMENTS FOR CROPS(MAN HOURS)
645
646                                              JAN  FEB  MAR  APR  MAY  JUN   JUL   AUG  SEP  OCT   NOV   DEC
647   BASMATI.PMW. BULLOCK.   STANDARD. STANDARD                         29.1  88.78 65.9 5.6  5.6   47.9  17.6
648   BASMATI.PCW. BULLOCK.   STANDARD. STANDARD                         29.1  88.78 65.9 5.6  5.6   47.9  17.6
649   BASMATI.PSW. BULLOCK.   STANDARD. STANDARD                         26.9  94.9  74.7 6.2  3.1   50.9  19.4
650   BASMATI.PRW. BULLOCK.   STANDARD. STANDARD                         44.2  91.8  63.9 8.4  5.6   41.9  25.0
651   BASMATI.PMW. SEMI-MECH.STANDARD. STANDARD                          8.4   71.6  65.9 5.6  5.6   47.4  16.1
652   BASMATI.PCW. SEMI-MECH.STANDARD. STANDARD                          8.4   71.6  65.9 5.6  5.6   47.4  16.1
653   BASMATI.PSW. SEMI-MECH.STANDARD. STANDARD                          5.6   74.4  74.7 6.2  3.1   50.4  16.9
654   BASMATI.PRW. SEMI-MECH.STANDARD. STANDARD                          11.0  67.7  63.9 8.4  5.6   41.4  22.5
655
656   RAB-FOD.SRWN.BULLOCK.   STANDARD. HEAVY    27.9 23.4                                        9.1  19.7  19.2  24.8
657   RAB-FOD.SRWS.BULLOCK.   STANDARD. HEAVY    27.4 24.8                                       10.7  19.5  16.8  27.9
658   RAB-FOD.SRWN.BULLOCK.   STANDARD. LIGHT    27.9 23.4 22.5                                   9.1  19.7  19.2  24.8
659   RAB-FOD.SRWS.BULLOCK.   STANDARD. LIGHT    27.4 24.8 19.7                                  10.7  19.5  16.8  27.9
660   RAB-FOD.NWFP.BULLOCK.   STANDARD. STANDARD 42.8 42.1 32.6 24.6 10.9                          8.0  14.3  17.8  35.6
661   RAB-FOD.PMW. BULLOCK.   STANDARD. STANDARD 29.3 37.1 36.6 28.1  9.7                              13.35 16.5  34.5
662   RAB-FOD.PCW. BULLOCK.   STANDARD. STANDARD 32.3 41.5 41.8 29.4 18.2                         10.1 15.5  23.5  29.1
663   RAB-FOD.PSW. BULLOCK.   STANDARD. STANDARD 31.9 39.8 41.7 28.6 13.4                              8.7  15.7  28.7
664   RAB-FOD.PRW. BULLOCK.   STANDARD. STANDARD 31.4 32.0 34.5 23.4 21.9                              4.5  14.4  27.1
665   RAB-FOD.SCWN.BULLOCK.   STANDARD. STANDARD 28.0 24.8 24.0 14.3                              9.6  21.5  19.2  24.5
666   RAB-FOD.SRWN.BULLOCK.   STANDARD. STANDARD 27.9 23.4 22.5 15.0                              9.1  19.7  19.2  24.8
667   RAB-FOD.SCWS.BULLOCK.   STANDARD. STANDARD 29.0 23.1 19.8 17.5  8.6                        10.3  19.8  17.5  27.8
668   RAB-FOD.SRWS.BULLOCK.   STANDARD. STANDARD 27.4 24.8 19.7 18.8  9.3                        10.7  19.5  16.8  27.9
669   RAB-FOD.SRWN.SEMI-MECH.STANDARD. HEAVY     27.9 23.4                                        1.0   8.7  19.2  24.8
670   RAB-FOD.SRWS.SEMI-MECH.STANDARD. HEAVY     27.4 24.8                                        1.2   9.3  16.8  27.9
671   RAB-FOD.SRWN.SEMI-MECH.STANDARD. LIGHT     27.9 23.4 22.5                                   1.0   8.7  19.2  24.8
672   RAB-FOD.SRWS.SEMI-MECH.STANDARD. LIGHT     27.4 24.8 19.7                                   1.2   9.3  16.8  27.9
673   RAB-FOD.NWFP.SEMI-MECH.STANDARD. STANDARD  42.2 43.5 31.2 22.0  9.5                         2.05  8.02 16.29 33.0
674   RAB-FOD.PMW. SEMI-MECH.STANDARD. STANDARD  28.6 36.4 35.6 27.5  8.3                              2.17  7.96 24.08
675   RAB-FOD.PCW. SEMI-MECH.STANDARD. STANDARD  31.1 39.8 40.3 28.7 17.6                         7.5  13.2  21.6  28.4
676   RAB-FOD.PSW. SEMI-MECH.STANDARD. STANDARD  30.5 38.4 40.3 27.1 11.9                              7.3  12.7  26.4
677   RAB-FOD.PRW. SEMI-MECH.STANDARD. STANDARD  29.6 30.4 29.6 22.5 20.7                              3.4  12.8  25.7
```

INDUS BASIN MODEL REVISED (IBMR) FILENAME=WSISD41
CROP DATA

07/25/91 17:17:54
GAMS 2.21 IBM CMS

#	Crop	Region	Power	Type	Std	c1	c2	c3	c4	c5	c6	c7	c8	c9	c10	c11	c12	c13	c14
678	RAB-FOD.	SCWN.	SEMI-MECH.	STANDARD.	STANDARD	26.5	21.4	22.8	12.7						1.2	8.8	17.5	22.8
679	RAB-FOD.	SRWN.	SEMI-MECH.	STANDARD.	STANDARD	27.9	23.4	22.5	15.0						1.0	8.7	19.2	24.8
680	RAB-FOD.	SCWS.	SEMI-MECH.	STANDARD.	STANDARD	29.0	23.1	19.8	17.5	8.6					1.2	8.9	17.5	27.8
681	RAB-FOD.	SRWS.	SEMI-MECH.	STANDARD.	STANDARD	27.4	24.8	19.7	18.8	9.3					1.2	9.3	16.8	27.9
682																		
683	COTTON.	PCW.	BULLOCK.	EL-PLANT.	STANDARD					26.8	6.7	20.5	2.5	7.6	13.2	41.3	57.4	22.1
684	COTTON.	PMW.	BULLOCK.	STANDARD.	STANDARD						16.85	9.6		2.9	12.0	50.74	70.5	26.5
685	COTTON.	PCW.	BULLOCK.	STANDARD.	STANDARD						35.59	18.4	2.5	7.6	13.2	41.3	57.4	22.1
686	COTTON.	PSW.	BULLOCK.	STANDARD.	STANDARD						10.2	19.32	2.5	3.9	11.0	40.1	52.9	16.3
687	COTTON.	PRW.	BULLOCK.	STANDARD.	STANDARD						22.3	20.28		6.7	10.6	40.1	57.1	21.0
688	COTTON.	SCWN.	BULLOCK.	STANDARD.	STANDARD					7.4	23.5	7.0	2.0	2.0	17.5	39.4	66.2	
689	COTTON.	SCWS.	BULLOCK.	STANDARD.	STANDARD					31.2	2.0	7.0	2.0	2.0	38.6	80.4		
690	COTTON.	SRWS.	BULLOCK.	STANDARD.	STANDARD					9.5	23.8	7.0	2.0	2.0	28.5	84.1		
691	COTTON.	PCW.	SEMI-MECH.	LA-PLANT.	STANDARD							14.4	2.5	7.6	12.7	40.8	56.9	21.6
692	COTTON.	PMW.	SEMI-MECH.	STANDARD.	STANDARD						13.36	5.3		2.9	11.5	50.24	70.0	26.0
693	COTTON.	PCW.	SEMI-MECH.	STANDARD.	STANDARD						4.99	9.42	2.5	7.6	12.7	40.8	56.9	21.6
694	COTTON.	PSW.	SEMI-MECH.	STANDARD.	STANDARD						4.41	9.28	2.5	3.9	10.5	39.6	52.4	15.8
695	COTTON.	PRW.	SEMI-MECH.	STANDARD.	STANDARD						1.45	9.04		6.7	10.1	39.6	56.6	20.5
696	COTTON.	SCWN.	SEMI-MECH.	STANDARD.	STANDARD						13.2				15.8	37.4	62.2	
697	COTTON.	SCWS.	SEMI-MECH.	STANDARD.	STANDARD					12.1		7.0	2.0	2.0	36.6	76.4		
698	COTTON.	SRWS.	SEMI-MECH.	STANDARD.	STANDARD					12.4	7.0	2.0	2.0	28.5	83.2			
699																		
700	GRAM.	NWFP.	BULLOCK.	STANDARD.	STANDARD	3.4			23.2						8.9	18.0		
701	GRAM.	PMW.	BULLOCK.	STANDARD.	STANDARD			21.5							10.7	9.1	1.9	2.5
702	GRAM.	PCW.	BULLOCK.	STANDARD.	STANDARD			21.7							10.8	8.9	0.7	2.5
703	GRAM.	PSW.	BULLOCK.	STANDARD.	STANDARD	2.5		22.3							10.2	8.6	0.8	
704	GRAM.	PRW.	BULLOCK.	STANDARD.	STANDARD	2.5		22.6							10.3	8.5	0.7	
705	GRAM.	SCWN.	BULLOCK.	STANDARD.	STANDARD			23.5	9.2						6.4	12.0	1.0	1.0
706	GRAM.	SCWS.	BULLOCK.	STANDARD.	STANDARD			23.5	9.2						6.4	12.0	1.0	1.0
707	GRAM.	SRWN.	BULLOCK.	STANDARD.	STANDARD			24.8	10.3						7.1	13.4	1.0	1.0
708	GRAM.	SRWS.	BULLOCK.	STANDARD.	STANDARD			24.8	10.3						7.1	13.4	1.0	1.0
709	GRAM.	NWFP.	SEMI-MECH.	STANDARD.	STANDARD	3.4			20.5						1.87	6.0		
710	GRAM.	PMW.	SEMI-MECH.	STANDARD.	STANDARD			19.0							1.63	4.64	0.9	2.5
711	GRAM.	PCW.	SEMI-MECH.	STANDARD.	STANDARD			20.0							1.62	4.41	0.7	2.5
712	GRAM.	PSW.	SEMI-MECH.	STANDARD.	STANDARD	2.5		21.4								1.54	4.39	0.8
713	GRAM.	PRW.	SEMI-MECH.	STANDARD.	STANDARD	2.5		21.9								1.45	4.19	0.7
714	GRAM.	SCWN.	SEMI-MECH.	STANDARD.	STANDARD			21.4	5.5						0.81	3.4	1.0	1.0
715	GRAM.	SCWS.	SEMI-MECH.	STANDARD.	STANDARD			21.4	5.5						0.81	3.4	1.0	1.0
716	GRAM.	SRWN.	SEMI-MECH.	STANDARD.	STANDARD			22.7	9.1						0.96	3.6	1.0	1.0
717	GRAM.	SRWS.	SEMI-MECH.	STANDARD.	STANDARD			22.7	9.1						0.96	3.6	1.0	1.0
718																		
719	IRRI.	PMW.	BULLOCK.	STANDARD.	STANDARD						22.9	122.3	35.9	8.4	6.4	43.9	20.4	
720	IRRI.	PCW.	BULLOCK.	STANDARD.	STANDARD						22.9	122.3	35.9	8.4	6.4	43.9	20.4	
721	IRRI.	PSW.	BULLOCK.	STANDARD.	STANDARD						22.3	135.7	37.8	6.2	6.2	51.7	22.2	
722	IRRI.	PRW.	BULLOCK.	STANDARD.	STANDARD						122.9	86.7	15.4	8.2	6.8	41.0	24.0	
723	IRRI.	SCWN.	BULLOCK.	STANDARD.	STANDARD					13.6	81.0	13.6	32.0	13.6	39.1	20.4		
724	IRRI.	SRWN.	BULLOCK.	STANDARD.	STANDARD					3.0	12.7	91.4	31.5	12.8	13.1	39.0	14.0	
725	IRRI.	SCWS.	BULLOCK.	STANDARD.	STANDARD					9.1	10.3	57.9	13.4	35.3	13.3	41.4	16.7	
726	IRRI.	SRWS.	BULLOCK.	STANDARD.	STANDARD					19.0	58.5	13.2	34.5	13.1	40.3	13.5	13.5	
727	IRRI.	PMW.	SEMI-MECH.	STANDARD.	STANDARD						5.3	105.5	29.9	8.4	6.4	43.4	17.9	
728	IRRI.	PCW.	SEMI-MECH.	STANDARD.	STANDARD						5.3	105.5	29.9	8.4	6.4	43.4	17.9	
729	IRRI.	PSW.	SEMI-MECH.	STANDARD.	STANDARD						7.9	119.2	28.8	6.2	6.2	51.2	19.7	
730	IRRI.	PRW.	SEMI-MECH.	STANDARD.	STANDARD						71.0	80.4	13.7	8.2	6.8	41.5	21.5	
731	IRRI.	SCWN.	SEMI-MECH.	STANDARD.	STANDARD					3.3	4.7	13.6	32.0	13.6	36.6	17.9		
732	IRRI.	SRWN.	SEMI-MECH.	STANDARD.	STANDARD					3.0	8.7	13.4	31.5	12.8	13.1	36.6	14.0	
733	IRRI.	SCWS.	SEMI-MECH.	STANDARD.	STANDARD					2.3	2.9	43.5	13.4	35.3	13.3	38.8	14.2	
734	IRRI.	SRWS.	SEMI-MECH.	STANDARD.	STANDARD					5.0	43.7	13.2	34.5	13.1	37.6	13.5	13.5	
735																		
736	MAIZE.	NWFP.	BULLOCK.	STANDARD.	STANDARD						43.4	38.6	0.7	2.7	46.1			
737	MAIZE.	PCW.	BULLOCK.	STANDARD.	STANDARD						10.8	4.5	16.7	44.2	2.5	26.8	27.0	
738	MAIZE.	PSW.	BULLOCK.	STANDARD.	STANDARD						10.2	4.1	15.9	42.3	2.0	23.0	23.8	
739	MAIZE.	PRW.	BULLOCK.	STANDARD.	STANDARD						10.2	4.1	15.9	42.3	2.0	23.0	23.8	
740	MAIZE.	SCWN.	BULLOCK.	STANDARD.	STANDARD						18.65	27.4	3.2	3.2	3.2	57.4		
741	MAIZE.	SCWS.	BULLOCK.	STANDARD.	STANDARD					17.9	26.8	3.0	3.0	3.0	3.0	59.3		
742	MAIZE.	NWFP.	SEMI-MECH.	STANDARD.	STANDARD						14.6	38.6	0.7	2.7	43.8			

INDUS BASIN MODEL REVISED (IBMR) FILENAME=WSISD41
CROP DATA

```
743  MAIZE.  PCW. SEMI-MECH.STANDARD. STANDARD                                    2.41  8.36 44.2    2.5  26.0  26.8
744  MAIZE.  PSW. SEMI-MECH.STANDARD. STANDARD                                    2.24  8.06 42.3    2.5  22.0  22.5
745  MAIZE.  PRW. SEMI-MECH.STANDARD. STANDARD                                    2.24  8.06 42.3    2.5  22.0  22.5
746  MAIZE.  SCWN.SEMI-MECH.STANDARD. STANDARD                         3.1        5.2   3.2   3.2    3.2  57.4
747  MAIZE.  SCWS.SEMI-MECH.STANDARD. STANDARD                  3.4    5.8        3.0   3.0   3.0    3.0  59.3
748
749  MUS+RAP.NWFP.BULLOCK.    STANDARD. STANDARD                                  8.9  10.5  18.0   14.6  23.0  23.0
750  MUS+RAP.PMW. BULLOCK.    STANDARD. STANDARD  25.7  15.0  20.5                                  19.3   9.0  23.0
751  MUS+RAP.PCW. BULLOCK.    STANDARD. STANDARD  21.75 20.5   6.0                                  10.8  13.6  13.9
752  MUS+RAP.PSW. BULLOCK.    STANDARD. STANDARD  20.5   4.0   3.0                            10.2  12.8  13.2  23.0
753  MUS+RAP.PRW. BULLOCK.    STANDARD. STANDARD   3.5   3.0                                        19.1  15.7  25.5  23.0
754  MUS+RAP.SCWN.BULLOCK.    STANDARD. STANDARD   1.5   1.5   1.5                                  15.1  12.3   1.5   1.5
755  MUS+RAP.SRWN.BULLOCK.    STANDARD. STANDARD   1.5   1.5   1.5                                  15.2  12.7   1.5   1.5
756  MUS+RAP.SRWS.BULLOCK.    STANDARD. STANDARD   1.5   1.5   1.5                                  15.2  12.7   1.5   1.5
757  MUS+RAP.SCWS.BULLOCK.    STANDARD. STANDARD   1.5   1.5   1.5                                  13.5   7.6   1.5   1.5
758  MUS+RAP.NWFP.SEMI-MECH.STANDARD. STANDARD                               1.87  3.45  6.0   14.6  23.0  23.0
759  MUS+RAP.PMW. SEMI-MECH.STANDARD. STANDARD    25.7  13.5  18.0                                   3.8   5.5  23.0
760  MUS+RAP.PCW. SEMI-MECH.STANDARD. STANDARD    21.75 20.25  6.0                                   1.62  5.08 13.9
761  MUS+RAP.PSW. SEMI-MECH.STANDARD. STANDARD    20.5   3.5   2.5                            1.54  4.9   13.2  23.0
762  MUS+RAP.PRW. SEMI-MECH.STANDARD. STANDARD     4.0   2.0                                         5.94 15.7  25.5  23.0
763  MUS+RAP.SCWN.SEMI-MECH.STANDARD. STANDARD     1.0   1.0   1.0                                   3.7   2.8   1.0   1.0
764  MUS+RAP.SRWN.SEMI-MECH.STANDARD. STANDARD     1.0   1.0   1.0                                   3.1   2.5   1.0   1.0
765  MUS+RAP.SRWS.SEMI-MECH.STANDARD. STANDARD     1.0   1.0   1.0                                   3.1   2.5   1.0   1.0
766  MUS+RAP.SCWS.SEMI-MECH.STANDARD. STANDARD     1.0   1.0   1.0                                   3.3   2.9   1.0   1.0
767
768  SC-GUR. NWFP.BULLOCK.    STANDARD. STANDARD  154.6      147 57.4  23.4   5.4   5.74  3.4   3.4  24.0  54.0    216 208.2
769  SC-GUR.(PMW,PCW,PSW,PRW).
770                BULLOCK.   STANDARD. STANDARD        159 80.4  54.4  22.5   4.75  5.05  3.0   3.0   3.0   1.5  150.3 148.5
771  SC-GUR. SCWN.BULLOCK.    STANDARD. STANDARD  207.6 149.4 53.2 20.8   3.2   8.9   3.2   3.2   3.2   3.2  183.4 195.6
772  SC-GUR. SRWN.BULLOCK.    STANDARD. STANDARD  213.7 145.2 54.4 21.7   3.2   9.0   3.2   3.2   3.2   3.2  187.3    198
773  SC-GUR. SCWS.BULLOCK.    STANDARD. STANDARD  207.1 163.5 53.4 31.5   3.2   9.0   3.2   3.2   3.2   3.2  187.4 191.5
774  SC-GUR. SRWS.BULLOCK.    STANDARD. STANDARD  208.2 161.4 49.5 37.2   3.2   9.0   3.2   3.2   3.2   3.2  183.2 197.5
775  SC-GUR. NWFP.SEMI-MECH.STANDARD. STANDARD    143.1 144.1 50.15 23.4  5.4   5.74  3.4   3.4  17.5  54.0    216 201.9
776  SC-GUR.(PMW,PCW,PSW,PRW).
777           SEMI-MECH.STANDARD. STANDARD       151.2 80.62 48.92 22.5  4.75  5.05  3.0   3.0   3.0   1.5  148.3 142.3
778  SC-GUR. SCWN.SEMI-MECH.STANDARD. STANDARD    201.2 143.4 48.5 16.4   3.2   8.9   3.2   3.2   3.2   3.2  179.2 193.4
779  SC-GUR. SRWN.SEMI-MECH.STANDARD. STANDARD    209.4 139.8 51.4 18.6   3.2   9.0   3.2   3.2   3.2   3.2  184.3 193.7
780  SC-GUR. SCWS.SEMI-MECH.STANDARD. STANDARD    201.6 156.5 47.4 28.2   3.2   9.0   3.2   3.2   3.2   3.2  181.5 186.4
781  SC-GUR. SRWS.SEMI-MECH.STANDARD. STANDARD    203.5 153.4 44.5 32.6   3.2   9.0   3.2   3.2   3.2   3.2  179.8 191.6
782
783  SC-MILL.NWFP.BULLOCK.    STANDARD. STANDARD  120   115   95.0 65.0   5.4   5.74  3.4   3.4  23.4  54.0  107   125
784  SC-MILL.PMW. BULLOCK.    STANDARD. STANDARD   87.0  85.2 74.0 64.5  44.5   5.05  3.0   3.0   3.0   1.5   74.0  85.5
785  SC-MILL.PCW. BULLOCK.    STANDARD. STANDARD   90.0  85.0 95.0 72.1  30.0   5.05  3.0   3.0   3.0   1.5   85.0  95.0
786  SC-MILL.PSW. BULLOCK.    STANDARD. STANDARD   88.0  85.0 95.0 70.0  40.0   5.05  3.0   3.0   3.0   1.5   85.0  85.0
787  SC-MILL.PRW. BULLOCK.    STANDARD. STANDARD   90.0  95.0 90.9 65.0  38.5   5.05  3.0   3.0   3.0   1.5   84.0  85.5
788  SC-MILL.SCWN.BULLOCK.    STANDARD. STANDARD   87.5  85.0 92.1 47.5   3.2   8.4  35.4   3.2   3.2   3.2   93.1  87.3
789  SC-MILL.SRWN.BULLOCK.    STANDARD. STANDARD   89.3  87.1 94.2 43.1   3.2   9.2  37.8   3.2   3.2   3.2   94.2  89.1
790  SC-MILL.SCWS.BULLOCK.    STANDARD. STANDARD   98.4  95.3 102.4 46.2  3.2   9.8  39.4   3.2   3.2   3.2   92.8  93.7
791  SC-MILL.SRWS.BULLOCK.    STANDARD. STANDARD   94.7  96.5 103.4 41.4  3.2   8.9  41.3   3.2   3.2   3.2   91.7  89.8
792  SC-MILL.NWFP.SEMI-MECH.STANDARD. STANDARD    116   116   90.6 58.0   5.4   5.74  3.4   3.4  22.0  51.0  102   120
793  SC-MILL.PMW. SEMI-MECH.STANDARD. STANDARD     84.0  81.5 20.0 59.0  59.0  38.5   5.05  3.0   3.0   1.5   69.5  80.5
794  SC-MILL.PCW. SEMI-MECH.STANDARD. STANDARD     86.5  81.5 90.0 64.0  34.5   5.05  3.0   3.0   3.0   1.5   80.5  90.5
795  SC-MILL.PSW. SEMI-MECH.STANDARD. STANDARD     86.0  82.0 90.5 64.5  35.0   5.05  3.0   3.0   3.0   1.5   81.5  84.0
796  SC-MILL.PRW. SEMI-MECH.STANDARD. STANDARD     87.0  92.5 87.0 62.5  35.0   5.05  3.0   3.0   3.0   1.5   80.0  81.5
797  SC-MILL.SCWN.SEMI-MECH.STANDARD. STANDARD     83.5  81.9 89.2 43.4   3.2   8.4  35.4   3.2   3.2   3.2   89.1  83.8
798  SC-MILL.SRWN.SEMI-MECH.STANDARD. STANDARD     84.1  82.3 90.1 44.5   3.2   9.2  35.7   3.2   3.2   3.2   91.2  84.5
799  SC-MILL.SCWS.SEMI-MECH.STANDARD. STANDARD     96.5  92.7 99.4 46.2   3.2   9.8  39.4   3.2   3.2   3.2   89.8  91.1
800  SC-MILL.SRWS.SEMI-MECH.STANDARD. STANDARD     92.8  93.7 101.8 39.1  3.2   8.9  40.5   3.2   3.2   3.2   89.5  84.7
801
802  KHA-FOD.SRWS.BULLOCK.    LA-PLANT. STANDARD                      13.4   3.0   3.0   8.3   3.0   3.0
803  KHA-FOD.NWFP.BULLOCK.    STANDARD. STANDARD              18.5   2.5   3.0  17.5   2.0  18.0   3.0   1.5
804  KHA-FOD.PMW. BULLOCK.    STANDARD. STANDARD              16.5   3.0   3.5  17.0   4.5  18.5   2.5   0.5
805  KHA-FOD.PCW. BULLOCK.    STANDARD. STANDARD              18.5   2.0   3.0  18.7   4.0  18.0   3.0   1.0
806  KHA-FOD.PSW. BULLOCK.    STANDARD. STANDARD              18.5   1.8   2.5  25.5   4.5  27.5   2.5   1.2
807  KHA-FOD.PRW. BULLOCK.    STANDARD. STANDARD              17.0   1.5   2.5  20.0   4.0  18.5   3.0   1.5
```

INDUS BASIN MODEL REVISED (IBMR) FILENAME=WSISD41
CROP DATA

```
808  KHA-FOD.SCWN.BULLOCK.   STANDARD. STANDARD                    10.7  3.0   7.3  3.0  3.0  7.3  3.0  3.0
809  KHA-FOD.SCWS.BULLOCK.   STANDARD. STANDARD                    13.4  3.0   8.2  3.0  3.0  8.3  3.0  3.0
810  KHA-FOD.SRWN.BULLOCK.   STANDARD. STANDARD                    14.3  3.0  11.4  3.0  3.0  7.5  3.0  3.0
811  KHA-FOD.SRWS.BULLOCK.   STANDARD. STANDARD                    14.3  3.0  11.4  3.0  3.0  7.5  3.0  3.0
812  KHA-FOD.SRWS.SEMI-MECH.LA-PLANT. STANDARD                                 4.0  2.1  0.8  2.3  1.2  1.2
813  KHA-FOD.NWFP.SEMI-MECH.STANDARD. STANDARD                      7.0  1.0   2.5  6.0  3.5  7.0  3.0  1.5
814  KHA-FOD.PMW. SEMI-MECH.STANDARD. STANDARD                      7.5  1.5   1.0  7.0  1.0  8.0  2.0  1.0
815  KHA-FOD.PCW. SEMI-MECH.STANDARD. STANDARD                      6.0  1.3   4.5  6.5  4.2  5.5  2.0  1.5
816  KHA-FOD.PSW. SEMI-MECH.STANDARD. STANDARD                      5.5  1.2   1.7  7.0  4.5  4.0  2.0  1.5
817  KHA-FOD.PRW. SEMI-MECH.STANDARD. STANDARD                      6.0  1.5   2.0  8.5  3.8  5.5  2.2  1.5
818  KHA-FOD.SCWN.SEMI-MECH.STANDARD. STANDARD                      2.1  1.2   2.7  1.2  1.2  2.7  1.2  1.2
819  KHA-FOD.SCWS.SEMI-MECH.STANDARD. STANDARD                      2.0  1.2   2.8  1.2  1.2  2.9  1.2  1.2
820  KHA-FOD.SRWN.SEMI-MECH.STANDARD. STANDARD                      2.2  1.2   2.1  1.2  1.2  2.3  1.2  1.2
821  KHA-FOD.SRWS.SEMI-MECH.STANDARD. STANDARD                      2.2  1.2   2.1  1.2  1.2  2.3  1.2  1.2
822
823  WHEAT. NWFP.BULLOCK.  LA-PLANT. HEAVY     5.8  3.7   3.7  38.3  14.2                              43.6   3.7
824  WHEAT. PMW. BULLOCK.  LA-PLANT. HEAVY     3.4  2.9   1.7  36.3  12.1                              31.9   2.9
825  WHEAT. PCW. BULLOCK.  LA-PLANT. HEAVY     4.3  3.9   3.9  33.2  12.3                              45.2   3.9
826  WHEAT. PSW. BULLOCK.  LA-PLANT. HEAVY     4.4  2.7   2.4  35.9  15.3                              41.4   4.0
827  WHEAT. PRW. BULLOCK.  LA-PLANT. HEAVY     4.1  2.4   2.4  36.1  15.4                              17.2  19.1
828  WHEAT. SCWN.BULLOCK.  LA-PLANT. HEAVY     3.2  3.2        24.0                                    34.0   3.2
829  WHEAT. SRWN.BULLOCK.  LA-PLANT. HEAVY     3.2  3.2        25.4                                    37.5   3.2
830  WHEAT. SCWS.BULLOCK.  LA-PLANT. HEAVY     3.2  3.2  14.0   7.3                                    36.3   3.2
831  WHEAT. SRWS.BULLOCK.  LA-PLANT. HEAVY     3.2  3.2  10.7   7.1                                    36.2   3.2
832  WHEAT. NWFP.BULLOCK.  LA-PLANT. JANUARY   5.8  3.7   3.7  38.3  14.2                              43.6   3.7
833  WHEAT. PMW. BULLOCK.  LA-PLANT. JANUARY   3.4  2.9   1.7  36.3  12.1                              31.9   2.9
834  WHEAT. PCW. BULLOCK.  LA-PLANT. JANUARY   4.3  3.9   3.9  33.2  12.3                              45.2   3.9
835  WHEAT. PSW. BULLOCK.  LA-PLANT. JANUARY   4.4  2.7   2.4  35.9  15.3                              41.4   4.0
836  WHEAT. PRW. BULLOCK.  LA-PLANT. JANUARY   4.1  2.4   2.4  36.1  15.4                              17.2  19.1
837  WHEAT. SCWN.BULLOCK.  LA-PLANT. JANUARY   3.2  3.2        24.0                                    34.0   3.2
838  WHEAT. SRWN.BULLOCK.  LA-PLANT. JANUARY   3.2  3.2        25.4                                    37.5   3.2
839  WHEAT. SCWS.BULLOCK.  LA-PLANT. JANUARY   3.2  3.2  14.0   7.3                                    36.3   3.2
840  WHEAT. SRWS.BULLOCK.  LA-PLANT. JANUARY   3.2  3.2  10.7   7.1                                    36.2   3.2
841  WHEAT. NWFP.BULLOCK.  LA-PLANT. LIGHT     5.8  3.7   3.7  50.1  18.5                              43.6   3.7
842  WHEAT. PMW. BULLOCK.  LA-PLANT. LIGHT     3.4  2.9   1.7  47.5  15.8                              31.9   2.9
843  WHEAT. PCW. BULLOCK.  LA-PLANT. LIGHT     4.3  3.9   3.9  43.4  16.1                              45.2   3.9
844  WHEAT. PSW. BULLOCK.  LA-PLANT. LIGHT     4.4  2.7   2.4  47.0  20.1                              41.4   4.0
845  WHEAT. PRW. BULLOCK.  LA-PLANT. LIGHT     4.1  2.4   2.4  47.1  20.1                              17.2  19.1
846  WHEAT. SCWN.BULLOCK.  LA-PLANT. LIGHT     3.2  3.2        31.4                                    34.0   3.2
847  WHEAT. SRWN.BULLOCK.  LA-PLANT. LIGHT     3.2  3.2        33.2                                    37.5   3.2
848  WHEAT. SCWS.BULLOCK.  LA-PLANT. LIGHT     3.2  3.2  18.3   9.6                                    36.3   3.2
849  WHEAT. SRWS.BULLOCK.  LA-PLANT. LIGHT     3.2  3.2  13.9   9.3                                    36.2   3.2
850  WHEAT. NWFP.BULLOCK.  LA-PLANT. STANDARD  5.8  3.7   3.7  59.0  21.8                              43.6   3.7
851  WHEAT. PMW. BULLOCK.  LA-PLANT. STANDARD  3.4  2.9   1.7  55.9  18.6                              31.9   2.9
852  WHEAT. PCW. BULLOCK.  LA-PLANT. STANDARD  4.3  3.9   3.9  51.4  18.9                              45.2   3.9
853  WHEAT. PSW. BULLOCK.  LA-PLANT. STANDARD  4.4  2.7   2.4  55.3  23.6                              41.4   4.0
854  WHEAT. PRW. BULLOCK.  LA-PLANT. STANDARD  4.1  2.4   2.4  55.5  23.7                              17.2  19.1
855  WHEAT. SCWN.BULLOCK.  LA-PLANT. STANDARD  3.2  3.2        37.0                                    34.0   3.2
856  WHEAT. SRWN.BULLOCK.  LA-PLANT. STANDARD  3.2  3.2        39.1                                    37.5   3.2
857  WHEAT. SCWS.BULLOCK.  LA-PLANT. STANDARD  3.2  3.2  21.5  11.3                                    36.3   3.2
858  WHEAT. SRWS.BULLOCK.  LA-PLANT. STANDARD  3.2  3.2  16.4  10.9                                    36.2   3.2
859  WHEAT. NWFP.BULLOCK.  QK-HARV.  HEAVY     5.8  3.7   3.7  59.8                              22.3  21.2   3.7
860  WHEAT. PMW. BULLOCK.  QK-HARV.  HEAVY     3.4  2.9   1.7  55.1                              15.9  16.0   2.9
861  WHEAT. PCW. BULLOCK.  QK-HARV.  HEAVY     4.3  3.9   3.9  57.3                              18.6  23.9   3.9
862  WHEAT. PSW. BULLOCK.  QK-HARV.  HEAVY     4.4  2.7   2.4  58.4                              18.3  23.1   4.0
863  WHEAT. PRW. BULLOCK.  QK-HARV.  HEAVY     4.1  2.4   2.4  58.6                                    33.9   2.4
864  WHEAT. SCWN.BULLOCK.  QK-HARV.  HEAVY     3.2  3.2  19.2   8.3                              18.9  15.1   3.2
865  WHEAT. SRWN.BULLOCK.  QK-HARV.  HEAVY     3.2  3.2  19.3   9.6                              20.2  17.3   3.2
866  WHEAT. SCWS.BULLOCK.  QK-HARV.  HEAVY     3.2  3.2  24.2                                    16.2  20.1   3.2
867  WHEAT. SRWS.BULLOCK.  QK-HARV.  HEAVY     3.2  3.2  22.3                                    16.5  19.7   3.2
868  WHEAT. NWFP.BULLOCK.  QK-HARV.  JANUARY   5.8  3.7   3.7  59.8                              22.3  21.2   3.7
869  WHEAT. PMW. BULLOCK.  QK-HARV.  JANUARY   3.4  2.9   1.7  55.1                              15.9  16.0   2.9
870  WHEAT. PCW. BULLOCK.  QK-HARV.  JANUARY   4.3  3.9   3.9  57.3                              18.6  23.9   3.9
871  WHEAT. PSW. BULLOCK.  QK-HARV.  JANUARY   4.4  2.7   2.4  58.4                              18.3  23.1   4.0
872  WHEAT. PRW. BULLOCK.  QK-HARV.  JANUARY   4.1  2.4   2.4  58.6                                    33.9   2.4
```

INDUS BASIN MODEL REVISED (IBMR) FILENAME=WSISD41   07/25/91 17:17:54
CROP DATA   GAMS 2.21 IBM CMS

#	Crop	Region	Method	Season	v1	v2	v3	v4	v5		v6	v7	v8
873	WHEAT.	SCWN.BULLOCK.	QK-HARV.	JANUARY	3.2	3.2	19.2	8.3			18.9	15.1	3.2
874	WHEAT.	SRWN.BULLOCK.	QK-HARV.	JANUARY	3.2	3.2	19.3	9.6			20.2	17.3	3.2
875	WHEAT.	SCWS.BULLOCK.	QK-HARV.	JANUARY	3.2	3.2	24.2				16.2	20.1	3.2
876	WHEAT.	SRWS.BULLOCK.	QK-HARV.	JANUARY	3.2	3.2	22.3				16.5	19.7	3.2
877	WHEAT.	NWFP.BULLOCK.	QK-HARV.	LIGHT	5.8	3.7	3.7	78.2			22.3	21.2	3.7
878	WHEAT.	PMW. BULLOCK.	QK-HARV.	LIGHT	3.4	2.9	1.7	72.0			15.9	16.0	2.9
879	WHEAT.	PCW. BULLOCK.	QK-HARV.	LIGHT	4.3	3.9	3.9	74.9			18.6	23.9	3.9
880	WHEAT.	PSW. BULLOCK.	QK-HARV.	LIGHT	4.4	2.7	2.4	76.3			18.3	23.1	4.0
881	WHEAT.	PRW. BULLOCK.	QK-HARV.	LIGHT	4.1	2.4	2.4	76.6				33.9	2.4
882	WHEAT.	SCWN.BULLOCK.	QK-HARV.	LIGHT	3.2	3.2	25.1	10.8			18.9	15.1	3.2
883	WHEAT.	SRWN.BULLOCK.	QK-HARV.	LIGHT	3.2	3.2	25.2	12.6			20.2	17.3	3.2
884	WHEAT.	SCWS.BULLOCK.	QK-HARV.	LIGHT	3.2	3.2	31.7				16.2	20.1	3.2
885	WHEAT.	SRWS.BULLOCK.	QK-HARV.	LIGHT	3.2	3.2	29.3				16.5	19.7	3.2
886	WHEAT.	NWFP.BULLOCK.	QK-HARV.	STANDARD	5.8	3.7	3.7	91.9			22.3	21.2	3.7
887	WHEAT.	PMW. BULLOCK.	QK-HARV.	STANDARD	3.4	2.9	1.7	84.8			15.9	16.0	2.9
888	WHEAT.	PCW. BULLOCK.	QK-HARV.	STANDARD	4.3	3.9	3.9	88.1			18.6	23.9	3.9
889	WHEAT.	PSW. BULLOCK.	QK-HARV.	STANDARD	4.4	2.7	2.4	89.8			18.3	23.1	4.0
890	WHEAT.	PRW. BULLOCK.	QK-HARV.	STANDARD	4.1	2.4	2.4	90.1				33.9	2.4
891	WHEAT.	SCWN.BULLOCK.	QK-HARV.	STANDARD	3.2	3.2	29.5	12.7			18.9	15.1	3.2
892	WHEAT.	SRWN.BULLOCK.	QK-HARV.	STANDARD	3.2	3.2	29.7	14.8			20.2	17.3	3.2
893	WHEAT.	SCWS.BULLOCK.	QK-HARV.	STANDARD	3.2	3.2	37.3				16.2	20.1	3.2
894	WHEAT.	SRWS.BULLOCK.	QK-HARV.	STANDARD	3.2	3.2	34.3				16.5	19.7	3.2
895	WHEAT.	NWFP.BULLOCK.	STANDARD.	HEAVY	5.8	3.7	3.7	43.6	16.1		22.3	21.2	3.7
896	WHEAT.	PMW. BULLOCK.	STANDARD.	HEAVY	3.4	2.9	1.7	41.3	13.8		15.9	16.0	2.9
897	WHEAT.	PCW. BULLOCK.	STANDARD.	HEAVY	4.3	3.9	3.9	41.9	15.5		18.6	23.9	3.9
898	WHEAT.	PSW. BULLOCK.	STANDARD.	HEAVY	4.4	2.7	2.4	40.9	17.5		18.3	23.1	4.0
899	WHEAT.	PRW. BULLOCK.	STANDARD.	HEAVY	4.1	2.4	2.4	41.0	17.6			33.9	2.4
900	WHEAT.	SCWN.BULLOCK.	STANDARD.	HEAVY	3.2	3.2		27.4			18.9	15.1	3.2
901	WHEAT.	SRWN.BULLOCK.	STANDARD.	HEAVY	3.2	3.2		28.9			20.2	17.3	3.2
902	WHEAT.	SCWS.BULLOCK.	STANDARD.	HEAVY	3.2	3.2	15.9	8.3			16.2	20.1	3.2
903	WHEAT.	SRWS.BULLOCK.	STANDARD.	HEAVY	3.2	3.2	14.2	8.0			16.5	19.7	3.2
904	WHEAT.	NWFP.BULLOCK.	STANDARD.	JANUARY	5.8	3.7	3.7	43.6	16.1		22.3	21.2	3.7
905	WHEAT.	PMW. BULLOCK.	STANDARD.	JANUARY	3.4	2.9	1.7	41.3	13.8		15.9	16.0	2.9
906	WHEAT.	PCW. BULLOCK.	STANDARD.	JANUARY	4.3	3.9	3.9	41.9	15.5		18.6	23.9	3.9
907	WHEAT.	PSW. BULLOCK.	STANDARD.	JANUARY	4.4	2.7	2.4	40.9	17.5		18.3	23.1	4.0
908	WHEAT.	PRW. BULLOCK.	STANDARD.	JANUARY	4.1	2.4	2.4	41.0	17.6			33.9	2.4
909	WHEAT.	SCWN.BULLOCK.	STANDARD.	JANUARY	3.2	3.2		27.4			18.9	15.1	3.2
910	WHEAT.	SRWN.BULLOCK.	STANDARD.	JANUARY	3.2	3.2		18.9			20.2	17.3	3.2
911	WHEAT.	SCWS.BULLOCK.	STANDARD.	JANUARY	3.2	3.2	15.9	8.3			16.2	20.1	3.2
912	WHEAT.	SRWS.BULLOCK.	STANDARD.	JANUARY	3.2	3.2	14.2	8.0			16.5	19.7	3.2
913	WHEAT.	NWFP.BULLOCK.	STANDARD.	LIGHT	5.8	3.7	3.7	57.0	21.1		22.3	21.2	3.7
914	WHEAT.	PMW. BULLOCK.	STANDARD.	LIGHT	3.4	2.9	1.7	54.1	18.0		15.9	16.0	2.9
915	WHEAT.	PCW. BULLOCK.	STANDARD.	LIGHT	4.3	3.9	3.9	54.7	20.2		18.6	23.9	3.9
916	WHEAT.	PSW. BULLOCK.	STANDARD.	LIGHT	4.4	2.7	2.4	53.5	22.9		18.3	23.1	4.0
917	WHEAT.	PRW. BULLOCK.	STANDARD.	LIGHT	4.1	2.4	2.4	53.6	23.0			33.9	2.4
918	WHEAT.	SCWN.BULLOCK.	STANDARD.	LIGHT	3.2	3.2		35.9			18.9	15.1	3.2
919	WHEAT.	SRWN.BULLOCK.	STANDARD.	LIGHT	3.2	3.2		37.8			20.2	17.3	3.2
920	WHEAT.	SCWS.BULLOCK.	STANDARD.	LIGHT	3.2	3.2	20.8	10.9			16.2	20.1	3.2
921	WHEAT.	SRWS.BULLOCK.	STANDARD.	LIGHT	3.2	3.2	18.6	10.7			16.5	19.7	3.2
922	WHEAT.	NWFP.BULLOCK.	STANDARD.	STANDARD	5.8	3.7	3.7	67.1	24.8		22.3	21.2	3.7
923	WHEAT.	PMW. BULLOCK.	STANDARD.	STANDARD	3.4	2.9	1.7	63.6	21.2		15.9	16.0	2.9
924	WHEAT.	PCW. BULLOCK.	STANDARD.	STANDARD	4.3	3.9	3.9	64.4	23.8		18.6	23.9	3.9
925	WHEAT.	PSW. BULLOCK.	STANDARD.	STANDARD	4.4	2.7	2.4	62.9	26.9		18.3	23.1	4.0
926	WHEAT.	PRW. BULLOCK.	STANDARD.	STANDARD	4.1	2.4	2.4	63.1	27.0			33.9	2.4
927	WHEAT.	SCWN.BULLOCK.	STANDARD.	STANDARD	3.2	3.2		42.2			18.9	15.1	3.2
928	WHEAT.	SRWN.BULLOCK.	STANDARD.	STANDARD	3.2	3.2		44.5			20.2	17.3	3.2
929	WHEAT.	SCWS.BULLOCK.	STANDARD.	STANDARD	3.2	3.2	24.5	12.8			16.2	20.1	3.2
930	WHEAT.	SRWS.BULLOCK.	STANDARD.	STANDARD	3.2	3.2	21.9	12.4			16.5	19.7	3.2
931	WHEAT.	NWFP.SEMI-MECH.LA-PLANT.		HEAVY	5.8	3.7	3.7	37.2	12.4			13.8	3.7
932	WHEAT.	PMW. SEMI-MECH.LA-PLANT.		HEAVY	3.4	2.9	1.7	33.1	11.1			2.3	10.7
933	WHEAT.	PCW. SEMI-MECH.LA-PLANT.		HEAVY	4.3	3.9	3.9	34.3	11.2			8.0	11.5
934	WHEAT.	PSW. SEMI-MECH.LA-PLANT.		HEAVY	4.4	2.7	2.4	32.4	13.9			3.4	15.6
935	WHEAT.	PRW. SEMI-MECH.LA-PLANT.		HEAVY	4.1	2.4	2.4	32.3	13.8				11.9
936	WHEAT.	SCWN.SEMI-MECH.LA-PLANT.		HEAVY	3.2	3.2		22.4				3.1	9.1
937	WHEAT.	SRWN.SEMI-MECH.LA-PLANT.		HEAVY	3.2	3.2		25.4				2.9	9.6

INDUS BASIN MODEL REVISED (IBMR) FILENAME=WSISD41
CROP DATA

#	Crop	Type	Season	v1	v2	v3	v4	v5		v6	v7	v8
938	WHEAT.	SCWS.SEMI-MECH.LA-PLANT.	HEAVY	3.2	3.2	14.0	7.3				3.2	11.0
939	WHEAT.	SRWS.SEMI-MECH.LA-PLANT.	HEAVY	3.2	3.2	10.7	7.1				3.2	11.0
940	WHEAT.	NWFP.SEMI-MECH.LA-PLANT.	JANUARY	5.8	3.7	3.7	37.2	12.4			13.8	3.7
941	WHEAT.	PMW. SEMI-MECH.LA-PLANT.	JANUARY	3.4	2.9	1.7	33.1	11.1			2.3	10.7
942	WHEAT.	PCW. SEMI-MECH.LA-PLANT.	JANUARY	4.3	3.9	3.9	34.3	11.2			8.0	11.5
943	WHEAT.	PSW. SEMI-MECH.LA-PLANT.	JANUARY	4.4	2.7	2.4	32.4	13.9			3.4	15.6
944	WHEAT.	PRW. SEMI-MECH.LA-PLANT.	JANUARY	4.1	2.4	2.4	32.3	13.8				11.9
945	WHEAT.	SCWN.SEMI-MECH.LA-PLANT.	JANUARY	3.2	3.2		22.9				3.1	9.1
946	WHEAT.	SRWN.SEMI-MECH.LA-PLANT.	JANUARY	3.2	3.2		25.4				2.9	9.6
947	WHEAT.	SCWS.SEMI-MECH.LA-PLANT.	JANUARY	3.2	3.2	14.0	7.3				3.2	11.0
948	WHEAT.	SRWS.SEMI-MECH.LA-PLANT.	JANUARY	3.2	3.2	10.7	7.1				3.2	11.0
949	WHEAT.	NWFP.SEMI-MECH.LA-PLANT.	LIGHT	5.8	3.7	3.7	48.6	16.2			13.8	3.7
950	WHEAT.	PMW. SEMI-MECH.LA-PLANT.	LIGHT	3.4	2.9	1.7	43.3	14.4			2.3	10.7
951	WHEAT.	PCW. SEMI-MECH.LA-PLANT.	LIGHT	4.3	3.9	3.9	44.9	14.6			8.0	11.5
952	WHEAT.	PSW. SEMI-MECH.LA-PLANT.	LIGHT	4.4	2.7	2.4	42.4	18.2			3.4	15.6
953	WHEAT.	PRW. SEMI-MECH.LA-PLANT.	LIGHT	4.1	2.4	2.4	42.2	18.1				11.9
954	WHEAT.	SCWN.SEMI-MECH.LA-PLANT.	LIGHT	3.2	3.2		31.4				3.1	9.1
955	WHEAT.	SRWN.SEMI-MECH.LA-PLANT.	LIGHT	3.2	3.2		33.2				2.9	9.6
956	WHEAT.	SCWS.SEMI-MECH.LA-PLANT.	LIGHT	3.2	3.2	18.3	9.6				3.2	11.0
957	WHEAT.	SRWS.SEMI-MECH.LA-PLANT.	LIGHT	3.2	3.2	13.9	9.3				3.2	11.0
958	WHEAT.	NWFP.SEMI-MECH.LA-PLANT.	STANDARD	5.8	3.7	3.7	57.2	19.1			13.8	3.7
959	WHEAT.	PMW. SEMI-MECH.LA-PLANT.	STANDARD	3.4	2.9	1.7	51.0	17.0			2.3	10.7
960	WHEAT.	PCW. SEMI-MECH.LA-PLANT.	STANDARD	4.3	3.9	3.9	52.8	17.2			8.0	11.5
961	WHEAT.	PSW. SEMI-MECH.LA-PLANT.	STANDARD	4.4	2.7	2.4	49.4	21.4			3.4	15.6
962	WHEAT.	PRW. SEMI-MECH.LA-PLANT.	STANDARD	4.1	2.4	2.4	49.7	21.3				11.9
963	WHEAT.	SCWN.SEMI-MECH.LA-PLANT.	STANDARD	3.2	3.2		37.0				3.1	9.1
964	WHEAT.	SRWN.SEMI-MECH.LA-PLANT.	STANDARD	3.2	3.2		39.1				2.9	9.6
965	WHEAT.	SCWS.SEMI-MECH.LA-PLANT.	STANDARD	3.2	3.2	21.5	11.3				3.2	11.0
966	WHEAT.	SRWS.SEMI-MECH.LA-PLANT.	STANDARD	3.2	3.2	16.4	10.9				3.2	11.0
967	WHEAT.	NWFP.SEMI-MECH.QK-HARV.	HEAVY	5.8	3.7	3.7	56.4			4.1	9.7	3.7
968	WHEAT.	PMW. SEMI-MECH.QK-HARV.	HEAVY	3.4	2.9	1.7	50.2			2.3	7.8	2.9
969	WHEAT.	PCW. SEMI-MECH.QK-HARV.	HEAVY	4.3	3.9	3.9	44.0				15.6	3.9
970	WHEAT.	PSW. SEMI-MECH.QK-HARV.	HEAVY	4.4	2.7	2.4	58.4			3.4	11.6	4.0
971	WHEAT.	PRW. SEMI-MECH.QK-HARV.	HEAVY	4.1	2.4	2.4	52.4				9.5	2.4
972	WHEAT.	SCWN.SEMI-MECH.QK-HARV.	HEAVY	3.2	3.2	19.2	8.2				9.0	3.2
973	WHEAT.	SRWN.SEMI-MECH.QK-HARV.	HEAVY	3.2	3.2	19.3	9.6				9.3	3.2
974	WHEAT.	SCWS.SEMI-MECH.QK-HARV.	HEAVY	3.2	3.2	24.2					11.0	3.2
975	WHEAT.	SRWS.SEMI-MECH.QK-HARV.	HEAVY	3.2	3.2	22.3					11.0	3.2
976	WHEAT.	NWFP.SEMI-MECH.QK-HARV.	JANUARY	5.8	3.7	3.7	56.4			4.1	9.7	3.7
977	WHEAT.	PMW. SEMI-MECH.QK-HARV.	JANUARY	3.4	2.9	1.7	50.2			2.3	7.8	2.9
978	WHEAT.	PCW. SEMI-MECH.QK-HARV.	JANUARY	4.3	3.9	3.9	44.0				15.6	3.9
979	WHEAT.	PSW. SEMI-MECH.QK-HARV.	JANUARY	4.4	2.7	2.4	58.4			3.4	11.6	4.0
980	WHEAT.	PRW. SEMI-MECH.QK-HARV.	JANUARY	4.1	2.4	2.4	52.4				9.5	2.4
981	WHEAT.	SCWN.SEMI-MECH.QK-HARV.	JANUARY	3.2	3.2	19.2	8.2				9.0	3.2
982	WHEAT.	SRWN.SEMI-MECH.QK-HARV.	JANUARY	3.2	3.2	19.3	9.6				9.3	3.2
983	WHEAT.	SCWS.SEMI-MECH.QK-HARV.	JANUARY	3.2	3.2	24.2					11.0	3.2
984	WHEAT.	SRWS.SEMI-MECH.QK-HARV.	JANUARY	3.2	3.2	22.3					11.0	3.2
985	WHEAT.	NWFP.SEMI-MECH.QK-HARV.	LIGHT	5.8	3.7	3.7	73.8			4.1	9.7	3.7
986	WHEAT.	PMW. SEMI-MECH.QK-HARV.	LIGHT	3.4	2.9	1.7	65.7			2.3	7.8	2.9
987	WHEAT.	PCW. SEMI-MECH.QK-HARV.	LIGHT	4.3	3.9	3.9	67.7				15.6	3.9
988	WHEAT.	PSW. SEMI-MECH.QK-HARV.	LIGHT	4.4	2.7	2.4	68.9			3.4	11.6	4.0
989	WHEAT.	PRW. SEMI-MECH.QK-HARV.	LIGHT	4.1	2.4	2.4	68.6				9.5	2.4
990	WHEAT.	SCWN.SEMI-MECH.QK-HARV.	LIGHT	3.2	3.2	25.1	10.8				9.0	3.2
991	WHEAT.	SRWN.SEMI-MECH.QK-HARV.	LIGHT	3.2	3.2	25.2	12.6				9.3	3.2
992	WHEAT.	SCWS.SEMI-MECH.QK-HARV.	LIGHT	3.2	3.2	31.7					11.0	3.2
993	WHEAT.	SRWS.SEMI-MECH.QK-HARV.	LIGHT	3.2	3.2	29.3					11.0	3.2
994	WHEAT.	NWFP.SEMI-MECH.QK-HARV.	STANDARD	5.8	3.7	3.7	86.8			4.1	9.7	3.7
995	WHEAT.	PMW. SEMI-MECH.QK-HARV.	STANDARD	3.4	2.9	1.7	77.3			2.3	7.8	2.9
996	WHEAT.	PCW. SEMI-MECH.QK-HARV.	STANDARD	4.3	3.9	3.9	79.6				15.6	3.9
997	WHEAT.	PSW. SEMI-MECH.QK-HARV.	STANDARD	4.4	2.7	2.4	81.1			3.4	11.6	4.0
998	WHEAT.	PRW. SEMI-MECH.QK-HARV.	STANDARD	4.1	2.4	2.4	80.7				9.5	2.4
999	WHEAT.	SCWN.SEMI-MECH.QK-HARV.	STANDARD	3.2	3.2	29.5	12.7				9.0	3.2
1000	WHEAT.	SRWN.SEMI-MECH.QK-HARV.	STANDARD	3.2	3.2	29.7	14.8				9.3	3.2
1001	WHEAT.	SCWS.SEMI-MECH.QK-HARV.	STANDARD	3.2	3.2	37.3					11.0	3.2
1002	WHEAT.	SRWS.SEMI-MECH.QK-HARV.	STANDARD	3.2	3.2	34.3					11.0	3.2

```
INDUS BASIN MODEL REVISED (IBMR) FILENAME=WSISD41                                           07/25/91 17:17:54
CROP DATA                                                                                   GAMS 2.21 IBM CMS


1003    WHEAT. NWFP.SEMI-MECH.STANDARD. HEAVY       5.8   3.7   3.7  42.3  14.1                   4.1   9.7   3.7
1004    WHEAT. PMW. SEMI-MECH.STANDARD. HEAVY       3.4   2.9   1.7  37.7  12.5                   2.3   7.8   2.9
1005    WHEAT. PCW. SEMI-MECH.STANDARD. HEAVY       4.3   3.9   3.9  39.1  12.7                        15.6   3.9
1006    WHEAT. PSW. SEMI-MECH.STANDARD. HEAVY       4.4   2.7   2.4  36.9  15.8                   3.4  11.6   4.0
1007    WHEAT. PRW. SEMI-MECH.STANDARD. HEAVY       4.1   2.4   2.4  36.7  15.7                         9.5   2.4
1008    WHEAT. SCWN.SEMI-MECH.STANDARD. HEAVY       3.2   3.2        27.4                               9.0   3.2
1009    WHEAT. SRWN.SEMI-MECH.STANDARD. HEAVY       3.2   3.2        28.9                               9.3   3.2
1010    WHEAT. SCWS.SEMI-MECH.STANDARD. HEAVY       3.2   3.2  15.9   8.3                              11.0   3.2
1011    WHEAT. SRWS.SEMI-MECH.STANDARD. HEAVY       3.2   3.2  14.2   8.0                              11.0   3.2
1012    WHEAT. NWFP.SEMI-MECH.STANDARD. JANUARY     5.8   3.7   3.7  42.3  14.1                   4.1   9.7   3.7
1013    WHEAT. PMW. SEMI-MECH.STANDARD. JANUARY     3.4   2.9   1.7  37.7  12.5                   2.3   7.8   2.9
1014    WHEAT. PCW. SEMI-MECH.STANDARD. JANUARY     4.3   3.9   3.9  39.1  12.7                        15.6   3.9
1015    WHEAT. PSW. SEMI-MECH.STANDARD. JANUARY     4.4   2.7   2.4  36.9  15.8                   3.4  11.6   4.0
1016    WHEAT. PRW. SEMI-MECH.STANDARD. JANUARY     4.1   2.4   2.4  36.7  15.7                         9.5   2.4
1017    WHEAT. SCWN.SEMI-MECH.STANDARD. JANUARY     3.2   3.2        27.4                               9.0   3.2
1018    WHEAT. SRWN.SEMI-MECH.STANDARD. JANUARY     3.2   3.2        28.9                               9.3   3.2
1019    WHEAT. SCWS.SEMI-MECH.STANDARD. JANUARY     3.2   3.2  15.9   8.3                              11.0   3.2
1020    WHEAT. SRWS.SEMI-MECH.STANDARD. JANUARY     3.2   3.2  14.2   8.0                              11.0   3.2
1021    WHEAT. NWFP.SEMI-MECH.STANDARD. LIGHT       5.8   3.7   3.7  55.3  18.4                   4.1   9.7   3.7
1022    WHEAT. PMW. SEMI-MECH.STANDARD. LIGHT       3.4   2.9   1.7  49.3  16.4                   2.3   7.8   2.9
1023    WHEAT. PCW. SEMI-MECH.STANDARD. LIGHT       4.3   3.9   3.9  51.1  16.6                        15.6   3.9
1024    WHEAT. PSW. SEMI-MECH.STANDARD. LIGHT       4.4   2.7   2.4  48.3  20.7                   3.4  11.6   4.0
1025    WHEAT. PRW. SEMI-MECH.STANDARD. LIGHT       4.1   2.4   2.4  48.0  20.6                         9.5   2.4
1026    WHEAT. SCWN.SEMI-MECH.STANDARD. LIGHT       3.2   3.2        35.9                               9.0   3.2
1027    WHEAT. SRWN.SEMI-MECH.STANDARD. LIGHT       3.2   3.2        37.8                               9.3   3.2
1028    WHEAT. SCWS.SEMI-MECH.STANDARD. LIGHT       3.2   3.2  20.8  10.9                              11.0   3.2
1029    WHEAT. SRWS.SEMI-MECH.STANDARD. LIGHT       3.2   3.2  18.6  10.7                              11.0   3.2
1030    WHEAT. NWFP.SEMI-MECH.STANDARD. STANDARD    5.8   3.7   3.7  65.1  21.7                   4.1   9.7   3.7
1031    WHEAT. PMW. SEMI-MECH.STANDARD. STANDARD    3.4   2.9   1.7  58.0  19.3                   2.3   7.8   2.9
1032    WHEAT. PCW. SEMI-MECH.STANDARD. STANDARD    4.3   3.9   3.9  60.1  19.5                        15.6   3.9
1033    WHEAT. PSW. SEMI-MECH.STANDARD. STANDARD    4.4   2.7   2.4  56.8  24.3                   3.4  11.6   4.0
1034    WHEAT. PRW. SEMI-MECH.STANDARD. STANDARD    4.1   2.4   2.4  56.5  24.2                         9.5   2.4
1035    WHEAT. SCWN.SEMI-MECH.STANDARD. STANDARD    3.2   3.2        42.2                               9.0   3.2
1036    WHEAT. SRWN.SEMI-MECH.STANDARD. STANDARD    3.2   3.2        44.5                               9.3   3.2
1037    WHEAT. SCWS.SEMI-MECH.STANDARD. STANDARD    3.2   3.2  24.5  12.8                              11.0   3.2
1038    WHEAT. SRWS.SEMI-MECH.STANDARD. STANDARD    3.2   3.2  21.9  12.4                              11.0   3.2
1039
1040    ORCHARD.(NWFP,PMW,PCW,PSW,PRW).
1041    (BULLOCK,SEMI-MECH).STANDARD.STANDARD        41  41.8  17.3  14.5  14.5  22.   22.   50.  29.  39.  56.5  28.3
1042    ORCHARD.(SCWN,SCWS,SRWN,SRWS).
1043    (BULLOCK,SEMI-MECH).STANDARD.STANDARD        31  31.   31.   31.   25.   22.   22.   22.  31.  31.  31.   31.
1044
1045    POTATOES.(SCWN,SCWS,SRWN,SRWS).
1046            SEMI-MECH.STANDARD.STANDARD         100  100                            50   50  100   50
1047    POTATOES.(NWFP,PMW,PSW,PCW,PRW).SEMI-MECH.
1048            STANDARD.STANDARD                    70   24         150                                              8
1049    ONIONS.(NWFP,PMW,PSW,PCW,PRW).SEMI-MECH.
1050            STANDARD.STANDARD                   160    8    32    32   160
1051    ONIONS.(SCWN,SCWS,SRWN,SRWS).SEMI-MECH.
1052            STANDARD.STANDARD                   160                                      160    8   32   32
1053    CHILLI.(NWFP,PMW,PCW,PSW,PRW).SEMI-MECH.
1054            STANDARD.STANDARD                               100     2    20   10    20   10   20    8    8
1055    CHILLI.(SCWN,SCWS,SRWN,SRWS).SEMI-MECH.
1056            STANDARD.STANDARD                    16   32    28    28                                             100
1058       TABLE WATER(C,Z,T,S,W,M) WATER REQUIREMENTS(ACRE FEET PER ACRE)
1059
1060                                                      JAN  FEB  MAR  APR  MAY  JUN  JUL  AUG  SEP  OCT  NOV  DEC
1061    BASMATI.(PMW,PCW,PSW).(BULLOCK,SEMI-MECH).STANDARD.STANDARD              .37  .84  .93  .93  .47
1062    BASMATI.PRW.(BULLOCK,SEMI-MECH).STANDARD.STANDARD                        .56  .84  .93  .93  .56
1063
1064    RAB-FOD.(PSW,PRW).(BULLOCK,SEMI-MECH).STANDARD.STANDARD   .20  .30  .50  .60  .35            .25  .15  .15
1065    RAB-FOD.(SCWS,SRWS).(BULLOCK,SEMI-MECH).STANDARD.STANDARD .27  .37  .53  .64            .21  .32  .27  .21
1066    RAB-FOD.SRWN.(BULLOCK,SEMI-MECH).STANDARD.HEAVY           .20                           .20  .25  .15  .15
1067    RAB-FOD.SRWS.(BULLOCK,SEMI-MECH).STANDARD.HEAVY           .27                           .21  .32  .27  .21
1068    RAB-FOD.SRWN.(BULLOCK,SEMI-MECH).STANDARD.LIGHT           .20  .30                      .20  .25  .15  .15
```

INDUS BASIN MODEL REVISED (IBMR) FILENAME=WSISD41
CROP DATA

07/25/91 17:17:54
GAMS 2.21 IBM CMS

#	Description													
1069	RAB-FOD.SRWS.(BULLOCK,SEMI-MECH).STANDARD.LIGHT	.27	.37							.21	.32	.27	.21
1070	RAB-FOD.NWFP.(BULLOCK,SEMI-MECH).STANDARD.STANDARD	.20	.25	.40	.45	.15				.20	.25	.15	.15
1071	RAB-FOD.PMW.(BULLOCK,SEMI-MECH).STANDARD.STANDARD	.20	.25	.40	.50	.10					.25	.15	.15
1072	RAB-FOD.PCW.(BULLOCK,SEMI-MECH).STANDARD.STANDARD	.20	.30	.45	.55	.25				.20	.25	.15	.15
1073	RAB-FOD.SCWN.(BULLOCK,SEMI-MECH).STANDARD.STANDARD	.21	.32	.48	.59					.21	.27	.16	.16
1074	RAB-FOD.SRWN.(BULLOCK,SEMI-MECH).STANDARD.STANDARD	.20	.30	.45	.55					.20	.25	.15	.15
1076	COTTON. PCW. BULLOCK. EL-PLANT. STANDARD					.19	.09	.37	.33	.37	.47	.33	.09
1077	COTTON.(PMW,PSW,PRW).(BULLOCK,SEMI-MECH).STANDARD.STANDARD						.19	.23	.28	.37	.37	.23	.09
1078	COTTON.PCW.(BULLOCK,SEMI-MECH).STANDARD.STANDARD						.19	.28	.33	.37	.47	.33	.09
1079	COTTON. SCWN.BULLOCK. STANDARD. STANDARD					.10	.20	.30	.40	.40	.40	.30	
1080	COTTON.SCWS.(BULLOCK,SEMI-MECH).STANDARD.STANDARD					.20	.20	.30	.40	.40	.40	.20	
1081	COTTON.SRWS.(BULLOCK,SEMI-MECH).STANDARD.STANDARD					.20	.20	.30	.40	.40	.40	.15	
1082	COTTON. PCW. SEMI-MECH.LA-PLANT. STANDARD							.37	.33	.37	.47	.33	.09
1083	COTTON. SCWN.SEMI-MECH.STANDARD. STANDARD						.25	.35	.40	.40	.40	.30	
1085	GRAM.NWFP.(BULLOCK,SEMI-MECH).STANDARD.STANDARD	.10								.20	.10	.10	.05
1086	GRAM.PMW.(BULLOCK,SEMI-MECH).STANDARD.STANDARD	.05	.05							.20	.10	.10	.15
1087	GRAM.PCW.(BULLOCK,SEMI-MECH).STANDARD.STANDARD	.10	.05							.25	.15	.10	.10
1088	GRAM.(PSW,PRW).(BULLOCK,SEMI-MECH).STANDARD.STANDARD	.10	.10							.20	.15	.10	
1089	GRAM.(SCWN,SCWS).(BULLOCK,SEMI-MECH).STANDARD.STANDARD	.11	.21							.21	.21	.11	
1090	GRAM.(SRWN,SRWS).(BULLOCK,SEMI-MECH).STANDARD.STANDARD	.10	.20							.20	.20	.10	
1092	IRRI.PMW.(BULLOCK,SEMI-MECH).STANDARD.STANDARD						.37	.84	.93	.93	.47		
1093	IRRI.PCW.(BULLOCK,SEMI-MECH).STANDARD.STANDARD						.37	.84	.93	.93	.47		
1094	IRRI.PSW.(BULLOCK,SEMI-MECH).STANDARD.STANDARD						.33	.79	.89	.89	.47		
1095	IRRI.PRW.(BULLOCK,SEMI-MECH).STANDARD.STANDARD						.56	.84	.93	.93	.65		
1096	IRRI.SCWN.(BULLOCK,SEMI-MECH).STANDARD.STANDARD						.36	.90	1.01	1.01	.52		
1097	IRRI.SRWN.(BULLOCK,SEMI-MECH).STANDARD.STANDARD						.78	.96	.96	.96	.45		
1098	IRRI.SCWS.(BULLOCK,SEMI-MECH).STANDARD.STANDARD					.20	.85	1.00	1.00	.75			
1099	IRRI.SRWS.(BULLOCK,SEMI-MECH).STANDARD.STANDARD					.20	1.15	1.15	1.15	.75			
1101	MAIZE.NWFP.(BULLOCK,SEMI-MECH).STANDARD.STANDARD						.15	.35	.30	.25	.20		
1102	MAIZE.(PCW,PSW,PRW).BULLOCK. STANDARD.STANDARD						.10	.10	.30	.30	.40	.20	
1103	MAIZE.(PCW,PSW,PRW).SEMI-MECH. STANDARD.STANDARD							.15	.30	.30	.40	.25	
1104	MAIZE.SCWN.(BULLOCK,SEMI-MECH).STANDARD.STANDARD						.21	.32	.32	.43	.43	.21	
1105	MAIZE.SCWS.(BULLOCK,SEMI-MECH).STANDARD.STANDARD					.21	.21	.21	.32	.43	.32	.21	
1107	MUS+RAP.NWFP.(BULLOCK,SEMI-MECH).STANDARD.STANDARD									.20	.20	.25	.25	.25
1108	MUS+RAP.PMW.(BULLOCK,SEMI-MECH).STANDARD.STANDARD	.20	.20								.25	.20	.20
1109	MUS+RAP.PCW.(BULLOCK,SEMI-MECH).STANDARD.STANDARD	.20	.15								.30	.20	.20
1110	MUS+RAP.PSW.(BULLOCK,SEMI-MECH).STANDARD.STANDARD	.20	.20							.20	.20	.20	.20
1111	MUS+RAP.PRW.(BULLOCK,SEMI-MECH).STANDARD.STANDARD	.20								.20	.20	.20	.20
1112	MUS+RAP.SCWN.(BULLOCK,SEMI-MECH).STANDARD.STANDARD	.21	.21							.21	.21	.21	.21
1113	MUS+RAP.SRWN.(BULLOCK,SEMI-MECH).STANDARD.STANDARD	.20	.20							.20	.20	.20	.20
1114	MUS+RAP.SRWS.(BULLOCK,SEMI-MECH).STANDARD.STANDARD	.20	.20							.20	.20	.20	.20
1115	MUS+RAP.SCWS.(BULLOCK,SEMI-MECH).STANDARD.STANDARD	.11	.11							.27	.21	.21	.27
1117	SC-GUR.NWFP.(BULLOCK,SEMI-MECH).STANDARD.STANDARD	.19	.19	.28	.37	.37	.37	.28	.47	.47	.37	.19	.28
1118	SC-GUR. PMW. (BULLOCK,SEMI-MECH).STANDARD.STANDARD	.14	.09	.23	.33	.33	.28	.23	.37	.37	.19	.09	.19
1119	SC-GUR. PCW. (BULLOCK,SEMI-MECH).STANDARD.STANDARD	.09	.14	.23	.33	.33	.33	.37	.47	.47	.28	.37	.19
1120	SC-GUR. PSW. (BULLOCK,SEMI-MECH).STANDARD.STANDARD	.09	.09	.23	.33	.33	.33	.37	.47	.47	.23	.23	.09
1121	SC-GUR. PRW. (BULLOCK,SEMI-MECH).STANDARD.STANDARD	.14	.19	.28	.37	.37	.28	.33	.37	.37	.28	.19	.19
1122	SC-GUR. SCWN.(BULLOCK,SEMI-MECH).STANDARD.STANDARD	.25	.30	.40	.40	.40	.55	.60	.60	.50	.40	.30	.25
1123	SC-GUR. SRWN.(BULLOCK,SEMI-MECH).STANDARD.STANDARD	.23	.28	.37	.37	.37	.51	.56	.56	.47	.37	.28	.23
1124	SC-GUR. SCWS.(BULLOCK,SEMI-MECH).STANDARD.STANDARD	.25	.35	.40	.40	.40	.55	.65	.65	.60	.50	.35	.25
1125	SC-GUR. SRWS.(BULLOCK,SEMI-MECH).STANDARD.STANDARD	.25	.35	.40	.40	.50	.55	.65	.65	.60	.50	.40	.20
1127	SC-MILL.NWFP.(BULLOCK,SEMI-MECH).STANDARD.STANDARD	.19	.19	.28	.33	.33	.37	.37	.47	.47	.37	.28	.19
1128	SC-MILL.PMW.(BULLOCK,SEMI-MECH).STANDARD.STANDARD	.14	.09	.23	.33	.33	.28	.23	.37	.37	.19	.09	.19
1129	SC-MILL.PCW.(BULLOCK,SEMI-MECH).STANDARD.STANDARD	.09	.14	.23	.33	.33	.33	.37	.47	.47	.28	.37	.19
1130	SC-MILL.PSW. (BULLOCK,SEMI-MECH).STANDARD.STANDARD	.09	.09	.23	.33	.33	.33	.37	.47	.47	.23	.23	.09
1131	SC-MILL.PRW. (BULLOCK,SEMI-MECH).STANDARD.STANDARD	.14	.19	.28	.37	.37	.28	.33	.37	.37	.28	.19	.19
1132	SC-MILL.SCWN.(BULLOCK,SEMI-MECH).STANDARD.STANDARD	.25	.30	.40	.40	.40	.55	.60	.60	.50	.40	.30	.25
1133	SC-MILL.SRWN.(BULLOCK,SEMI-MECH).STANDARD.STANDARD	.23	.28	.37	.37	.37	.51	.56	.56	.47	.37	.28	.23

```
INDUS BASIN MODEL REVISED (IBMR) FILENAME=WSISD41                        07/25/91 17:17:54
CROP DATA                                                                GAMS 2.21 IBM CMS


1134  SC-MILL.SCWS.(BULLOCK,SEMI-MECH).STANDARD.STANDARD      .25 .35 .40 .40 .40 .55 .65 .65 .60 .50 .35 .25
1135  SC-MILL.SRWS.(BULLOCK,SEMI-MECH).STANDARD.STANDARD      .25 .35 .40 .40 .50 .55 .65 .65 .60 .50 .40 .20
1136
1137  KHA-FOD.SRWS.(BULLOCK,SEMI-MECH).LA-PLANT.STANDARD                      .21 .27 .32 .32 .16 .11
1138  KHA-FOD.NWFP.(BULLOCK,SEMI-MECH).STANDARD.STANDARD                  .05 .15 .20 .20 .25 .25 .05
1139  KHA-FOD.PMW. (BULLOCK,SEMI-MECH).STANDARD.STANDARD              .05 .05 .20 .20 .30 .25 .15
1140  KHA-FOD.PCW. (BULLOCK,SEMI-MECH).STANDARD.STANDARD                  .10 .15 .25 .30 .40 .30 .20
1141  KHA-FOD.PSW. (BULLOCK,SEMI-MECH).STANDARD.STANDARD                  .15 .25 .35 .35 .30 .25 .20
1142  KHA-FOD.PRW. (BULLOCK,SEMI-MECH).STANDARD.STANDARD                  .10 .20 .30 .30 .30 .30 .30
1143  KHA-FOD.SCWN.(BULLOCK,SEMI-MECH).STANDARD.STANDARD                  .11 .11 .21 .21 .32 .32 .21 .11
1144  KHA-FOD.SCWS.(BULLOCK,SEMI-MECH).STANDARD.STANDARD                  .11 .21 .21 .27 .32 .32 .16 .05
1145  KHA-FOD.SRWN.(BULLOCK,SEMI-MECH).STANDARD.STANDARD                  .21 .32 .27 .27 .32     .21 .32
1146  KHA-FOD.SRWS.(BULLOCK,SEMI-MECH).STANDARD.STANDARD                  .21 .32 .27 .27 .32     .21 .32
1147
1148  WHEAT. NWFP.BULLOCK.  LA-PLANT. HEAVY                   .28                                      .25 .25
1149  WHEAT. PMW. BULLOCK.  LA-PLANT. HEAVY                   .22                                      .28 .16
1150  WHEAT.(PCW,PSW,SRWN).BULLOCK.LA-PLANT.HEAVY             .28                                      .28 .19
1151  WHEAT. PRW. BULLOCK.  LA-PLANT. HEAVY                   .28                                      .09 .28
1152  WHEAT.(SCWN,SRWS).BULLOCK.LA-PLANT.HEAVY                .30                                      .30 .20
1153  WHEAT. SCWS.BULLOCK.  LA-PLANT. HEAVY                   .30                                      .30 .35
1154  WHEAT. NWFP.BULLOCK.  LA-PLANT. JANUARY                     .28                                  .25 .25
1155  WHEAT. PMW. BULLOCK.  LA-PLANT. JANUARY                     .28                                  .28 .16
1156  WHEAT.(PCW,PSW,SRWN).BULLOCK.LA-PLANT.JANUARY               .28                                  .28 .19
1157  WHEAT. PRW. BULLOCK.  LA-PLANT. JANUARY                     .28                                  .09 .28
1158  WHEAT.(SCWN,SRWS).BULLOCK.LA-PLANT.JANUARY                  .30                                  .30 .20
1159  WHEAT. SCWS.BULLOCK.  LA-PLANT. JANUARY                     .30                                  .30 .35
1160  WHEAT. NWFP.BULLOCK.  LA-PLANT. LIGHT                   .22 .35                                  .25 .25
1161  WHEAT. PMW. BULLOCK.  LA-PLANT. LIGHT                   .22 .28                                  .28 .16
1162  WHEAT.(PCW,PSW).      BULLOCK.LA-PLANT.LIGHT            .23 .37                                  .28 .19
1163  WHEAT. SRWN.BULLOCK.  LA-PLANT. LIGHT                   .19 .33                                  .28 .19
1164  WHEAT. PRW. BULLOCK.  LA-PLANT. LIGHT                   .23 .37                                  .09 .28
1165  WHEAT. SCWN.BULLOCK.  LA-PLANT. LIGHT                   .20 .40                                  .30 .20
1166  WHEAT. SCWS.BULLOCK.  LA-PLANT. LIGHT                   .25 .40                                  .30 .35
1167  WHEAT. SRWS.BULLOCK.  LA-PLANT. LIGHT                   .20 .35                                  .30 .20
1168  WHEAT. NWFP.BULLOCK.  LA-PLANT. STANDARD                .22 .35 .41                              .25 .25
1169  WHEAT. PMW. BULLOCK.  LA-PLANT. STANDARD                .22 .28 .35                              .28 .16
1170  WHEAT. PCW. BULLOCK.  LA-PLANT. STANDARD                .23 .37 .47                              .28 .19
1171  WHEAT. PSW. BULLOCK.  LA-PLANT. STANDARD                .23 .37 .42                              .28 .19
1172  WHEAT. PRW. BULLOCK.  LA-PLANT. STANDARD                .23 .37 .47                              .09 .28
1173  WHEAT. SCWN.BULLOCK.  LA-PLANT. STANDARD                .20 .40 .40                              .30 .20
1174  WHEAT. SRWN.BULLOCK.  LA-PLANT. STANDARD                .19 .33 .37                              .28 .19
1175  WHEAT. SCWS.BULLOCK.  LA-PLANT. STANDARD                .25 .40 .50                              .30 .35
1176  WHEAT. SRWS.BULLOCK.  LA-PLANT. STANDARD                .20 .35 .40                              .30 .20
1177
1178  WHEAT.NWFP.BULLOCK.(QK-HARV,STANDARD).HEAVY             .28                                  .19 .16 .16
1179  WHEAT.PMW. BULLOCK.(QK-HARV,STANDARD).HEAVY             .22                                  .19 .16 .16
1180  WHEAT.PCW. BULLOCK.(QK-HARV,STANDARD).HEAVY             .28                                  .19 .19 .19
1181  WHEAT.PSW. BULLOCK.(QK-HARV,STANDARD).HEAVY             .28                                  .19 .19 .19
1182  WHEAT.PRW. BULLOCK.(QK-HARV,STANDARD).HEAVY             .28                                      .19 .19
1183  WHEAT.SCWN.BULLOCK.(QK-HARV,STANDARD).HEAVY             .30                                  .20 .20 .20
1184  WHEAT.SRWN.BULLOCK.(QK-HARV,STANDARD).HEAVY             .28                                  .19 .19 .19
1185  WHEAT.SCWS.BULLOCK.(QK-HARV,STANDARD).HEAVY             .30                                  .20 .30 .25
1186  WHEAT.SRWS.BULLOCK.(QK-HARV,STANDARD).HEAVY             .30                                  .20 .20 .20
1187  WHEAT.NWFP.BULLOCK.(QK-HARV,STANDARD).JANUARY               .28                              .19 .16 .16
1188  WHEAT.PMW. BULLOCK.(QK-HARV,STANDARD).JANUARY               .28                              .19 .16 .16
1189  WHEAT.PCW. BULLOCK.(QK-HARV,STANDARD).JANUARY               .28                              .19 .19 .19
1190  WHEAT.PSW. BULLOCK.(QK-HARV,STANDARD).JANUARY               .28                              .19 .19 .19
1191  WHEAT.PRW. BULLOCK.(QK-HARV,STANDARD).JANUARY               .28                                  .19 .19
1192  WHEAT.SCWN.BULLOCK.(QK-HARV,STANDARD).JANUARY               .30                              .20 .20 .20
1193  WHEAT.SRWN.BULLOCK.(QK-HARV,STANDARD).JANUARY               .28                              .19 .19 .19
1194  WHEAT.SCWS.BULLOCK.(QK-HARV,STANDARD).JANUARY               .30                              .20 .30 .25
1195  WHEAT.SRWS.BULLOCK.(QK-HARV,STANDARD).JANUARY               .30                              .20 .20 .20
1196  WHEAT.NWFP.BULLOCK.(QK-HARV,STANDARD).LIGHT             .22 .35                              .19 .16 .16
1197  WHEAT.PMW. BULLOCK.(QK-HARV,STANDARD).LIGHT             .22 .28                              .19 .16 .16
1198  WHEAT.PCW. BULLOCK.(QK-HARV,STANDARD).LIGHT             .23 .37                              .19 .19 .19
```

INDUS BASIN MODEL REVISED (IBMR) FILENAME=WSISD41
CROP DATA

07/25/91 17:17:54
GAMS 2.21 IBM CMS

#	Description	v1	v2	v3		v4	v5	v6
1199	WHEAT.PSW. BULLOCK.(QK-HARV,STANDARD).LIGHT	.23	.37			.19	.19	.19
1200	WHEAT.PRW. BULLOCK.(QK-HARV,STANDARD).LIGHT	.23	.37				.19	.19
1201	WHEAT.SCWN.BULLOCK.(QK-HARV,STANDARD).LIGHT	.20	.40			.20	.20	.20
1202	WHEAT.SRWN.BULLOCK.(QK-HARV,STANDARD).LIGHT	.19	.33			.19	.19	.19
1203	WHEAT.SCWS.BULLOCK.(QK-HARV,STANDARD).LIGHT	.25	.40			.20	.30	.25
1204	WHEAT.SRWS.BULLOCK.(QK-HARV,STANDARD).LIGHT	.20	.35			.20	.20	.20
1205	WHEAT.NWFP.BULLOCK.(QK-HARV,STANDARD).STANDARD	.22	.35	.41		.19	.16	.16
1206	WHEAT.PMW. BULLOCK.(QK-HARV,STANDARD).STANDARD	.22	.28	.35		.19	.16	.16
1207	WHEAT.PCW. BULLOCK.(QK-HARV,STANDARD).STANDARD	.23	.37	.47		.19	.19	.19
1208	WHEAT.PSW. BULLOCK.(QK-HARV,STANDARD).STANDARD	.23	.37	.42		.19	.19	.19
1209	WHEAT.PRW. BULLOCK.(QK-HARV,STANDARD).STANDARD	.23	.37	.47			.19	.19
1210	WHEAT.SCWN.BULLOCK.(QK-HARV,STANDARD).STANDARD	.20	.40	.40		.20	.20	.20
1211	WHEAT.SRWN.BULLOCK.(QK-HARV,STANDARD).STANDARD	.19	.33	.37		.19	.19	.19
1212	WHEAT.SCWS.BULLOCK.(QK-HARV,STANDARD).STANDARD	.25	.40	.50		.20	.30	.25
1213	WHEAT.SRWS.BULLOCK.(QK-HARV,STANDARD).STANDARD	.20	.35	.40		.20	.20	.20
1214								
1215	WHEAT.NWFP.SEMI-MECH.(QK-HARV,STANDARD).HEAVY	.28				.19	.16	.16
1216	WHEAT.PMW. SEMI-MECH.(QK-HARV,STANDARD).HEAVY	.22				.19	.16	.16
1217	WHEAT.PCW. SEMI-MECH.(QK-HARV,STANDARD).HEAVY	.28					.28	.19
1218	WHEAT.PSW. SEMI-MECH.(QK-HARV,STANDARD).HEAVY	.28				.19	.19	.19
1219	WHEAT.PRW. SEMI-MECH.(QK-HARV,STANDARD).HEAVY	.28					.19	.19
1220	WHEAT.SRWN.SEMI-MECH.(QK-HARV,STANDARD).HEAVY	.28					.28	.19
1221	WHEAT.SRWS.SEMI-MECH.(QK-HARV,STANDARD).HEAVY	.30					.30	.30
1222	WHEAT.NWFP.SEMI-MECH.(QK-HARV,STANDARD).JANUARY		.28			.19	.16	.16
1223	WHEAT.PMW. SEMI-MECH.(QK-HARV,STANDARD).JANUARY		.28			.19	.16	.16
1224	WHEAT.PCW. SEMI-MECH.(QK-HARV,STANDARD).JANUARY		.28				.28	.19
1225	WHEAT.PSW. SEMI-MECH.(QK-HARV,STANDARD).JANUARY		.28			.19	.19	.19
1226	WHEAT.PRW. SEMI-MECH.(QK-HARV,STANDARD).JANUARY		.28				.19	.19
1227	WHEAT.SRWN.SEMI-MECH.(QK-HARV,STANDARD).JANUARY		.28				.28	.19
1228	WHEAT.SRWS.SEMI-MECH.(QK-HARV,STANDARD).JANUARY		.30				.30	.30
1229	WHEAT.NWFP.SEMI-MECH.(QK-HARV,STANDARD).LIGHT	.22	.35			.19	.16	.16
1230	WHEAT.PMW. SEMI-MECH.(QK-HARV,STANDARD).LIGHT	.22	.28			.19	.16	.16
1231	WHEAT.PCW. SEMI-MECH.(QK-HARV,STANDARD).LIGHT	.23	.37				.28	.19
1232	WHEAT.PSW. SEMI-MECH.(QK-HARV,STANDARD).LIGHT	.23	.37			.19	.19	.19
1233	WHEAT.PRW. SEMI-MECH.(QK-HARV,STANDARD).LIGHT	.23	.37				.19	.19
1234	WHEAT.SRWN.SEMI-MECH.(QK-HARV,STANDARD).LIGHT	.19	.33				.28	.19
1235	WHEAT.SRWS.SEMI-MECH.(QK-HARV,STANDARD).LIGHT	.20	.35				.30	.30
1236	WHEAT.NWFP.SEMI-MECH.(QK-HARV,STANDARD).STANDARD	.22	.35	.41		.19	.16	.16
1237	WHEAT.PMW. SEMI-MECH.(QK-HARV,STANDARD).STANDARD	.22	.28	.35		.19	.16	.16
1238	WHEAT.PCW. SEMI-MECH.(QK-HARV,STANDARD).STANDARD	.23	.37	.47			.28	.19
1239	WHEAT.PSW. SEMI-MECH.(QK-HARV,STANDARD).STANDARD	.23	.37	.42		.19	.19	.19
1240	WHEAT.PRW. SEMI-MECH.(QK-HARV,STANDARD).STANDARD	.23	.37	.47			.19	.19
1241								
1242	WHEAT. SCWN.SEMI-MECH.QK-HARV. HEAVY	.30					.30	.20
1243	WHEAT. SCWS.SEMI-MECH.QK-HARV. HEAVY	.30					.35	.30
1244	WHEAT. SCWN.SEMI-MECH.QK-HARV. JANUARY		.30				.30	.20
1245	WHEAT. SCWS.SEMI-MECH.QK-HARV. JANUARY		.30				.35	.30
1246	WHEAT. SCWN.SEMI-MECH.QK-HARV. LIGHT	.20	.40				.30	.20
1247	WHEAT. SCWS.SEMI-MECH.QK-HARV. LIGHT	.25	.40				.35	.30
1248	WHEAT. SCWN.SEMI-MECH.QK-HARV. STANDARD	.20	.40	.40			.30	.20
1249	WHEAT. SRWN.SEMI-MECH.QK-HARV. STANDARD	.19	.33	.37			.28	.19
1250	WHEAT. SCWS.SEMI-MECH.QK-HARV. STANDARD	.25	.40	.50			.35	.30
1251	WHEAT. SRWS.SEMI-MECH.QK-HARV. STANDARD	.20	.35	.40			.30	.30
1252	WHEAT. SCWN.SEMI-MECH.STANDARD. HEAVY	.30					.30	.25
1253	WHEAT. SCWS.SEMI-MECH.STANDARD. HEAVY	.30					.40	.30
1254	WHEAT. SCWN.SEMI-MECH.STANDARD. JANUARY		.30				.30	.25
1255	WHEAT. SCWS.SEMI-MECH.STANDARD. JANUARY		.30				.40	.30
1256	WHEAT. SCWN.SEMI-MECH.STANDARD. LIGHT	.25	.40				.30	.25
1257	WHEAT. SCWS.SEMI-MECH.STANDARD. LIGHT	.30	.40				.30	.30
1258	WHEAT. SCWN.SEMI-MECH.STANDARD. STANDARD	.20	.40	.50			.30	.20
1259	WHEAT. SRWN.SEMI-MECH.STANDARD. STANDARD	.19	.33	.47			.28	.19
1260	WHEAT. SCWS.SEMI-MECH.STANDARD. STANDARD	.25	.45	.55			.30	.25
1261	WHEAT. SRWS.SEMI-MECH.STANDARD. STANDARD	.20	.40	.55			.20	.20
1262								
1263	WHEAT. NWFP.SEMI-MECH.LA-PLANT. HEAVY	.28					.25	.25

INDUS BASIN MODEL REVISED (IBMR) FILENAME=WSISD41
CROP DATA

```
1264   WHEAT. PMW. SEMI-MECH.LA-PLANT. HEAVY              .22                                       .23 .23
1265   WHEAT. PCW. SEMI-MECH.LA-PLANT. HEAVY              .28                                       .23 .23
1266   WHEAT. PSW. SEMI-MECH.LA-PLANT. HEAVY              .28                                       .28 .19
1267   WHEAT. PRW. SEMI-MECH.LA-PLANT. HEAVY              .28                                       .09 .28
1268   WHEAT. SCWN.SEMI-MECH.LA-PLANT. HEAVY              .30                                       .30 .25
1269   WHEAT. SRWN.SEMI-MECH.LA-PLANT. HEAVY              .28                                       .28 .19
1270   WHEAT. SCWS.SEMI-MECH.LA-PLANT. HEAVY              .30                                       .35 .30
1271   WHEAT. SRWS.SEMI-MECH.LA-PLANT. HEAVY              .30                                       .20 .30
1272   WHEAT. NWFP.SEMI-MECH.LA-PLANT. JANUARY                .28                                   .25 .25
1273   WHEAT. PMW. SEMI-MECH.LA-PLANT. JANUARY                .28                                   .23 .23
1274   WHEAT. PCW. SEMI-MECH.LA-PLANT. JANUARY                .28                                   .23 .23
1275   WHEAT. PSW. SEMI-MECH.LA-PLANT. JANUARY                .28                                   .28 .19
1276   WHEAT. PRW. SEMI-MECH.LA-PLANT. JANUARY                .28                                   .09 .28
1277   WHEAT. SCWN.SEMI-MECH.LA-PLANT. JANUARY                .30                                   .30 .25
1278   WHEAT. SRWN.SEMI-MECH.LA-PLANT. JANUARY                .28                                   .28 .19
1279   WHEAT. SCWS.SEMI-MECH.LA-PLANT. JANUARY                .30                                   .35 .30
1280   WHEAT. SRWS.SEMI-MECH.LA-PLANT. JANUARY                .30                                   .20 .30
1281   WHEAT. NWFP.SEMI-MECH.LA-PLANT. LIGHT              .22 .35                                   .25 .25
1282   WHEAT. PMW. SEMI-MECH.LA-PLANT. LIGHT              .22 .28                                   .23 .23
1283   WHEAT. PCW. SEMI-MECH.LA-PLANT. LIGHT              .23 .37                                   .23 .23
1284   WHEAT. PSW. SEMI-MECH.LA-PLANT. LIGHT              .23 .37                                   .28 .19
1285   WHEAT. PRW. SEMI-MECH.LA-PLANT. LIGHT              .23 .37                                   .09 .28
1286   WHEAT. SCWN.SEMI-MECH.LA-PLANT. LIGHT              .20 .40                                   .30 .25
1287   WHEAT. SRWN.SEMI-MECH.LA-PLANT. LIGHT              .19 .33                                   .28 .19
1288   WHEAT. SCWS.SEMI-MECH.LA-PLANT. LIGHT              .25 .40                                   .35 .30
1289   WHEAT. SRWS.SEMI-MECH.LA-PLANT. LIGHT              .20 .35                                   .20 .30
1290   WHEAT. NWFP.SEMI-MECH.LA-PLANT. STANDARD           .22 .35 .41                               .25 .25
1291   WHEAT. PMW. SEMI-MECH.LA-PLANT. STANDARD           .22 .28 .35                               .23 .23
1292   WHEAT. PCW. SEMI-MECH.LA-PLANT. STANDARD           .23 .37 .47                               .23 .23
1293   WHEAT. PSW. SEMI-MECH.LA-PLANT. STANDARD           .23 .37 .42                               .28 .19
1294   WHEAT. PRW. SEMI-MECH.LA-PLANT. STANDARD           .23 .37 .47                               .09 .28
1295   WHEAT. SCWN.SEMI-MECH.LA-PLANT. STANDARD           .20 .40 .40                               .30 .25
1296   WHEAT. SRWN.SEMI-MECH.LA-PLANT. STANDARD           .19 .33 .37                               .28 .19
1297   WHEAT. SCWS.SEMI-MECH.LA-PLANT. STANDARD           .25 .40 .50                               .35 .30
1298   WHEAT. SRWS.SEMI-MECH.LA-PLANT. STANDARD           .20 .35 .40                               .20 .30
1299
1300   ORCHARD. NWFP.(BULLOCK,SEMI-MECH).STANDARD.STANDARD   .08 .10 .19 .33 .46 .50 .55 .58 .49 .37 .19 .09
1301   ORCHARD. PMW. (BULLOCK,SEMI-MECH).STANDARD.STANDARD   .10 .13 .23 .39 .53 .56 .60 .60 .59 .45 .23 .11
1302   ORCHARD. PCW. (BULLOCK,SEMI-MECH).STANDARD.STANDARD   .11 .14 .30 .44 .54 .59 .57 .61 .58 .47 .23 .12
1303   ORCHARD. PSW. (BULLOCK,SEMI-MECH).STANDARD.STANDARD   .08 .12 .24 .40 .49 .49 .46 .50 .51 .40 .18 .10
1304   ORCHARD. PRW. (BULLOCK,SEMI-MECH).STANDARD.STANDARD   .08 .12 .24 .40 .47 .47 .46 .50 .50 .39 .18 .10
1305   ORCHARD. SCWN.(BULLOCK,SEMI-MECH).STANDARD.STANDARD   .17 .17 .32 .45 .52 .54 .56 .60 .60 .45 .26 .14
1306   ORCHARD. SCWS.(BULLOCK,SEMI-MECH).STANDARD.STANDARD   .18 .20 .38 .54 .61 .56 .56 .61 .62 .56 .32 .20
1307   ORCHARD. SRWN.(BULLOCK,SEMI-MECH).STANDARD.STANDARD   .15 .16 .32 .45 .54 .52 .54 .59 .57 .47 .26 .15
1308   ORCHARD. SRWS.(BULLOCK,SEMI-MECH).STANDARD.STANDARD   .19 .21 .39 .57 .61 .54 .55 .59 .63 .58 .33 .21
1309
1310   POTATOES.(SCWN,SCWS,SRWN,SRWS).
1311            SEMI-MECH.STANDARD.STANDARD                                                .25 .667 .5 .25
1312   POTATOES.(NWFP,PMW,PCW,PSW,PRW).SEMI-MECH.
1313            STANDARD.STANDARD                        .25 .25 .5                                        .5
1314   ONIONS.(NWFP,PMW,PCW,PSW,PRW).SEMI-MECH.
1315            STANDARD.STANDARD                            .25 .25 .5 .5 25
1316   ONIONS.(SCWN,SCWS,SRWN,SRWS).SEMI-MECH.
1317            STANDARD.STANDARD                        .25                               .25 .25 .5  .5
1318   CHILLI.(NWFP,PMW,PSW,PCW,PRW).SEMI-MECH.
1319            STANDARD.STANDARD                                  .167 .25 .667 .667 .5 .5 .333 .333 .167
1320   CHILLI.(SCWN,SCWS,SRWN,SRWS).SEMI-MECH.
1321            STANDARD.STANDARD                        .5 .667 .667 .667                                 .25
1322
1323      TABLE TRACTOR(C,Z,T,S,W,M) TRACTOR REQUIREMENTS (TRACTOR HOURS PER ACRE)
1324
1325                                              JAN  FEB  MAR  APR  MAY  JUN  JUL  AUG  SEP  OCT  NOV  DEC
1326   BASMATI.PMW. SEMI-MECH.STANDARD.STANDARD                            2.70 1.70                1.50 0.50
1327   BASMATI.PCW. SEMI-MECH.STANDARD.STANDARD                            2.70 1.70                1.50 0.50
1328   BASMATI.PSW. SEMI-MECH.STANDARD.STANDARD                            3.30 1.80                1.50 0.50
```

INDUS BASIN MODEL REVISED (IBMR) FILENAME=WSISD41   07/25/91 17:17:54
CROP DATA   GAMS 2.21 IBM CMS

#	Activity													
1329	BASMATI.PRW. SEMI-MECH.STANDARD.STANDARD							3.10	2.00				1.50	0.50
1330	RAB-FOD.SRWN.SEMI-MECH.STANDARD.HEAVY										1.00	2.80		
1331	RAB-FOD.SRWS.SEMI-MECH.STANDARD.HEAVY										1.50	2.50		
1332	RAB-FOD.SRWN.SEMI-MECH.STANDARD.LIGHT										1.00	2.80		
1333	RAB-FOD.SRWS.SEMI-MECH.STANDARD.LIGHT										1.50	2.50		
1334	RAB-FOD.NWFP.SEMI-MECH.STANDARD.STANDARD	1.50	1.50	1.50	1.50	1.50					2.05	1.62	1.50	1.50
1335	RAB-FOD.PMW. SEMI-MECH.STANDARD.STANDARD	1.50	1.50	1.50	1.50	1.50						1.63	0.78	1.50
1336	RAB-FOD.PCW. SEMI-MECH.STANDARD.STANDARD	1.50	1.50	1.50	1.50	1.50					1.26	2.01	2.29	1.50
1337	RAB-FOD.PSW. SEMI-MECH.STANDARD.STANDARD	3.00	3.00	3.00	1.50	1.50						1.91	1.23	1.50
1338	RAB-FOD.PRW. SEMI-MECH.STANDARD.STANDARD	3.00	3.00	3.00	1.50	1.50						1.49	1.05	1.57
1339	RAB-FOD.SCWN.SEMI-MECH.STANDARD.STANDARD										1.20	2.60		
1340	RAB-FOD.SRWN.SEMI-MECH.STANDARD.STANDARD										1.00	2.80		
1341	RAB-FOD.SCWS.SEMI-MECH.STANDARD.STANDARD										1.50	2.40		
1342	RAB-FOD.SRWS.SEMI-MECH.STANDARD.STANDARD										1.50	2.50		
1343	COTTON. PCW. SEMI-MECH.LA-PLANT.STANDARD							4.70			0.50	0.50	0.50	0.50
1344	COTTON. PMW. SEMI-MECH.STANDARD.STANDARD						3.06	0.64			0.50	0.50	0.50	0.50
1345	COTTON. PCW. SEMI-MECH.STANDARD.STANDARD						2.49	2.22			0.50	0.50	0.50	0.50
1346	COTTON. PSW. SEMI-MECH.STANDARD.STANDARD						1.66	1.98			0.50	0.50	0.50	0.50
1347	COTTON. PRW. SEMI-MECH.STANDARD.STANDARD						1.45	1.84			0.50	0.50	0.50	0.50
1348	COTTON. SCWN.SEMI-MECH.STANDARD.STANDARD						12.10				1.00	1.00	2.00	
1349	COTTON. SCWS.SEMI-MECH.STANDARD.STANDARD					12.00					1.00	2.00		
1350	COTTON. SRWS.SEMI-MECH.STANDARD.STANDARD						12.40				1.00	2.00		
1351	GRAM. NWFP.SEMI-MECH.STANDARD.STANDARD					3.75					1.37	2.00		
1352	GRAM. PMW. SEMI-MECH.STANDARD.STANDARD				3.75						1.27	0.78	0.90	2.50
1353	GRAM. PCW. SEMI-MECH.STANDARD.STANDARD				3.75						1.30	0.79		
1354	GRAM. PSW. SEMI-MECH.STANDARD.STANDARD				3.75							1.25	0.70	
1355	GRAM. PRW. SEMI-MECH.STANDARD.STANDARD				3.75							1.14	0.68	
1356	GRAM.(SCWN,SCWS).SEMI-MECH.STANDARD.STANDARD				3.50						0.81	1.29		
1357	GRAM.(SRWN,SRWS).SEMI-MECH.STANDARD.STANDARD				3.80						0.96	1.39		
1358														
1359	IRRI. PMW. SEMI-MECH.STANDARD.STANDARD							2.60	2.20				1.00	0.50
1360	IRRI. PCW. SEMI-MECH.STANDARD.STANDARD							2.60	2.20				1.00	0.50
1361	IRRI. PSW. SEMI-MECH.STANDARD.STANDARD							2.30	1.90				1.00	0.50
1362	IRRI. PRW. SEMI-MECH.STANDARD.STANDARD							4.90	1.50				1.00	0.50
1363	IRRI. SCWN.SEMI-MECH.STANDARD.STANDARD						1.70	2.30				1.00	0.50	
1364	IRRI. SRWN.SEMI-MECH.STANDARD.STANDARD							4.30					1.00	
1365	IRRI. SCWS.SEMI-MECH.STANDARD.STANDARD						2.10	2.30	3.50			1.00	0.50	
1366	IRRI. SRWS.SEMI-MECH.STANDARD.STANDARD							4.20	2.50			1.00		
1367														
1368	MAIZE. NWFP.SEMI-MECH.STANDARD.STANDARD							5.40				2.50		
1369	MAIZE. PCW. SEMI-MECH.STANDARD.STANDARD								1.77	1.58				2.50
1370	MAIZE. PSW. SEMI-MECH.STANDARD.STANDARD								1.68	1.40				2.50
1371	MAIZE. PRW. SEMI-MECH.STANDARD.STANDARD								1.68	1.40				2.50
1372	MAIZE. SCWN.SEMI-MECH.STANDARD.STANDARD							2.60	4.30					
1373	MAIZE. SCWS.SEMI-MECH.STANDARD.STANDARD						2.70	4.50						
1374														
1375	MUS+RAP.NWFP.SEMI-MECH.STANDARD.STANDARD								1.37	0.75	2.00	1.00	1.00	1.00
1376	MUS+RAP.PMW. SEMI-MECH.STANDARD.STANDARD	1.00	1.00	1.00								2.47	1.00	1.00
1377	MUS+RAP.PCW. SEMI-MECH.STANDARD.STANDARD	1.00	1.00	1.00								1.30	1.26	1.00
1378	MUS+RAP.PSW. SEMI-MECH.STANDARD.STANDARD	1.00	1.00	1.00							1.25	1.11	1.00	1.00
1379	MUS+RAP.PRW. SEMI-MECH.STANDARD.STANDARD	1.00	1.00								1.82	1.00	1.00	1.00
1380	MUS+RAP.SCWN.SEMI-MECH.STANDARD.STANDARD	0.50	0.50	0.50							2.80	1.90	0.50	0.50
1381	MUS+RAP.SRWN.SEMI-MECH.STANDARD.STANDARD	0.50	0.50	0.50							2.40	1.80	0.50	0.50
1382	MUS+RAP.SRWS.SEMI-MECH.STANDARD.STANDARD	0.50	0.50	0.50							2.40	1.80	0.50	0.50
1383	MUS+RAP.SCWS.SEMI-MECH.STANDARD.STANDARD	0.50	0.50	0.50							2.40	1.70	0.50	0.50
1384														
1385	SC-GUR. NWFP.SEMI-MECH.STANDARD.STANDARD	2.50	0.37	1.25										0.98
1386	SC-GUR.(PMW,PCW,PSW,PRW).SEMI-MECH.													
1387	STANDARD.STANDARD	2.41	0.62	0.72										0.25
1388	SC-GUR. SCWN.SEMI-MECH.STANDARD.STANDARD	1.70	1.30	2.10										
1389	SC-GUR. SRWN.SEMI-MECH.STANDARD.STANDARD	1.80	1.20	2.10										
1390	SC-GUR. SCWS.SEMI-MECH.STANDARD.STANDARD	1.90	1.30	2.40										
1391	SC-GUR. SRWS.SEMI-MECH.STANDARD.STANDARD	2.00	1.50	2.30										
1392	SC-MILL.NWFP.SEMI-MECH.STANDARD.STANDARD	13.00	13.00	10.00	7.50						6.00	5.70	12.00	13.00
1393	SC-MILL.(PMW,PCW,PSW). SEMI-MECH.													

INDUS BASIN MODEL REVISED (IBMR) FILENAME=WSISD41   07/25/91 17:17:54
CROP DATA   GAMS 2.21 IBM CMS

```
1394                     STANDARD.STANDARD        6.40 6.60 5.22 4.50 4.00                              5.75 5.75
1395     SC-MILL.PRW. SEMI-MECH.STANDARD.STANDARD 6.50 7.00 6.22 4.50 3.00                              5.70 5.75
1396     SC-MILL.SCWN.SEMI-MECH.STANDARD.STANDARD 1.10 1.00 2.00
1397     SC-MILL.SRWN.SEMI-MECH.STANDARD.STANDARD 1.10 1.00 2.00
1398     SC-MILL.SCWS.SEMI-MECH.STANDARD.STANDARD 0.40 1.50 2.00
1399     SC-MILL.SRWS.SEMI-MECH.STANDARD.STANDARD 0.50 1.70 1.80
1400     KHA-FOD.SRWS.SEMI-MECH.LA-PLANT.STANDARD                1.20 0.50 0.50 0.90 0.50 0.50
1401     KHA-FOD.NWFP.SEMI-MECH.STANDARD.STANDARD      3.00 0.50 1.00 0.50 1.00 1.00 1.00 0.50
1402     KHA-FOD.PMW. SEMI-MECH.STANDARD.STANDARD      3.50 0.50 2.00 1.00 2.00 1.50 1.00 0.50
1403     KHA-FOD.PCW. SEMI-MECH.STANDARD.STANDARD      2.50 0.50 1.80 1.00 2.20 1.50 1.00 0.50
1404     KHA-FOD.PSW. SEMI-MECH.STANDARD.STANDARD      2.00 0.50 1.50 1.00 2.00 1.20 0.80 0.50
1405     KHA-FOD.PRW. SEMI-MECH.STANDARD.STANDARD      3.00 0.80 1.50 1.20 2.20 1.50 1.00 0.50
1406     KHA-FOD.SCWN.SEMI-MECH.STANDARD.STANDARD      1.10 0.50 0.90 0.50 0.50 0.90 0.50 0.50
1407     KHA-FOD.SCWS.SEMI-MECH.STANDARD.STANDARD      1.20 0.50 0.90 0.50 0.50 0.90 0.50 0.50
1408     KHA-FOD.SRWN.SEMI-MECH.STANDARD.STANDARD      1.30 0.50 0.90 0.50 0.50 0.80 0.50 0.50
1409     KHA-FOD.SRWS.SEMI-MECH.STANDARD.STANDARD      1.30 0.50 0.90 0.50 0.50 0.80 0.50 0.50
1410     WHEAT.  NWFP.SEMI-MECH.LA-PLANT.HEAVY              0.40 0.40                                    8.60
1411
1412     WHEAT.  PMW. SEMI-MECH.LA-PLANT.HEAVY              0.40 0.50                                    2.30 3.00
1413     WHEAT.  PCW. SEMI-MECH.LA-PLANT.HEAVY              0.40 0.40                                    4.00 3.60
1414     WHEAT.  PRW. SEMI-MECH.LA-PLANT.HEAVY              0.80 0.80                                         3.10
1415     WHEAT.  SCWN.SEMI-MECH.LA-PLANT.HEAVY                                                           3.10 3.30
1416     WHEAT.  SRWN.SEMI-MECH.LA-PLANT.HEAVY                                                           2.90 3.00
1417     WHEAT. (SCWS,SRWS).SEMI-MECH.LA-PLANT.
1418            (HEAVY,JANUARY,LIGHT,STANDARD)                                                           3.20 4.20
1419     WHEAT.  NWFP.SEMI-MECH.LA-PLANT.JANUARY            0.40 0.40                                    8.60
1420     WHEAT.  PMW. SEMI-MECH.LA-PLANT.JANUARY            0.40 0.50                                    2.30 3.00
1421     WHEAT.  PCW. SEMI-MECH.LA-PLANT.JANUARY            0.40 0.40                                    4.00 3.60
1422     WHEAT.  PRW. SEMI-MECH.LA-PLANT.JANUARY            0.80 0.80                                         3.10
1423     WHEAT.  SCWN.SEMI-MECH.LA-PLANT.JANUARY                                                         3.10 3.30
1424     WHEAT.  SRWN.SEMI-MECH.LA-PLANT.JANUARY                                                         2.90 3.00
1425     WHEAT.  NWFP.SEMI-MECH.LA-PLANT.LIGHT              0.50 0.60                                    8.60
1426     WHEAT.  PMW. SEMI-MECH.LA-PLANT.LIGHT              0.60 0.60                                    2.30 3.00
1427     WHEAT.  PCW. SEMI-MECH.LA-PLANT.LIGHT              0.50 0.60                                    4.00 3.60
1428     WHEAT.  PRW. SEMI-MECH.LA-PLANT.LIGHT              1.10 1.10                                         3.10
1429     WHEAT.  SCWN.SEMI-MECH.LA-PLANT.LIGHT                                                           3.10 3.30
1430     WHEAT.  SRWN.SEMI-MECH.LA-PLANT.LIGHT                                                           2.90 3.00
1431     WHEAT.  NWFP.SEMI-MECH.LA-PLANT.STANDARD           0.60 0.70                                    8.60
1432     WHEAT.  PMW. SEMI-MECH.LA-PLANT.STANDARD           0.70 0.70                                    2.30 3.00
1433     WHEAT.  PCW. SEMI-MECH.LA-PLANT.STANDARD           0.60 0.70                                    4.00 3.60
1434     WHEAT.  PRW. SEMI-MECH.LA-PLANT.STANDARD           1.30 1.30                                         3.10
1435     WHEAT.  SCWN.SEMI-MECH.LA-PLANT.STANDARD                                                        3.10 3.30
1436     WHEAT.  SRWN.SEMI-MECH.LA-PLANT.STANDARD                                                        2.90 3.00
1437
1438     WHEAT.  NWFP.SEMI-MECH.QK-HARV. HEAVY              1.00                                         4.10 4.50
1439     WHEAT.  PMW. SEMI-MECH.QK-HARV. HEAVY              1.00                                         2.30 3.00
1440     WHEAT.  PCW. SEMI-MECH.QK-HARV. HEAVY              1.00                                              7.60
1441     WHEAT.  PSW. SEMI-MECH.QK-HARV. HEAVY              0.90                                         3.40 3.00
1442     WHEAT.  PRW. SEMI-MECH.QK-HARV. HEAVY              1.00                                              3.10
1443     WHEAT.SCWN.SEMI-MECH.(QK-HARV,STANDARD).
1444           (HEAVY,JANUARY,LIGHT,STANDARD)                                                                 6.40
1445     WHEAT.SRWN.SEMI-MECH.(QK-HARV,STANDARD).
1446           (HEAVY,JANUARY,LIGHT,STANDARD)                                                                 5.9
1447     WHEAT. (SCWS,SRWS).SEMI-MECH.
1448           (QK-HARV,STANDARD).
1449           (HEAVY,JANUARY,LIGHT,STANDARD)                                                                 7.40
1450     WHEAT.  NWFP.SEMI-MECH.QK-HARV. JANUARY            1.00                                         4.10 4.50
1451     WHEAT.  PMW. SEMI-MECH.QK-HARV. JANUARY            1.00                                         2.30 3.00
1452     WHEAT.  PCW. SEMI-MECH.QK-HARV. JANUARY            1.00                                              7.60
1453     WHEAT.  PSW. SEMI-MECH.QK-HARV. JANUARY            0.90                                         3.40 3.00
1454     WHEAT.  PRW. SEMI-MECH.QK-HARV. JANUARY            1.00                                              3.10
1455     WHEAT.  NWFP.SEMI-MECH.QK-HARV. LIGHT              1.30                                         4.10 4.50
1456     WHEAT.  PMW. SEMI-MECH.QK-HARV. LIGHT              1.40                                         2.30 3.00
1457     WHEAT.  PCW. SEMI-MECH.QK-HARV. LIGHT              1.30                                              7.60
1458     WHEAT.  PSW. SEMI-MECH.QK-HARV. LIGHT              1.20                                         3.40 3.00
```

INDUS BASIN MODEL REVISED (IBMR) FILENAME=WSISD41
CROP DATA

```
1459     WHEAT. PRW. SEMI-MECH.QK-HARV. LIGHT              1.30                                  3.10
1460     WHEAT. NWFP.SEMI-MECH.QK-HARV. STANDARD           1.50                          4.10   4.50
1461     WHEAT. PMW. SEMI-MECH.QK-HARV. STANDARD           1.60                          2.30   3.00
1462     WHEAT. PCW. SEMI-MECH.QK-HARV. STANDARD           1.50                                  7.60
1463     WHEAT. PSW. SEMI-MECH.QK-HARV. STANDARD           1.40                          3.40   3.00
1464     WHEAT. PRW. SEMI-MECH.QK-HARV. STANDARD           1.50                                  3.10
1465     WHEAT. NWFP.SEMI-MECH.STANDARD.HEAVY              0.50  0.50                    4.10   4.50
1466     WHEAT. PMW. SEMI-MECH.STANDARD.HEAVY              0.50  0.50                    2.30   3.00
1467     WHEAT. PCW. SEMI-MECH.STANDARD.HEAVY              0.50  0.50                           7.60
1468     WHEAT. PSW. SEMI-MECH.STANDARD.HEAVY              0.40  0.50                    3.40   3.00
1469     WHEAT. PRW. SEMI-MECH.STANDARD.HEAVY              0.50  0.50                           3.10
1470     WHEAT. NWFP.SEMI-MECH.STANDARD.JANUARY            0.50  0.50                    4.10   4.50
1471     WHEAT. PMW. SEMI-MECH.STANDARD.JANUARY            0.50  0.50                    2.30   3.00
1472     WHEAT. PCW. SEMI-MECH.STANDARD.JANUARY            0.50  0.50                           7.60
1473     WHEAT. PSW. SEMI-MECH.STANDARD.JANUARY            0.40  0.50                    3.40   3.00
1474     WHEAT. PRW. SEMI-MECH.STANDARD.JANUARY            0.50  0.50                           3.10
1475     WHEAT. NWFP.SEMI-MECH.STANDARD.LIGHT              0.60  0.70                    4.10   4.50
1476     WHEAT. PMW. SEMI-MECH.STANDARD.LIGHT              0.70  0.70                    2.30   3.00
1477     WHEAT. PCW. SEMI-MECH.STANDARD.LIGHT              0.60  0.70                           7.60
1478     WHEAT. PSW. SEMI-MECH.STANDARD.LIGHT              0.60  0.60                    3.40   3.00
1479     WHEAT. PRW. SEMI-MECH.STANDARD.LIGHT              0.60  0.70                           3.10
1480     WHEAT. NWFP.SEMI-MECH.STANDARD.STANDARD           0.70  0.80                    4.10   4.50
1481     WHEAT. PMW. SEMI-MECH.STANDARD.STANDARD           0.80  0.80                    2.30   3.00
1482     WHEAT. PCW. SEMI-MECH.STANDARD.STANDARD           0.70  0.80                           7.60
1483     WHEAT. PSW. SEMI-MECH.STANDARD.STANDARD           0.70  0.70                    3.40   3.00
1484     WHEAT. PRW. SEMI-MECH.STANDARD.STANDARD           0.70  0.80                           3.10
1485     ORCHARD.(NWFP,PMW,PCW,PSW,PRW).SEMI-MECH.
1486             STANDARD.STANDARD          .1   .2   .1   .1   .1   .1   .1   .1   .1   .2   .3
1487     ORCHARD.(SCWN,SCWS,SRWN,SRWS).SEMI-MECH.
1488             STANDARD.STANDARD          .1   .3   .1   .1   .1   .2   .1   .1   .1   .1   .3
1489     POTATOES.(SCWN,SCWS,SRWN,SRWS).
1490             SEMI-MECH.STANDARD.STANDARD                               2.   2.
1491     POTATOES.(NWFP,PMW,PCW,PSW,PRW).SEMI-MECH.
1492             STANDARD.STANDARD          2.                                                   2.
1493     ONIONS.(NWFP,PMW,PCW,PSW,PRW).SEMI-MECH.
1494             STANDARD.STANDARD          4.
1495     ONIONS.(SCWN,SCWS,SRWN,SRWS). SEMI-MECH.
1496             STANDARD.STANDARD                                         4.
1497     CHILLI.(NWFP,PMW,PCW,PSW,PRW).SEMI-MECH.
1498             STANDARD.STANDARD               1.
1499     CHILLI.(SCWN,SCWS,SRWN,SRWS). SEMI-MECH.
1500             STANDARD.STANDARD                                                               1.
1501
1502     TABLE SYLDS(C,Z,T,S,W,CI) STRAW YIELD AND SEED DATA
1503                                         STRAW-YLD    SEED
1504     *                                   (PROPORTION  (KG)
1505     *                                   OF YIELD)
1506     BASMATI.(PMW,PCW,PRW).(BULLOCK,SEMI-MECH).
1507             STANDARD.STANDARD           2.33         6.4
1508     BASMATI.PSW. (BULLOCK,SEMI-MECH).
1509             STANDARD. STANDARD          2.12         6.4
1510
1511     RAB-FOD.(SRWN,SRWS).(BULLOCK,SEMI-MECH).
1512             STANDARD.(LIGHT,HEAVY)      1            6.0
1513     RAB-FOD.(NWFP,PMW,PCW,PSW,PRW).(BULLOCK,SEMI-MECH).
1514             STANDARD.STANDARD           1            2.0
1515     RAB-FOD.(SCWN,SRWN,SCWS,SRWS).(BULLOCK,SEMI-MECH).
1516             STANDARD. STANDARD          1            6.0
1517
1518     COTTON. PCW. BULLOCK.         EL-PLANT. STANDARD       9.
1519     COTTON. PCW. SEMI-MECH.       LA-PLANT. STANDARD       9.
1520     COTTON. (PMW,PCW,PSW,PRW,SCWN,SCWS,SRWN,SRWS).
1521             (BULLOCK,SEMI-MECH).STANDARD. STANDARD         9.
1522
1523     GRAM.   NWFP.(BULLOCK,SEMI-MECH).STANDARD. STANDARD  1.70   10.
```

INDUS BASIN MODEL REVISED (IBMR) FILENAME=WSISD41
CROP DATA

```
1524    GRAM.  (PMW,PCW,PSW,PRW).
1525           (BULLOCK,SEMI-MECH).STANDARD. STANDARD    1.52     14.
1526    GRAM.  (SCWN,SCWS).(BULLOCK,SEMI-MECH).
1527                  STANDARD. STANDARD                 1.50     12.8
1528    GRAM.  (SRWN,SRWS).(BULLOCK,SEMI-MECH).
1529                  STANDARD. STANDARD                 2.00     12.0
1530
1531    IRRI.  (PMW,PCW,PSW,PRW).(BULLOCK,SEMI-MECH).
1532                  STANDARD.STANDARD                  1.80      7.1
1533    IRRI.  SCWN.(BULLOCK,SEMI-MECH).STANDARD. STANDARD  1.34   4.5
1534    IRRI.  (SRWN,SCWS,SRWS).(BULLOCK,SEMI-MECH).
1535                  STANDARD. STANDARD                 1.32      4.5
1536
1537    MAIZE. NWFP.(BULLOCK,SEMI-MECH).STANDARD. STANDARD  2.09  14.2
1538    MAIZE. PCW. (BULLOCK,SEMI-MECH).STANDARD. STANDARD  3.00   6.1
1539    MAIZE. (PSW,PRW). (BULLOCK,SEMI-MECH).
1540                  STANDARD. STANDARD                 2.50      6.1
1541    MAIZE. (SCWN,SCWS).
1542           (BULLOCK,SEMI-MECH).STANDARD. STANDARD    2.80      8.1
1543
1544    MUS+RAP.NWFP.(BULLOCK,SEMI-MECH).STANDARD. STANDARD 0.50   1.8
1545    MUS+RAP.(PMW,PCW,PSW,PRW,SCWN,SRWN,SRWS,SCWS).
1546           (BULLOCK,SEMI-MECH).STANDARD. STANDARD    0.65      2.4
1547
1548    SC-GUR. NWFP.(BULLOCK,SEMI-MECH).STANDARD. STANDARD 1.54   3238
1549    SC-GUR. PMW. (BULLOCK,SEMI-MECH).STANDARD. STANDARD 1.81   3238
1550    SC-GUR. PCW. (BULLOCK,SEMI-MECH).STANDARD. STANDARD 1.66   3238
1551    SC-GUR. PSW. (BULLOCK,SEMI-MECH).STANDARD. STANDARD 1.54   3238
1552    SC-GUR. PRW. (BULLOCK,SEMI-MECH).STANDARD. STANDARD 2.17   3238
1553    SC-GUR. SCWN.(BULLOCK,SEMI-MECH).STANDARD. STANDARD 1.81   3238
1554    SC-GUR. SRWN.(BULLOCK,SEMI-MECH).STANDARD. STANDARD 3.08   3238
1555    SC-GUR. SCWS.(BULLOCK,SEMI-MECH).STANDARD. STANDARD 2.17   3238
1556    SC-GUR. SRWS.(BULLOCK,SEMI-MECH).STANDARD. STANDARD 2.05   3238
1557
1558    SC-MILL.NWFP.(BULLOCK,SEMI-MECH).STANDARD. STANDARD 0.12   3238
1559    SC-MILL.PMW. (BULLOCK,SEMI-MECH).STANDARD. STANDARD 0.14   3238
1560    SC-MILL.PCW. (BULLOCK,SEMI-MECH).STANDARD. STANDARD 0.13   3238
1561    SC-MILL.PSW. (BULLOCK,SEMI-MECH).STANDARD. STANDARD 0.10   3238
1562    SC-MILL.PRW. (BULLOCK,SEMI-MECH).STANDARD. STANDARD 0.17   3238
1563    SC-MILL.SCWN.(BULLOCK,SEMI-MECH).STANDARD. STANDARD 0.14   3238
1564    SC-MILL.SRWN.(BULLOCK,SEMI-MECH).STANDARD. STANDARD 0.25   3238
1565    SC-MILL.SCWS.(BULLOCK,SEMI-MECH).STANDARD. STANDARD 0.17   3238
1566    SC-MILL.SRWS.(BULLOCK,SEMI-MECH).STANDARD. STANDARD 0.16   3238
1567
1568    KHA-FOD.SRWS.(BULLOCK,SEMI-MECH).LA-PLANT. STANDARD 1       5.7
1569    KHA-FOD.NWFP.(BULLOCK,SEMI-MECH).STANDARD. STANDARD 1      20.2
1570    KHA-FOD.(PMW,PCW,PSW,PRW).
1571           (BULLOCK,SEMI-MECH).STANDARD. STANDARD    1        27.3
1572    KHA-FOD.(SCWN,SCWS,SRWN,SRWS).
1573           (BULLOCK,SEMI-MECH).STANDARD. STANDARD    1         5.7
1574
1575    WHEAT.NWFP.(BULLOCK,SEMI-MECH).
1576           (STANDARD,LA-PLANT,QK-HARV).
1577           (STANDARD,LIGHT,HEAVY,JANUARY)            1.3      40.1
1578    WHEAT.PMW. (BULLOCK,SEMI-MECH).
1579           (STANDARD,LA-PLANT,QK-HARV).
1580           (STANDARD,LIGHT,HEAVY,JANUARY)            1.3      34.8
1581    WHEAT.(PCW,PSW). (BULLOCK,SEMI-MECH).
1582           (STANDARD,LA-PLANT,QK-HARV).
1583           (STANDARD,LIGHT,HEAVY,JANUARY)            1.5      34.8
1584    WHEAT.PRW. (BULLOCK,SEMI-MECH).
1585           (STANDARD,LA-PLANT,QK-HARV).
1586           (STANDARD,LIGHT,HEAVY,JANUARY)            1.6      34.8
1587    WHEAT.(SCWN,SRWN,SCWS,SRWS).(BULLOCK,SEMI-MECH).
1588           (STANDARD,LA-PLANT,QK-HARV).
```

```
1589                  (STANDARD,LIGHT,HEAVY,JANUARY)             1.5       49.8
1590
1591
1592     ORCHARD.(NWFP,PMW,PCW,PSW,PRW,SCWN,SCWS,SRWN,SRWS).
1593            (BULLOCK,SEMI-MECH).STANDARD.STANDARD                        1
1594
1595     POTATOES.(SCWN,SCWS,SRWN,SRWS).
1596            SEMI-MECH.STANDARD.STANDARD                                1200
1597     POTATOES.(NWFP,PMW,PCW,PSW,PRW).
1598            SEMI-MECH.STANDARD.STANDARD                                 600
1599
1600     ONIONS.(NWFP,PMW,PCW,PSW,PRW,SCWN,SCWS,SRWN,SRWS).
1601            SEMI-MECH.STANDARD.STANDARD                                   3.
1602
1603     CHILLI.(NWFP,PMW,PCW,PSW,PRW,SCWN,SCWS,SRWN,SRWS).
1604            SEMI-MECH.STANDARD.STANDARD                                   5.
1605
1606     TABLE FERT(P2,C,Z)  FERTILIZER APPLICATIONS  (KG PER ACRE)
1607                       NWFP   PCW   PMW   PRW   PSW   SCWN  SCWS  SRWN  SRWS
1608
1609     NITROGEN.BASMATI          26.6  26.6  21.9  23.4
1610     NITROGEN.IRRI             39.4  39.4  23.3  26.8  68.6  61.5  48.9  41.7
1611     NITROGEN.COTTON   26.7    42.3  30.0  30.0  19.7  55.0  54.9  39.6  39.6
1612     NITROGEN.MAIZE    27.0    27.1  27.1  23.6  19.5  42.0  42.0  42.0  42.0
1613     NITROGEN.KHA-FOD  21.0    25.3  18.1  19.2  18.4  49.0  49.0  49.0  49.0
1614     NITROGEN.WHEAT    46.8    40.9  36.5  32.2  33.3  54.9  53.5  29.5  39.8
1615     NITROGEN.RAB-FOD  10.0    25.3  18.1  19.2  18.4  28.0  28.0  28.0  28.0
1616     NITROGEN.SC-MILL  83.4    44.8  63.2  33.9  33.9  65.1  65.1  65.1  65.1
1617     NITROGEN.SC-GUR   24.0    19.0  19.0  19.0  19.0  28.0  28.0  28.0  28.0
1618     NITROGEN.ONIONS   60.6    48.0  48.0  48.0  48.0  70.7  70.7  70.7  70.7
1619     NITROGEN.POTATOES 48.6    38.5  38.5  38.5  38.5  56.7  56.7  56.7  56.7
1620     NITROGEN.MUS+RAP  33.6    30.8  30.8  30.8  30.8  45.4  45.4  45.4  45.4
1621     NITROGEN.CHILLI   48.6    38.5  38.5  38.5  38.5  56.7  56.7  56.7  56.7
1622     NITROGEN.ORCHARD  60.0    47.5  47.5  47.5  47.5  70.0  70.0  70.0  70.0
1623
1624     PHOSPHATE.BASMATI         13.3  13.3  11.4  10.6   0.0   0.0   0.0   0.0
1625     PHOSPHATE.IRRI             8.0   8.0   6.8   6.4  15.7  16.6  13.5  12.5
1626     PHOSPHATE.COTTON  26.2    11.3   5.2   7.0   7.0  12.8  14.4  13.0  13.0
1627     PHOSPHATE.MAIZE   17.6     8.8   6.9   8.6   8.6   7.7   7.7   7.7   7.7
1628     PHOSPHATE.KHA-FOD 25.9    11.4  11.4  11.4  11.4  15.3  15.3  15.3  15.3
1629     PHOSPHATE.WHEAT   12.3    12.8   8.8  11.4  10.9  15.1  15.1  11.8  17.3
1630     PHOSPHATE.RAB-FOD  5.4     7.6   5.4   5.8   5.6   7.7   7.7   7.7   7.7
1631     PHOSPHATE.SC-MILL 46.8    14.0  19.8  10.6  10.6  18.5  18.5  18.5  18.5
1632     PHOSPHATE.SC-GUR  10.4     5.7   5.7   5.7   5.7   7.7   7.7   7.7   7.7
1633     PHOSPHATE.ONIONS  26.2     7.0   7.0   7.0   7.0   9.3   9.3   9.3   9.3
1634     PHOSPHATE.POTATOES 42.0   18.5  18.5  18.5  18.5  24.9  24.9  24.9  24.9
1635     PHOSPHATE.MUS+RAP 25.9    13.3  13.3  13.3  13.3  17.9  17.9  17.9  17.9
1636     PHOSPHATE.CHILLI  26.2    11.6  11.6  11.6  11.6  15.5  15.5  15.5  15.5
1637     PHOSPHATE.ORCHARD 13.0     5.7   5.7   5.7   5.7   7.7   7.7   7.7   7.7
1638
1639     PARAMETER FERTGR(C)  FERTILIZER APPLICATION GROWTH RATE   PERCENT
1640
1641       /(BASMATI, IRRI,    COTTON,   RAB-FOD
1642         GRAM   , MAIZE,  MUS+RAP,  KHA-FOD
1643         SC-GUR , SC-MILL,WHEAT   , ORCHARD)         3,
1644        (POTATOES, ONIONS,CHILLI)                    2.4/
1645         ;
1646
1647     *&&Z   ZONE4XXXXX(Z,C,P2) = FERT(P2,C,Z) ;
1648
1649
1650     FERT(P2,C,Z) = FERT(P2,C,Z)*SUM(IS$ISR(IS),
1651                    (1+FERTGR(C)/100)**(ORD(IS)+1979-BASEYEAR)) ;
1652
1653
```

```
1654    PARAMETER
1655      NATYIELD(C) NATIONAL CROP YIELDS 1988 FOR STANDARD TECHNOLOGIES (KGS)
1656        /    BASMATI       457
1657             IRRI          880
1658             COTTON        695
1659             RAB-FOD     15000
1660             GRAM          183
1661             MAIZE         534
1662             MUS+RAP       307
1663             KHA-FOD     10000
1664             SC-GUR       1270
1665             SC-MILL     15870
1666             WHEAT         780
1667             ORCHARD      3400
1668             POTATOES     3980
1669             ONIONS       4615
1670             CHILLI        567 /
1671
1672
1673    TABLE YLDPRPV(C,PV) PROVINCE YIELDS PROPORTION OF NATIONAL 1987-88
1674  *                    NOTE:  USED 3-YEAR AVERAGE FROM LATEST ASP
1675
1676                  NWFP    PUNJAB    SIND
1677
1678      WHEAT       0.926   0.978    1.103
1679      BASMATI     0       1        0
1680      IRRI        0       0.859    1.001
1681      COTTON      0       1.153    0.861
1682      SC-MILL     1.090   0.919    1.190
1683      SC-GUR      1.090   0.919    1.190
1684      MAIZE       1.006   1.029    0.408
1685      MUS+RAP     0.622   1.192    0.842
1686      GRAM        0.911   0.937    1.465
1687      RAB-FOD      .400   1.000    1.000
1688      KHA-FOD      .770   1.000    1.300
1689      ORCHARD     0.936   1.070    1
1690      POTATOES     .973    .960     .873
1691      ONIONS      1.172   1.085    0.843
1692      CHILLI      0.786   1.167    0.952
1693
1694
1695    TABLE YLDPRZS(C,Z) ZONES YIELDS AS PROPORTION OF PROVINCE-STANDARD TECHNOLOGIES
1696
1697              NWFP   PCW    PMW    PRW    PSW    SCWN   SCWS   SRWN   SRWS
1698   WHEAT      1     0.96   0.84   1.06   1.13   0.96   1.07   0.82   1.30
1699   BASMATI    1      .83   0.00   1.26   0.79   0.00   0.00   0.00   0.00
1700   IRRI       1     0.83   0.00   1.26   0.00   1.06   0.91   1.10   0.90
1701   COTTON     1     1.10   0.71   0.70   0.80   1.09   0.96   0.78   0.78
1702   SC-MILL    1     1.04   0.93   0.93   0.98   0.95   1.10   0.64   0.87
1703   SC-GUR     1     1.04   0.93   0.93   0.98   0.95   1.10   0.64   0.87
1704   MAIZE      1     1.03   0.76   1.08   0.97   1.07   1.06   0.87   1.00
1705   MUS+RAP    1     1.03   0.76   1.08   0.97   1.07   1.06   0.87   1.00
1706   GRAM       1     1.03   0.76   1.08   0.97   1.07   1.06   0.87   1.00
1707   RAB-FOD    1     1.23   0.81   1.17   1.23   1.     1.     1.     1.
1708   KHA-FOD    1      .83   0.54   0.82   0.82   1.     1.     1.     1.00
1709   ORCHARD    1     1.     1.     1.     1.0    1.     1.     1.     1.
1710   POTATOES   1     1.03   0.76   1.08   0.97   1.07   1.06   0.87   1.00
1711   ONIONS     1     1.03   0.76   1.08   0.97   1.07   1.06   0.87   1.00
1712   CHILLI     1     1.03   0.76   1.08   0.97   1.07   1.06   0.87   1.00
1713
1714    PARAMETER YLDPRZO(C,S,W) YIELDS AS PROPORTION OF STANDARD TECHNOLOGIES
1715       /
1716      RAB-FOD.STANDARD.HEAVY        0.70
1717      RAB-FOD.STANDARD.LIGHT        0.80
```

```
1718        KHA-FOD.LA-PLANT.STANDARD       1.05
1719
1720        WHEAT.   QK-HARV .STANDARD                      1.00
1721        WHEAT.   LA-PLANT.(HEAVY,JANUARY)               0.57
1722        WHEAT.   LA-PLANT.LIGHT                         0.75
1723        WHEAT.   LA-PLANT.STANDARD                      0.88
1724        WHEAT.   (QK-HARV,STANDARD).(HEAVY,JANUARY)     0.65
1725        WHEAT.   (QK-HARV,STANDARD).LIGHT               0.85
1726        /
1727   TABLE GROWTHCY(C,Z) GROWTH RATE OF CROP YIELDS FROM 1988 BASE (PERCENT)
1728
1729               (NWFP,PMW,PCW,PSW,PRW,SCWN,SCWS,SRWN,SRWS)
1730   COTTON          5.00
1731   MAIZE           0.73
1732   SC-GUR          2.29
1733   SC-MILL         2.29
1734   WHEAT           0.41
1735   ONIONS          1.66
1736
1737       TABLE WEEDY(Z,SEA,C) WEED YIELDS BY CROP (TONNS PER ACER)
1738                           BASMATI  IRRI  COTTON  GRAM  MUS+RAP  MAIZE  SC-GUR  SC-MILL  WHEAT  ORCHARD
1739   (NWFP,PMW,PCW,PSW,PRW).RABI                            .3      .3              .3      .3      .4      .3
1740   (SCWN,SCWS,SRWN,SRWS). RABI                            .3      .3              .3      .3      .4      .3
1741   (NWFP,PMW,PCW,PSW,PRW).KHARIF    1       1       1                      1       1       1               1
1742   (SCWN,SCWS,SRWN,SRWS). KHARIF    1       1                              1       1       1               1
1743
1744       TABLE GRAZ(Z,SEA)  GRAZING FROM SLACK LAND (TONNS PER ACRE )
1745                                    RABI    KHARIF
1746   PRW                               .15     .5
1747   PMW                               .1      .2
1748   (PSW,PCW)                         .2      .3
1749   (SCWN,SCWS,SRWN,SRWS)             .2      .5
1750
1751   PARAMETER
1752      YIELD(C,T,S,W,Z) YIELD BY ZONE CROP TECHNOLOGY IN METRIC TONNS
1753      GROWTHCYF(C,Z)   GROWTH FACTOR FOR CROP YIELDS USING GROWTHCY ;
1754
1755      YIELD(C,T,"STANDARD","STANDARD",Z) =         SUM(PV$PVZ(PV,Z),
1756                        NATYIELD(C)/1000 *YLDPRPV(C,PV) *YLDPRZS(C,Z) );
1757
1758      GROWTHCY(C,Z)$((GROWTHCY(C,Z) GT 3)$ISR("2000")) = 3.0 ;
1759      YIELD(C,T,S,W,Z)$YLDPRZO(C,S,W) =
1760                  YIELD(C,T,"STANDARD","STANDARD",Z)*YLDPRZO(C,S,W) ;
1761
1762      GROWTHCYF(C,Z) = SUM(IS$ISR(IS),
1763                      (1+GROWTHCY(C,Z)/100)**(ORD(IS)+1979-BASEYEAR));
1764      YIELD(C,T,S,W,Z) = YIELD(C,T,S,W,Z) * GROWTHCYF(C,Z) ;
1765   DISPLAY BASEYEAR, GROWTHCYF, FERT , GROWTHCY ;
1766   * DISPLAY YIELD;
       *  REPORT ON INPUT DATA
          SET IC CROP INPUTS /LAND, LABOR, BULLOCK, WATER, TRACTOR/
          PARAMETER REP1,REP2 ;
          REP1(Z,C,T,S,W,"LAND",M)    = LAND(C,Z,T,S,W,M)     ;
          REP1(Z,C,T,S,W,"BULLOCK",M) = BULLOCK(C,Z,T,S,W,M)  ;
          REP1(Z,C,T,S,W,"LABOR",M)   = LABOR(C,Z,T,S,W,M)    ;
          REP1(Z,C,T,S,W,"WATER",M)   = WATER(C,Z,T,S,W,M)    ;
          REP1(Z,C,T,S,W,"TRACTOR",M) = TRACTOR(C,Z,T,S,W,M)  ;

          REP1(Z,C,T,S,W,IC,"TOTAL")  = SUM(M, REP1(Z,C,T,S,W,IC,M)) ;
          REP2(Z,C,T,S,W,IC)          = REP1(Z,C,T,S,W,IC,"TOTAL")   ;
          REP2(Z,C,T,S,W,CI)          = sylds(C,Z,T,S,W,CI)          ;
          REP2(Z,C,T,S,W,p2)$tech(z,c,t,s,w) = fert(p2,c,z)          ;
          REP2(Z,C,T,S,W,"YIELD")     = Yield(c,t,s,w,z)             ;
          OPTIONS REP1:2:4:1, REP2:2:4:1 ; display rep1, rep2 ;
```

INDUS BASIN MODEL REVISED (IBMR) FILENAME=WSISD41
LIVESTOCK DATA

```
1786       TABLE IOLIVE(A,Z,*) LIVESTOCK INPUT OUTPUT COEFFICIENTS BY ZONES
1787  *      MEAT AND MILK YEILDS ARE OF 1985.
1788                            TDN      DP      LABOR     COW-MILK  BUFF-MILK   MEAT    FIX-COST
1789  *                       (METRIC TONS/ SEASON)  MAN HRS              LITERS       KGS
1790  *                                             PER MONTH  (- - - PER YEAR   - - -)
1791     COW. NWFP             .68     .061      25.1       359                  13.4      15
1792     COW. PMW              .68     .061      25.1       284                  13.4      15
1793     COW. PCW              .68     .061      25.1       303                  13.4      15
1794     COW. PSW              .68     .061      25.1       268                  13.4      15
1795     COW. PRW              .68     .061      25.1       332                  13.4      15
1796     COW. SCWN             .68     .046      23.5       460                  10.5      17
1797     COW. SRWN             .68     .083      23.5       383                  15.4      17
1798     COW. SCWS             .68     .046      23.5       350                  10.5      17
1799     COW. SRWS             .68     .083      23.5       468                  15.4      17
1800
1801     BULLOCK.(NWFP,PMW,PCW,PSW,PRW)  .635  .058   15.1                       14.3      50
1802     BULLOCK.(SCWN,SRWN,SCWS,SRWS)   .635  .058   14.8                       14.1      50
1803
1804     BUFFALO.NWFP         1.04     .095      33.6                  602       17.3      50
1805     BUFFALO.PMW          1.04     .095      33.6                  653       17.3      50
1806     BUFFALO.PCW          1.04     .095      33.6                  652       17.3      50
1807     BUFFALO.PSW          1.04     .095      33.6                  643       17.3      50
1808     BUFFALO.PRW          1.04     .095      33.6                  590       17.3      50
1809     BUFFALO.SCWN         1.04     .073      31.4                  946       16.2      50
1810     BUFFALO.SRWN         1.04     .128      31.4                  750       22.5      50
1811     BUFFALO.SCWS         1.04     .073      31.4                  773       16.2      50
1812     BUFFALO.SRWS         1.04     .128      31.4                 1018       22.5      50
1813
1814       TABLE SCONV(NT,SEA,C)   TDN AND DP CONVERSION FACTOR FROM CROP STRAW
1815
1816              BASMATI COTTON RAB-FOD GRAM  IRRI  KHA-FOD MAIZE MUS+RAP SC-MILL SC-GUR WHEAT
1817     TDN.KHARIF                       .6          .14          .5                      .6
1818     DP. KHARIF                       .1          .019         .005                    .005
1819     TDN.RABI    .5     .14           .5                .5             .17     .17
1820     DP. RABI   .005    .025          .005             .1             .005    .005
1821     ;
1822     SCALARS    REPCO    REPRODUCTIVE COEFFICIENT                      / 2.5 /
1823                GR       REQUIRED PROPORTION OF GREEN FODDER IN TOTAL FODDER / 0.3 /
1824                GROWTHQ  GROWTH RATE OF MILK AND MEAT YIELDS (PERCENT)  / 2.5 /
1825     PARAMETERS BP(M)    DRAFT POWER AVAILABLE PER BULLOCK(HOURS PER MONTH)
1826                ;
1827          BP(M)=96;            BP("MAY")=77;         BP("JUN")=77;
1828  *&&Z   ZONE2XXXXX(Z,A,SET1) = IOLIVE(A,Z,SET1) ;
1829       IOLIVE(A,Z,Q) = IOLIVE(A,Z,Q) * SUM(IS$ISR(IS), (1+GROWTHQ/100)**(ORD(IS)+1979 -BASEYEAR) ) ;
1830       OPTIONS IOLIVE:3;   DISPLAY IOLIVE ;
```

```
1832    SET CNL   IRRIGATION CANALS IN THE INDUS RIVER IRRIGATION SYSTEM/
1833        01-UD          UPPER DIPALPUR
1834        02-CBD         CENTRAL BARI DOAB CANAL
1835        03-RAY         RAYA CANAL
1836        04-UC          UPPER CHENAB CANAL
1837        05-MR          MARALA RAVI CANAL
1838        06-SAD         SADIQIA CANAL
1839        07-FOR         FORDWAH CANAL
1840        08-PAK         UPPER PAKPATTAN+U-BAHAWAL+QAIM+U-MAILSI CANAL
1841        09-LD          LOWER DIPALPUR CANAL
1842        10-LBD         LOWER BARI DOAB CANAL
1843        11-JHA         JHANG CANAL (LCC)
1844        12-GUG         GUGERA BRANCH CANAL (LCC)
1845        13-UJ          UPPER JEHLUM CANAL
1846        14-LJ          LOWER JEHLUM CANAL
1847        15-BAH         BAHAWAL CANAL
1848        16-MAI         LOWER MAILSI+ LOWER PAKPATTAN CANAL
1849        17-SID         SIDHNAI CANAL
1850        18-HAV         HAVELI CANAL
1851        19-RAN         RANGPUR CANAL
1852        20-PAN         PANJNAD CANAL
1853        21-ABB         ABBASIA CANAL
1854        22-USW         UPPER SWAT CANAL
1855        22A-PHL        PEHUR HIGH LEVEL CANAL (IN YEAR 2000)
1856        23-LSW         LOWER SWAT CANAL
1857        24-WAR         WARSAK CANAL
1858        25-KAB         KABUL RIVER CANAL
1859        26-THA         THAL CANAL  (+ GREATER THAL CANAL IN YEAR 2000)
1860        27-PAH         PAHARPUR CANAL (OR C.R.B.C. IN YEAR 1988 & 2000)
1861        28-MUZ         MUZAFFARGARH CANAL
1862        29-DGK         DERA GHAZI KHAN CANAL
1863        31-P+D         PAT PLUS DESERT CANAL
1864        32-BEG         BEGARI CANAL
1865        33-GHO         GHOTKI CANAL
1866        34-NW          NORTH WEST CANAL
1867        35-RIC         RICE CANAL
1868        36-DAD         DADU CANAL
1869        37-KW          KHAIRPUR WEST CANAL
1870        38-KE          KHAIRPUR EAST CANAL
1871        39-ROH         ROHRI CANAL
1872        41-NAR         NARA CANAL  (+ MAKHI FRASH CANAL IN YEAR 2000)
1873        42-KAL         KALRI CANAL
1874        43-LCH         LINED CHANNEL
1875        44-FUL         FULELI CANAL
1876        45-PIN         PINYARI CANAL/
1877    PVCNL(PV,CNL)   PROVINCE TO CANALS MAP /
1878        NWFP.  (22-USW, 22A-PHL,23-LSW, 24-WAR, 25-KAB)
1879        PUNJAB.(01-UD,  02-CBD, 03-RAY, 04-UC,  05-MR,  06-SAD, 07-FOR,
1880                08-PAK, 09-LD,  10-LBD, 11-JHA, 12-GUG, 13-UJ,  14-LJ,
1881                15-BAH, 16-MAI, 17-SID, 18-HAV, 19-RAN, 20-PAN, 21-ABB,
1882                26-THA, 27-PAH, 28-MUZ, 29-DGK)
1883  *     SIND CANALS INCLUDING BALUCHISTAN
1884        SIND.  (31-P+D, 32-BEG, 33-GHO, 34-NW,  35-RIC, 36-DAD, 37-KW,
1885                38-KE,  39-ROH, 41-NAR, 42-KAL, 43-LCH, 44-FUL, 45-PIN)/
1886    GWFG(CNL,SA,G) SUBAREA IDENTIFICATION BY THE GROUNDWATER QUALITY;
1887
1888
1889    TABLE  COMDEF(IS,DC,CNL) CANAL COMMAND CHARACTERISTICS
1890  *    CCA       --   CULTURABLE COMMANDED AREA OF THE CANAL( MILLIONS OF ACRES)
1891  *    CCAP      --   CANAL CAPACITY AT THE CANAL HEAD(MILLIONS OF ACRE FEET)
1892  *    CEFF      --   CANAL EFFICIENCY FROM BARRAGE TO THE WATER COURSE HEAD
1893  *    WCE-R     --   WATER COURSE COMMAND EFFICINECY IN RABI SEASON (WC&FLD)
1894  *    WCE-K     --   WATER COURSE COMMAND EFFICINECY IN KHARIF SEASON(WC&FLD)
```

```
INDUS BASIN MODEL REVISED (IBMR) FILENAME=WSISD41                              07/25/91  17:17:54
CANAL AND AGROCLIMATIC ZONE DATA                                               GAMS 2.21 IBM CMS


1895  *    FLDE      --   FIELD EFFICIENCY
1896  *
1897  *    NOTE: CCAP FOR 05-MR, 24-WAR, 25-KAB ARE SET EQUAL TO POST TARBELA AVERAGE DIVERSION. ORIGINAL CAPACITIES WERE
1898  *              .166    .044    .047
1899  *
1900  * 1988 PARAMETERS ARE DERIVED CONSIDERING THE FOLLOWING PROJECTS;
1901  * COMMAND WATER MANAGEMENT PROJECT  50% COMPLETION
1902  * SCARP VI, SCARP MARDAN, KHAIRPUR TILE, FOURTH DRAINAGE, OFWM I AND II
1903  * IRRIGATION SYSTEM REHABLITATION PHASE I, CHASMA RIGHT BANK.
1904
1905              01-UD   02-CBD   03-RAY   04-UC    05-MR    06-SAD   07-FOR   08-PAK   09-LD    10-LBD   11-JHA
1906
1907  1980.CCA    .36     .649     .424     1.017    .158     .969     .426     1.049    .615     1.67     1.168
1908  1980.CCAP   .131    .163     .11      .506     .282     .332     .204     .746     .244     .518     .41
1909  1980.CEFF   .70     .80      .80      .76      .80      .72      .70      .61      .80      .72      .70
1910  1980.WCE-R  .57     .57      .45      .57      .57      .51      .51      .57      .57      .57      .55
1911  1980.WCE-K  .57     .57      .45      .57      .57      .51      .51      .57      .57      .57      .55
1912  1980.FLDE   .90     .90      .90      .90      .90      .80      .80      .90      .90      .90      .90
1913
1914  1988.CCA    .36     .649     .424     1.017    .158     .969     .426     1.049    .615     1.67     1.168
1915  1988.CCAP   .131    .163     .11      .506     .282     .332     .204     .746     .244     .518     .41
1916  1988.CEFF   .703    .811     .800     .760     .800     .728     .704     .621     .810     .722     .706
1917  1988.WCE-R  .57     .573     .45      .57      .57      .522     .51      .575     .57      .57      .554
1918  1988.WCE-K  .57     .573     .45      .57      .57      .522     .51      .575     .57      .57      .554
1919  1988.FLDE   .90     .90      .90      .90      .90      .80      .80      .90      .90      .90      .90
1920
1921  1993.CCA    .36     .649     .424     1.017    .158     .969     .426     1.049    .615     1.67     1.168
1922  1993.CCAP   .131    .163     .11      .506     .282     .332     .204     .746     .244     .518     .41
1923  1993.CEFF   .705    .815     .808     .765     .800     .733     .708     .625     .810     .724     .713
1924  1993.WCE-R  .58     .585     .46      .58      .58      .543     .53      .59      .58      .58      .564
1925  1993.WCE-K  .58     .585     .46      .58      .58      .543     .53      .59      .58      .58      .564
1926  1993.FLDE   .90     .90      .90      .90      .90      .80      .80      .90      .90      .90      .90
1927
1928  2000.CCA    .36     .649     .424     1.017    .158     .969     .426     1.049    .615     1.67     1.168
1929  2000.CCAP   .131    .163     .11      .506     .282     .332     .204     .746     .244     .518     .41
1930  2000.CEFF   .703    .811     .800     .760     .800     .728     .704     .621     .810     .722     .706
1931  2000.WCE-R  .59     .585     .47      .59      .59      .543     .53      .59      .59      .59      .574
1932  2000.WCE-K  .59     .585     .47      .59      .59      .543     .53      .59      .59      .59      .574
1933  2000.FLDE   .90     .90      .90      .90      .90      .80      .80      .90      .90      .90      .90
1934
1935     +        12-GUG  13-UJ    14-LJ    15-BAH   16-MAI   17-SID   18-HAV   19-RAN   20-PAN   21-ABB   22-USW  22A-PHL
1936
1937  1980.CCA    1.866   .544     1.50     .605     .996     .869     .179     .344     1.348    .154     .279
1938  1980.CCAP   .433    .345     .464     .346     .362     .268     .068     .176     .671     .08      .117
1939  1980.CEFF   .74     .80      .64      .77      .77      .72      .77      .70      .72      .64      .75
1940  1980.WCE-R  .55     .51      .51      .57      .57      .57      .57      .46      .57      .57      .52
1941  1980.WCE-K  .55     .51      .51      .57      .57      .57      .57      .46      .57      .57      .52
1942  1980.FLDE   .90     .90      .90      .90      .90      .90      .85      .80      .90      .90      .90
1943
1944  1988.CCA    1.866   .544     1.50     .605     .996     .869     .179     .344     1.348    .154     .279
1945  1988.CCAP   .433    .345     .464     .346     .362     .268     .068     .176     .773     .08      .117
1946  1988.CEFF   .749    .800     .642     .782     .780     .721     .773     .700     .728     .640     .750
1947  1988.WCE-R  .553    .51      .51      .57      .57      .57      .57      .46      .59      .59      .53
1948  1988.WCE-K  .553    .51      .51      .57      .57      .57      .57      .46      .59      .59      .53
1949  1988.FLDE   .90     .90      .90      .90      .90      .90      .85      .80      .90      .90      .90
1950
1951  1993.CCA    1.866   .544     1.500    .605     .996     .869     .179     .344     1.348    .154     .279
1952  1993.CCAP   .433    .345     .464     .346     .362     .268     .068     .176     .773     .08      .132
1953  1993.CEFF   .751    .808     .644     .783     .783     .727     .776     .706     .73      .641     .757
1954  1993.WCE-R  .565    .52      .52      .58      .58      .58      .58      .48      .60      .60      .54
1955  1993.WCE-K  .565    .52      .52      .58      .58      .58      .58      .48      .60      .60      .54
1956  1993.FLDE   .90     .90      .90      .90      .90      .90      .85      .80      .90      .90      .90
1957
1958  2000.CCA    1.866   .544     1.500    .605     .996     .869     .179     .344     1.348    .154     .279    .0
1959  2000.CCAP   .433    .345     .464     .346     .362     .268     .068     .176     .773     .08      .117    .0
```

INDUS BASIN MODEL REVISED (IBMR) FILENAME=WSISD41
CANAL AND AGROCLIMATIC ZONE DATA

07/25/91 17:17:54
GAMS 2.21 IBM CMS

1960	2000.CEFF	.749	.800	.642	.782	.780	.721	.773	.700	.728	.640	.750	.0
1961	2000.WCE-R	.565	.53	.53	.59	.59	.59	.59	.49	.60	.60	.53	.0
1962	2000.WCE-K	.565	.53	.53	.59	.59	.59	.59	.49	.60	.60	.53	.0
1963	2000.FLDE	.90	.90	.90	.90	.90	.90	.85	.80	.90	.90	.90	.0
1964													
1965	+	23-LSW	24-WAR	25-KAB	26-THA	27-PAH	28-MUZ	29-DGK	31-P+D	32-BEG	33-GHO	34-NW	
1966													
1967	1980.CCA	.182	.119	.048	1.641	.104	.809	.909	1.075	1.002	.858	1.215	
1968	1980.CCAP	.06	.049	.047	.577	.037	.495	.529	.799	1.155	.648	.566	
1969	1980.CEFF	.75	.80	.72	.65	.73	.70	.70	.83	.82	.76	.80	
1970	1980.WCE-R	.52	.52	.52	.48	.51	.46	.54	.55	.55	.45	.55	
1971	1980.WCE-K	.52	.52	.52	.48	.51	.46	.54	.60	.65	.45	.60	
1972	1980.FLDE	.90	.90	.90	.80	.85	.80	.85	.85	.85	.85	.85	
1973													
1974	1988.CCA	.182	.119	.048	1.641	.570	.809	.909	1.075	1.002	.858	1.215	
1975	1988.CCAP	.120	.049	.047	.577	.330	.495	.529	.799	1.155	.648	.566	
1976	1988.CEFF	.750	.800	.720	.663	.730	.720	.703	.830	.820	.763	.800	
1977	1988.WCE-R	.57	.57	.53	.48	.51	.46	.54	.55	.55	.45	.55	
1978	1988.WCE-K	.57	.57	.53	.48	.51	.46	.54	.60	.65	.45	.60	
1979	1988.FLDE	.90	.90	.90	.80	.85	.80	.85	.85	.85	.85	.85	
1980													
1981	1993.CCA	.182	.119	.048	1.641	.570	.809	.909	1.075	1.002	.858	1.215	
1982	1993.CCAP	.120	.049	.047	.577	.330	.495	.529	.799	1.155	.648	.566	
1983	1993.CEFF	.751	.80	.723	.671	.73	.725	.707	.831	.82	.765	.80	
1984	1993.WCE-R	.59	.60	.54	.49	.56	.47	.55	.56	.56	.47	.56	
1985	1993.WCE-K	.59	.60	.54	.49	.56	.47	.55	.61	.66	.47	.61	
1986	1993.FLDE	.90	.90	.90	.80	.85	.80	.85	.85	.85	.85	.85	
1987													
1988	2000.CCA	.182	.119	.048	1.641	.570	.809	.909	1.075	1.002	.858	1.215	
1989	2000.CCAP	.120	.049	.047	.577	.330	.495	.529	.799	1.155	.648	.566	
1990	2000.CEFF	.750	.800	.720	.663	.730	.720	.703	.830	.820	.763	.801	
1991	2000.WCE-R	.60	.61	.55	.50	.57	.48	.56	.594	.57	.48	.57	
1992	2000.WCE-K	.60	.61	.55	.50	.57	.48	.56	.65	.67	.48	.62	
1993	2000.FLDE	.90	.90	.90	.80	.85	.80	.85	.85	.85	.85	.85	
1994													
1995	+	35-RIC	36-DAD	37-KW	38-KE	39-ROH	41-NAR	42-KAL	43-LCH	44-FUL	45-PIN		
1996													
1997	1980.CCA	.519	.584	.417	.373	2.561	2.176	.592	.502	.923	.758		
1998	1980.CCAP	.829	.319	.157	.207	.981	.873	.546	.205	.894	.831		
1999	1980.CEFF	.85	.80	.74	.73	.80	.80	.80	.80	.80	.82		
2000	1980.WCE-R	.55	.55	.45	.45	.45	.45	.55	.55	.55	.55		
2001	1980.WCE-K	.75	.60	.45	.45	.45	.45	.60	.61	.65	.62		
2002	1980.FLDE	.85	.85	.85	.85	.85	.85	.85	.85	.85	.85		
2003													
2004	1988.CCA	.519	.584	.417	.373	2.561	2.176	.592	.502	.923	.758		
2005	1988.CCAP	.829	.319	.157	.214	.981	.873	.546	.205	.894	.831		
2006	1988.CEFF	.850	.800	.748	.736	.807	.811	.800	.800	.800	.820		
2007	1988.WCE-R	.55	.55	.45	.455	.456	.45	.55	.55	.55	.55		
2008	1988.WCE-K	.75	.60	.45	.455	.456	.45	.60	.61	.65	.62		
2009	1988.FLDE	.85	.85	.85	.85	.85	.85	.85	.85	.85	.85		
2010													
2011	1993.CCA	.519	.584	.417	.373	2.561	2.176	.592	.502	.923	.758		
2012	1993.CCAP	.829	.319	.157	.214	.981	1.041	.546	.205	.894	.831		
2013	1993.CEFF	.854	.804	.75	.738	.81	.816	.80	.80	.80	.821		
2014	1993.WCE-R	.56	.56	.46	.465	.47	.469	.56	.56	.56	.56		
2015	1993.WCE-K	.74	.61	.46	.465	.47	.469	.61	.62	.65	.63		
2016	1993.FLDE	.85	.85	.85	.85	.85	.85	.85	.85	.85	.85		
2017													
2018	2000.CCA	.519	.584	.417	.373	2.561	2.176	.592	.502	.923	.758		
2019	2000.CCAP	.829	.319	.157	.214	.981	.873	.546	.205	.894	.831		
2020	2000.CEFF	.850	.800	.748	.736	.807	.811	.800	.800	.800	.820		
2021	2000.WCE-R	.57	.57	.47	.475	.49	.479	.57	.57	.57	.57		
2022	2000.WCE-K	.75	.62	.47	.475	.49	.479	.61	.62	.65	.63		
2023	2000.FLDE	.85	.85	.85	.85	.85	.85	.85	.85	.85	.85		
2024													

INDUS BASIN MODEL REVISED (IBMR) FILENAME=WSISD41
CANAL AND AGROCLIMATIC ZONE DATA
07/25/91 17:17:54
GAMS 2.21 IBM CMS

```
* Original data with Plan
*
```

I		01-UD	02-CBD	03-RAY	04-UC	05-MR	06-SAD	07-FOR	08-PAK	09-LD	10-LBD	11-JHA	I
I2000.CCA		.36	.649	.424	1.017	.158	.969	.426	1.049	.615	1.67	1.168	I
I2000.CCAP		.131	.163	.11	.506	.282	.332	.204	.746	.244	.518	.41	I
I2000.CEFF		.705	.815	.808	.765	.800	.733	.708	.625	.810	.724	.713	I
I2000.WCE-R		.59	.595	.47	.59	.59	.553	.54	.60	.59	.59	.574	I
I2000.WCE-K		.59	.595	.47	.59	.59	.553	.54	.60	.59	.59	.574	I
I2000.FLDE		.90	.90	.90	.90	.90	.80	.80	.90	.90	.90	.90	I

I	+	12-GUG	13-UJ	14-LJ	15-BAH	16-MAI	17-SID	18-HAV	19-RAN	20-PAN	21-ABB	22-USW	22A-PHL	I
I2000.CCA		1.866	.544	1.500	.605	.996	.869	.179	.344	1.348	.154	.192	.166	I
I2000.CCAP		.433	.345	.464	.346	.362	.268	.068	.176	.773	.08	.176	.0589	I
I2000.CEFF		.751	.808	.644	.783	.783	.727	.776	.706	.73	.641	.757	.757	I
I2000.WCE-R		.575	.53	.53	.59	.59	.59	.59	.49	.61	.61	.55	.55	I
I2000.WCE-K		.575	.53	.53	.59	.59	.59	.59	.49	.61	.61	.55	.55	I
I2000.FLDE		.90	.90	.90	.90	.90	.90	.85	.80	.90	.90	.90	.90	I

I	+	23-LSW	24-WAR	25-KAB	26-THA	27-PAH	28-MUZ	29-DGK	31-P+D	32-BEG	33-GHO	34-NW	I
I2000.CCA		.182	.119	.048	3.201	.904	.809	1.292	1.215	1.002	.858	1.215	I
I2000.CCAP		.120	.049	.047	1.202	.523	.495	0.692	1.011	1.155	.648	.587	I
I2000.CEFF		.751	.80	.723	.671	.73	.725	.707	.832	.82	.765	.804	I
I2000.WCE-R		.60	.61	.55	.50	.57	.48	.56	.594	.57	.48	.57	I
I2000.WCE-K		.60	.61	.55	.50	.57	.48	.56	.65	.67	.48	.62	I
I2000.FLDE		.90	.90	.90	.80	.85	.80	.85	.85	.85	.85	.85	I

I	+	35-RIC	36-DAD	37-KW	38-KE	39-ROH	41-NAR	42-KAL	43-LCH	44-FUL	45-PIN	I
I2000.CCA		.519	.584	.417	.373	2.561	2.176	.592	.502	.923	.758	I
I2000.CCAP		.829	.319	.157	.214	.998	1.041	.546	.205	.894	.831	I
I2000.CEFF		.854	.804	.75	.738	.811	.816	.80	.80	.80	.821	I
I2000.WCE-R		.57	.57	.47	.475	.49	.479	.57	.57	.57	.57	I
I2000.WCE-K		.75	.62	.47	.475	.49	.479	.61	.62	.65	.63	I
I2000.FLDE		.85	.85	.85	.85	.85	.85	.85	.85	.85	.85	I

```
*
* YEAR 2000 PARAMETERS ARE THE WSIPS BASIC PLAN. THE BASIC PLAN INCLUDE
* COMMAND WATER MANAGEMENT, IRRIGATION REHABLITATION I & II.
* ON-FARM MANGEMENT III, CHASMA RIGHT BANK STAGE III,
* FORDWAH SADIQIA SCARP, GREATER THAL, DGK SCARP, PUNJNAD ABBASIA,
* REMAINING FOURTH DRAINAGE, LBOD, MAKKHI FARESH LINK, NW REMODELLING
* NORTH DADU, GHOTKI FRESH GW, KOTRI SURFACE DRAIN,
* MARDAN SCARP, PEHUR HIGH LEVEL CANAL, SWABI SCARP,
* PAT FEEDER CANAL, KIRTHER BRANCH CANAL REMODELLING.
*
* PEHUR HIGH LEVEL CANAL 22A-PHL TAKES OFF ABOVE TARBELA, AND COMMANDS
* 79,000 ACRES OF NEW AREA AND 87,000 ACRES OF UPPER SWAT CANAL.
*
      TABLE PLAN1(CNL,DC)  WATER SECTOR INVESTMENT PLAN COEFFICIENTS

              CCA    CCAP    CEFF    WCE-R   WCE-K   FLDE

* COMMAND WATER MANAGEMENT REST OF THE 50%, SAD INCLUDE FORWAH SAD SCARP

  02-CBD                            .595    .595
  06-SAD                            .553    .553
  08-PAK                            .60     .60
  12-GUG                            .575    .575
```

```
2092  * FORWAH SADIQIA SCRAP
2093    07-FOR                               .54      .54
2094
2095  * CRBC STAGE III
2096    27-PAH         .590    .523
2097
2098  * GREATER THAL
2099    26-THA        3.201   1.202
2100
2101  * PANJNAD ABBASIA
2102    20-PAN                                .61      .61
2103    21-ABB                                .61      .61
2104
2105  * LBOD, INCLUDES CHOTIARI RESERVOIR
2106    41-NAR                1.041
2107
2108  * NW AND KITHAR REMODELLING
2109
2110    34-NW                  .587    .804
2111
2112  * MARDAN SCARP, PEHUR HIGH LEVEL AND SWABI SCARP
2113    22-USW         .192    .176    .757    .55      .55     .90
2114    22A-PHL        .166   .0589    .757    .55      .55     .90
2115
2116  * DGK SCARP , DAJAL BRANCH
2117    29-DGK        1.292    .692
2118
2119  * PAT AND DESERT CANAL
2120    31-P+D        1.215   1.011
2121
2122  * ROHRI REMODELLING
2123    39-ROH                 .998
2124
2125       TABLE PLAN2(DC,CNL)  IRRIGATION REHABLITATION PROJECT PHASE II
2126  * DOES NOT INCLUDE, USW, PHL, AND NW, THESE ARE IN PLAN1
2127
2128              01-UD   02-CBD   03-RAY   04-UC    05-MR    06-SAD   07-FOR   08-PAK   09-LD    10-LBD   11-JHA
2129    CEFF      .705    .815     .808     .765     .800     .733     .708     .625     .810     .724     .713
2130       +      12-GUG  13-UJ    14-LJ    15-BAH   16-MAI   17-SID   18-HAV   19-RAN   20-PAN   21-ABB
2131    CEFF      .751    .808     .644     .783     .783     .727     .776     .706     .73      .641
2132       +      23-LSW  24-WAR   25-KAB   26-THA   27-PAH   28-MUZ   29-DGK   31-P+D   32-BEG   33-GHO
2133    CEFF      .751    .80      .723     .671     .73      .725     .707     .832     .82      .765
2134       +      35-RIC  36-DAD   37-KW    38-KE    39-ROH   41-NAR   42-KAL   43-LCH   44-FUL   45-PIN
2135    CEFF      .854    .804     .75      .738     .811     .816     .80      .80      .80      .821
2136    ;
2137  * FOR YEAR 2000 CHECK WHETHER PLAN IS TO BE INCLUDED OR NOT
2138    COMDEF("2000",DC,CNL)$(INCP("WITH")$(PLAN1(CNL,DC) NE 0))
2139                   = PLAN1(CNL,DC);
2140    COMDEF("2000",DC,CNL)$(INCP("WITH")$(PLAN2(DC,CNL) NE 0))
2141                   = PLAN2(DC,CNL)  ;
2142  DISPLAY COMDEF;
2143
2144       TABLE SUBDEF(SA,CNL) SUB-AREA DEFINITION(PROPORTION OF CCA)  BY CANALS
2145
2146              01-UD   02-CBD   03-RAY   04-UC    05-MR    06-SAD   07-FOR   08-PAK   09-LD    10-LBD   11-JHA
2147    S1        1       .5       1        1        1        1        1        1        1        .5       .32
2148    S2                .5                                                                      .5       .49
2149    S3                                                                                                 .19
2150
2151       +      12-GUG  13-UJ    14-LJ    15-BAH   16-MAI   17-SID   18-HAV   19-RAN   20-PAN   21-ABB   22-USW   22A-PHL
2152    S1        .26     1        .64      .8       .65      1        1        1        .7       1        1        1
2153    S2        .53              .36      .2       .35                                 .3
2154    S3        .21
2155
2156       +      23-LSW  24-WAR   25-KAB   26-THA   27-PAH   28-MUZ   29-DGK   31-P+D   32-BEG   33-GHO   34-NW
```

```
INDUS BASIN MODEL REVISED (IBMR) FILENAME=WSISD41                                07/25/91 17:17:54
CANAL AND AGROCLIMATIC ZONE DATA                                                   GAMS 2.21 IBM CMS


2157     S1              1          1           1         .35       1        .25       1         1        .5        .5         1
2158     S2                                               .17                .75                           .5        .5
2159     S3                                               .30
2160     S4                                               .18
2161
2162             +    35-RIC      36-DAD      37-KW     38-KE    39-ROH    41-NAR    42-KAL    43-LCH    44-FUL    45-PIN
2163     S1           1           1           1         1        .39       .2        1         1         1         1
2164     S2                                                      .20       .8
2165     S3                                                      .16
2166     S4                                                      .25
2167
2168     SET ZSA(Z,CNL,SA)    CANAL-SUBAREA TO AGROCLIMATIC ZONE MAPPING /
2169
2170         NWFP. (22-USW.  S1,22A-PHL. S1, 23-LSW.  S1,     24-WAR.  S1,       25-KAB.  S1)
2171         PMW.  (26-THA.  (S1,S2,S3,S4), 27-PAH.   S1,     28-MUZ.  S1)
2172         PCW.  (01-UD.   S1,            02-CBD.   S2,     06-SAD.  S1,       07-FOR.  S1,    08-PAK. S1
2173                09-LD.   S1,            10-LBD.  (S1,S2), 15-BAH. (S1,S2),   16-MAI. (S1,S2),17-SID. S1
2174                19-RAN.  S1,            20-PAN.  (S1,S2), 21-ABB.  S1,       28-MUZ.  S2,    29-DGK. S1)
2175         PSW.  (11-JHA. (S2,S3),        12-GUG.  (S2,S3), 13-UJ.   S1,       14-LJ.  (S1,S2),18-HAV. S1)
2176         PRW.  (02-CBD.  S1,            03-RAY.   S1,     04-UC.   S1,       05-MR.   S1,    11-JHA. S1, 12-GUG.S1)
2177
2178         SCWN. (33-GHO. (S1,S2),        37-KW.    S1,     38-KE.   S1,       39-ROH. (S1,S2),41-NAR. S1)
2179         SRWN. (31-P+D.  S1,            32-BEG.  (S1,S2), 34-NW.   S1,       35-RIC.  S1,    36-DAD. S1)
2180         SCWS. (39-ROH. (S3,S4),        41-NAR.   S2  )
2181         SRWS. (42-KAL.  S1,            43-LCH.   S1,     44-FUL.  S1,       45-PIN.  S1 ) /
2182
2183         GWF(CNL,SA)    SUBAREAS WITH FRESH GROUND WATER /
2184         01-UD.   S1,   02-CBD. (S1,S2),03-RAY.  S1,  04-UC.   S1,    05-MR.   S1
2185         07-FOR.  S1,   08-PAK.  S1,    09-LD.   S1,  10-LBD. (S1,S2),11-JHA. (S1,S2)
2186         12-GUG. (S1,S2),13-UJ.  S1,    14-LJ.   S1,  15-BAH.  S1,    16-MAI.  S1
2187         17-SID.  S1,   19-RAN.  S1,    20-PAN.  S1,  21-ABB.  S1,
2188         22-USW.  S1,   22A-PHL. S1,    23-LSW.  S1,  24-WAR.  S1,    25-KAB.  S1,
2189         26-THA.(S1,S3),27-PAH.  S1,    28-MUZ.  S1,  29-DGK.  S1,    31-P+D.  S1, 32-BEG. S1, 33-GHO. S1
2190         37-KW.   S1,   39-ROH. (S1,S3) / ;
2191
2192         GWFG(CNL,SA,"SALINE")$SUBDEF(SA,CNL)  = YES;
2193         GWFG(CNL,SA,"SALINE")$GWF(CNL,SA)     = NO ;  GWFG(CNL,SA,"FRESH")$GWF(CNL,SA) = YES ;
2194      PARAMETER  CAREA(CNL,*)    CCA CLASSIFIED BY GROUNDWATER QUALITY FOR EACH CANAL ;
2195      LOOP(ISR,
2196      CAREA(CNL,G)        = SUM(SA$GWFG(CNL,SA,G), SUBDEF(SA,CNL)*COMDEF(ISR,"CCA",CNL) ) );
2197      CAREA(CNL,"TOTAL") = SUM(G,  CAREA(CNL,G) );
2198      DISPLAY CAREA ;
          *- REPORT AND CHECK ON ACZ DEFINITION
          SET ZSA1(CNL,SA,Z);
          PARAMETER REP3
                    REP4
          REP3(IS,CNL,SA)       = SUBDEF(SA,CNL)*COMDEF(IS,"CCA",CNL) ;
          REP3(IS,CNL,G)        = SUM(SA$GWFG(CNL,SA,G), SUBDEF(SA,CNL)*COMDEF(IS,"CCA",CNL) ) ;
          REP3(IS,CNL,"TOTAL")  = SUM(G, REP3(IS,CNL,G) )      ;
          REP3(IS,"TOTAL",T1)   = SUM(CNL, REP3(IS,CNL,T1))    ;
          REP3(IS,CNL,"CCA")    = COMDEF(IS,"CCA",CNL)         ;
          REP3(IS,"TOTAL","CCA")= SUM(CNL, COMDEF(IS,"CCA",CNL)) ;

          REP4(IS,Z,G)       = SUM((CNL,SA)$(ZSA(Z,CNL,SA)$GWFG(CNL,SA,G)), COMDEF(IS,"CCA",CNL)*SUBDEF(SA,CNL) );
          REP4(IS,Z,"TOTAL") = SUM((CNL,SA)$ZSA(Z,CNL,SA), COMDEF(IS,"CCA",CNL)*SUBDEF(SA,CNL) );
          REP4(IS,"TOTAL",T1)= SUM(Z, REP4(IS,Z,T1)   )   ;
          ZSA1(CNL,SA,Z)$ZSA(Z,CNL,SA) = YES ;
          DISPLAY ZSA1, GWF, GWFG, REP3, REP4 ;
2218
2219     TABLE EVAP(CNL,M)  PAN EVAPORATION(FEET)
2220              JAN      FEB      MAR      APR     MAY     JUN     JUL     AUG     SEP     OCT     NOV     DEC
2221     01-UD    .244     .314     .545     .786    .982    1.040   .913    .814    .729    .571    .357    .244
2222     02-CBD   .200     .314     .529     .771    .968    .998    .829    .743    .686    .543    .329    .229
2223     03-RAY   .200     .314     .514     .771    .951    .968    .812    .743    .671    .543    .326    .214
2224     04-UC    .200     .283     .514     .771    .940    .998    .829    .743    .671    .543    .326    .214
```

INDUS BASIN MODEL REVISED (IBMR) FILENAME=WSISD41
CANAL AND AGROCLIMATIC ZONE DATA

		JAN	FEB	MAR	APR	MAY	JUN	JUL	AUG	SEP	OCT	NOV	DEC
2225	05-MR	.200	.286	.529	.769	.942	.998	.829	.743	.671	.543	.326	.214
2226	06-SAD	.286	.414	.648	.814	1.139	1.254	1.012	.940	.743	.629	.402	.286
2227	07-FOR	.271	.400	.614	.786	1.083	1.196	.968	.893	.714	.564	.386	.286
2228	08-PAK	.271	.343	.588	.829	1.012	1.056	.970	.885	.771	.614	.400	.288
2229	09-LD	.243	.312	.586	.786	.969	1.056	.955	.871	.771	.586	.371	.286
2230	10-LBD	.243	.314	.557	.786	.982	1.040	.925	.938	.745	.581	.360	.243
2231	11-JHA	.229	.300	.514	.729	.990	1.012	.870	.757	.700	.486	.343	.217
2232	12-GUG	.214	.312	.526	.769	.965	1.008	.826	.740	.683	.555	.326	.226
2233	13-UJ	.229	.300	.514	.729	.967	1.012	.870	.760	.700	.571	.343	.214
2234	14-LJ	.257	.314	.557	.788	.982	1.026	.930	.843	.743	.600	.371	.260
2235	15-BAH	.271	.360	.657	.860	1.069	1.098	1.011	.899	.771	.657	.400	.271
2236	16-MAI	.271	.340	.629	.842	1.042	1.056	.940	.869	.757	.643	.400	.271
2237	17-SID	.257	.343	.588	.842	1.026	1.026	.940	.860	.757	.643	.414	.271
2238	18-HAV	.271	.329	.571	.800	.982	1.025	.940	.857	.757	.600	.386	.271
2239	19-RAN	.244	.314	.557	.786	.982	1.040	.913	.814	.729	.571	.357	.244
2240	20-PAN	.200	.314	.529	.771	.968	1.112	1.026	.913	.800	.543	.329	.457
2241	21-ABB	.271	.338	.671	.858	1.085	1.098	1.026	.913	.788	.671	.400	.288
2242	(22-USW,22A-PHL,												
2243	23-LSW,												
2244	24-WAR,25-KAB)	.214	.257	.429	.643	.913	1.025	.969	.870	.657	.529	.329	.200
2245	26-THA	.260	.331	.517	.760	1.057	1.143	1.057	.899	.721	.631	.402	.245
2246	27-PAH	.257	.329	.502	.729	1.094	1.225	1.154	.943	.829	.671	.357	.243
2247	28-MUZ	.271	.340	.600	.870	1.056	1.039	.968	.885	.771	.657	.414	.271
2248	29-DGK	.286	.357	.657	.885	1.082	1.098	1.011	.913	.798	.671	.414	.271
2249	31-P+D	.371	.457	.758	.956	1.142	1.126	1.026	.942	.830	.715	.486	.343
2250	(32-BEG,34-NW,												
2251	35-RIC)	.386	.474	.700	.870	1.068	.949	.943	.949	.757	.657	.457	.343
2252	33-GHO	.429	.429	.700	.871	1.040	1.098	.998	.900	.800	.629	.457	.329
2253	36-DAD	.357	.429	.700	.956	1.169	1.185	1.027	.940	.786	.700	.429	.357
2254	37-KW	.386	.414	.674	.900	1.083	1.112	.998	.899	.786	.643	.429	.343
2255	38-KE	.429	.443	.745	.926	1.152	1.154	1.025	.955	.829	.700	.486	.371
2256	39-ROH	.371	.414	.700	.940	1.154	1.139	.982	.926	.757	.698	.457	.371
2257	41-NAR	.457	.500	.843	1.054	1.212	1.155	.999	.914	.829	.786	.557	.443
2258	42-KAL	.529	.557	.857	.982	1.094	.970	.814	.757	.786	.800	.614	.529
2259	(43-LCH,44-FUL)	.471	.529	.870	1.095	1.210	1.111	.985	.886	.843	.814	.586	.471
2260	45-PIN	.557	.586	.899	1.044	1.140	1.012	.843	.800	.829	.843	.643	.557
2261													
2262	TABLE RAIN(CNL,M) RAIN (INCHES)												
2263		JAN	FEB	MAR	APR	MAY	JUN	JUL	AUG	SEP	OCT	NOV	DEC
2264	01-UD	.770	.520	.660	.470	.450	1.180	4.480	4.070	1.530	.160	.220	.360
2265	02-CBD	.878	.700	.712	.510	.506	1.364	4.868	4.502	1.798	.192	.172	.404
2266	03-RAY	1.180	1.120	1.040	.740	.710	1.820	5.610	6.450	2.460	.230	.140	.490
2267	04-UC	1.141	1.089	1.013	.724	.693	1.779	5.444	6.278	2.396	.224	.138	.478
2268	05-MR	1.180	1.120	1.040	.740	.710	1.820	5.610	6.450	2.460	.230	.140	.490
2269	(06-SAD,07-FOR)	.500	.520	.520	.360	.310	.720	3.500	3.000	.860	.090	.110	.240
2270	08-PAK	.505	.520	.523	.362	.313	.729	3.520	3.021	.873	.091	.112	.242
2271	09-LD	.689	.520	.618	.437	.408	1.042	4.186	3.749	1.329	.139	.187	.324
2272	10-LBD	.564	.522	.543	.380	.360	.778	3.545	3.130	1.018	.111	.122	.278
2273	11-JHA	.627	.686	.662	.509	.465	1.239	3.286	4.035	1.564	.139	.112	.315
2274	12-GUG	.548	.624	.608	.476	.430	1.156	2.954	3.690	1.436	.126	.108	.290
2275	13-UJ	1.210	1.170	1.240	.850	.690	1.500	5.320	5.610	2.030	.230	.160	.480
2276	14-LJ	.920	.889	.942	.646	.524	1.140	4.043	4.264	1.543	.175	.122	.365
2277	15-BAH	.230	.290	.380	.250	.170	.260	2.520	1.930	.190	.020	.120	.130
2278	16-MAI	.402	.434	.463	.316	.270	.553	3.020	2.530	.618	.068	.107	.208
2279	17-SID	.420	.435	.445	.305	.325	.495	2.570	2.265	.665	.085	.080	.245
2280	18-HAV	.390	.500	.500	.410	.360	.990	2.290	3.000	1.180	.100	.100	.240
2281	19-RAN	.370	.380	.400	.270	.330	.550	2.010	1.820	.540	.080	.060	.240
2282	20-PAN	.220	.252	.296	.184	.140	.184	1.716	1.444	.152	.010	.108	.120
2283	21-ABB	.221	.263	.326	.205	.143	.206	1.998	1.615	.163	.011	.102	.121
2284	(22-USW,22A-PHL												
2285	23-LSW,												
2286	24-WAR,25-KAB)	1.440	1.530	2.440	1.760	.770	.310	1.260	2.030	.810	.230	.310	.670
2287	26-THA	.610	.717	.882	.621	.448	.809	2.946	2.769	.944	.132	.130	.298
2288	27-PAH	.450	.670	.960	.690	.390	.610	2.290	1.900	.630	.110	.150	.240
2289	28-MUZ	.375	.400	.435	.298	.335	.555	2.028	1.825	.548	.083	.068	.240

INDUS BASIN MODEL REVISED (IBMR) FILENAME=WSISD41
CANAL AND AGROCLIMATIC ZONE DATA

```
2290   29-DGK              .260   .330   .270   .220   .180   .380   .740  1.230   .310   .030   .070   .210
2291   (31-P+D,32-BEG)     .199   .282   .204   .179   .100   .257   .447   .905   .187  0.000   .097   .179
2292   33-GHO              .210   .230   .250   .140   .120   .140  1.230  1.160   .130  0.000   .110   .110
2293   34-NW               .171   .336   .263   .171   .100   .242   .612   .994   .330  0.000   .035   .173
2294   35-RIC              .173   .341   .273   .173   .100   .251   .706  1.020   .346  0.000   .032   .163
2295   36-DAD              .170   .335   .270   .170   .100   .275  1.330  1.170   .400  0.000   .035   .130
2296   37-KW               .190   .240   .200   .140   .100   .250  1.310  1.210   .270  0.000   .090   .100
2297   38-KE               .150   .200   .200   .100   .100   .250  1.350  1.450   .350  0.000   .050   .100
2298   39-ROH              .138   .235   .188   .143   .150   .368  2.297  1.876   .564   .034   .051   .095
2299   41-NAR              .091   .151   .119   .103   .159   .397  3.003  2.097   .658   .071   .056   .060
2300   (42-KAL,43-LCH,
2301    44-FUL,45-PIN)     .140   .300   .120   .060   .140   .720  4.020  2.080   .700   .060   .060   .120
2302
```

INDUS BASIN MODEL REVISED (IBMR) FILENAME=WSISD41
HISTORIC CANAL DIVERSIONS(MAF)

07/25/91 17:17:54
GAMS 2.21 IBM CMS

```
2304
2305   TABLE DIVPOST(CNL,M1) AVERAGE(APRIL 1977 TO MARCH 1989) CANAL DIVERSIONS(MAF)
2306  *       SOURCE : INDUS BASIN IRRIGATION SYSTEM
2307  *                HISTORIC RIVERS AND CANALS DISCHARGE DATA,
2308  *                WATER RESOURCES MANAGEMENT, DIRECTORATE. WAPDA LAHORE  , JAN 1990
2309
2310            APR    MAY    JUN    JUL    AUG    SEP    OCT    NOV    DEC    JAN    FEB    MAR    RABI   KHARIF   ANNUAL
2311   01-UD   .0258  .1014  .1087  .1180  .1246  .1279  .0626  .0209  .0090  .0100  .0167  .0312   .1504   .6063   0.7567
2312   02-CBD  .1089  .1315  .1381  .1285  .1247  .1275  .1200  .1307  .1122  .0461  .0960  .1014   .6064   .7591   1.3656
2313   03-RAY  .0024  .0311  .0856  .0870  .0810  .0843  .0413  .0017  .0000  .0005  .0022  .0003   .0460   .3714   0.4173
2314   04-UC   .0873  .1707  .2711  .3085  .2789  .2729  .1886  .0816  .0356  .0428  .0622  .0567   .4673  1.3895   1.8568
2315   05-MR   .0157  .0959  .1188  .2098  .2528  .1690  .0384  .0116  .0107  .0109  .0195  .0098   .1009   .8620   0.9629
2316   06-SAD  .2675  .2827  .2848  .2740  .2773  .2776  .2347  .2822  .2759  .1032  .2334  .2834  1.4128  1.6640   3.0768
2317   07-FOR  .0244  .1208  .1471  .1405  .1470  .1590  .0934  .0162  .0144  .0041  .0137  .0137   .1554   .7388   0.8942
2318   08-PAK  .2125  .3274  .3473  .3416  .3763  .3692  .2633  .2119  .2430  .1068  .1815  .2515  1.2581  1.9743   3.2324
2319   09-LD   .0457  .1908  .2043  .2333  .2398  .2308  .1470  .0399  .0289  .0188  .0406  .0279   .3003  1.1447   1.4451
2320   10-LBD  .3972  .4829  .4672  .4517  .4924  .4638  .3921  .4234  .3791  .1436  .3531  .4091  2.1003  2.7551   4.8554
2321   11-JHA  .2886  .3394  .3361  .3146  .3178  .3366  .3250  .3049  .3995  .1317  .1973  .2762  1.5346  1.9329   3.4675
2322   12-GUG  .3267  .3868  .3839  .3518  .3653  .3856  .3730  .3468  .3396  .1461  .2193  .3121  1.7369  2.2000   3.9369
2323   13-UJ   .0812  .1167  .1286  .1443  .1382  .1300  .1175  .0913  .0743  .0407  .0487  .0510   .4235   .7389   1.1625
2324   14-LJ   .2392  .3118  .3151  .3021  .3020  .2961  .2993  .2465  .2147  .0964  .2133  .2078  1.2780  1.7663   3.0442
2325   15-BAH  .1149  .2062  .2339  .2501  .2401  .2430  .2010  .1200  .1229  .0907  .0859  .1445   .7649  1.2883   2.0532
2326   16-MAI  .1340  .2917  .3313  .3122  .3456  .3562  .2776  .1223  .1419  .1207  .1011  .1590   .9227  1.7711   2.6937
2327   17-SID  .1460  .2228  .2348  .2162  .2353  .2424  .2210  .1653  .1590  .0915  .1184  .1547   .9099  1.2975   2.2074
2328   18-HAV  .0329  .0471  .0484  .0449  .0479  .0520  .0475  .0240  .0294  .0217  .0273  .0307   .1806   .2732   0.4538
2329   19-RAN  .0307  .0751  .0845  .0759  .0747  .0838  .0651  .0152  .0164  .0173  .0105  .0227   .1472   .4246   0.5718
2330   20-PAN  .2709  .4425  .5353  .5382  .5574  .5616  .4861  .2192  .2018  .1229  .1727  .2205  1.4232  2.9059   4.3291
2331   21-ABB  .0417  .0527  .0600  .0570  .0627  .0629  .0569  .0395  .0351  .0204  .0297  .0356   .2171   .3370   0.5542
2332   22-USW  .1047  .1130  .1100  .1100  .1053  .1058  .1048  .0815  .0791  .0632  .0650  .0853   .4788   .6488   1.1276
2333   22A-PHL
2334   23-LSW  .0422  .0516  .0515  .0469  .0386  .0470  .0480  .0388  .0367  .0048  .0229  .0275   .1790   .2777   0.4567
2335   24-WAR  .0381  .0435  .0422  .0383  .0331  .0404  .0450  .0429  .0377  .0021  .0142  .0260   .1678   .2357   0.4036
2336   25-KAB  .0349  .0440  .0466  .0449  .0408  .0441  .0434  .0392  .0328  .0017  .0176  .0344   .1690   .2553   0.4243
2337   26-THA  .3770  .4203  .4077  .4096  .4140  .4199  .4112  .3997  .3717  .1131  .3187  .3646  1.9786  2.4458   4.4247
2338   27-PAH  .0285  .0355  .0387  .0379  .0325  .0356  .0372  .0326  .0328  .0073  .0278  .0307   .1683   .2086   0.3769
2339   28-MUZ  .0959  .3193  .3917  .3961  .3930  .3868  .2624  .1239  .0749  .0676  .0843  .0927   .7058  1.9829   2.6887
2340   29-DGK  .1841  .3498  .4124  .4074  .3602  .3955  .2956  .1723  .1148  .0880  .1295  .1437   .9440  2.1094   3.0534
2341   31-P+D  .0020  .0913  .5223  .5805  .4495  .4418  .2244  .0944  .0757  .1748  .0748  .0671   .7111  2.0873   2.7984
2342   32-BEG  .0017  .0857  .7770 1.0634  .6544  .5580  .1456  .0023  .0111  .1597  .0041  .0005   .3234  3.1402   3.4636
2343   33-GHO  .0114  .2297  .4291  .4089  .4450  .4472  .3862  .2106  .1031  .2035  .0819  .1221  1.1074  1.9712   3.0786
2344   34-NW   .0839  .1085  .3588  .5011  .4307  .3715  .3067  .2109  .1998  .1010  .2115  .1874  1.2173  1.8547   3.0720
2345   35-RIC  .0000  .1275  .6033  .5902  .4426  .3964  .1425  .0000  .0000  .0000  .0000  .0000   .1425  2.1599   2.3024
2346   36-DAD  .0704  .0856  .2014  .2988  .2650  .2396  .1664  .1581  .1448  .0644  .1519  .1419   .8576  1.1607   2.0183
2347   37-KW   .0937  .1044  .1136  .1181  .1166  .1385  .1224  .1131  .1136  .0357  .0952  .0957   .5756   .6849   1.2606
2348   38-KE   .1125  .1319  .1397  .1520  .1502  .1561  .1598  .1433  .1425  .0433  .1168  .1180   .7237   .8422   1.5659
2349   39-ROH  .6944  .8037  .7851  .8233  .8596  .8817  .8111  .7295  .7511  .2630  .7118  .7542  4.0205  4.8477   8.8683
2350   41-NAR  .5853  .6515  .6339  .6764  .6679  .6934  .6591  .5947  .5752  .2066  .5104  .5389  3.0849  3.9084   6.9933
2351   42-KAL  .1192  .1688  .3069  .4274  .3686  .4103  .2800  .1412  .0857  .1559  .0973  .0794   .8396  1.8012   2.6409
2352   43-LCH  .1079  .1452  .1628  .1662  .1388  .1496  .1338  .0967  .0643  .0627  .0756  .0745   .5076   .8705   1.3780
2353   44-FUL  .1121  .3900  .6827  .6503  .4310  .4283  .3038  .1480  .0808  .1524  .1062  .0637   .8649  2.6944   3.5593
2354   45-PIN  .0594  .1900  .4164  .5292  .4193  .3378  .2322  .1159  .0574  .1113  .0680  .0350   .6198  1.9481   2.5679
2355  ;
2356    DIVPOST("22A-PHL",M)$ISR("2000") =.0589 ;
```

```
INDUS BASIN MODEL REVISED (IBMR) FILENAME=WSISD41                    07/25/91 17:17:54
GOVT. TUBEWELL PUMPAGE AND DEPTH TO WATER TABLE                      GAMS 2.21 IBM CMS
```

2358 TABLE GWT(CNL,M) PUBLIC TUNEWELL PUMPAGE(KAF)

	JAN	FEB	MAR	APR	MAY	JUN	JUL	AUG	SEP	OCT	NOV	DEC
01-UD	0	0	1	0	1	0	1	1	1	0	0	0
02-CBD	2	2	3	2	2	2	6	7	8	2	2	2
03-RAY	15	29	41	45	40	50	30	17	48	43	19	7
04-UC	39	49	49	30	49	49	55	58	58	16	36	15
05-MR	3	6	20	17	9	11	8	6	12	9	5	2
09-LD	1	1	1	1	1	1	0	0	0	1	1	1
10-LBD	2	2	2	2	2	2	2	3	3	1	2	2
11-JHA	38	42	49	57	57	57	50	54	60	57	48	32
12-GUG	50	62	69	78	79	79	69	73	82	85	67	34
13-UJ	67	88	104	119	141	121	114	114	140	155	117	5
14-LJ	51	65	88	70	81	72	87	81	94	91	70	77
15-BAH	0	0	1	0	1	1	2	2	2	0	1	1
18-HAV	6	6	8	6	7	7	6	6	6	5	7	7
19-RAN	55	66	82	84	68	52	36	57	59	52	51	43
20-PAN	6	6	8	6	6	6	6	6	5	5	6	6
21-ABB	1	1	2	1	2	2	1	2	1	1	2	2
28-MUZ	114	114	159	147	109	89	80	90	94	45	79	119
32-BEG	2	3	4	2	2	1	0	1	1	2	2	2
37-KW	18	20	24	19	16	20	18	18	16	10	11	11
39-ROH	6	7	8	6	5	7	6	48	42	3	4	4

2382 TABLE DEP1(CNL,IS, *)   DEPTH TO WATER TABLE (FEET)
*   1980 DEPTH IS FROM IBM AND IRRGATION SUPPORTING REPORT OF RAP.
*   YEAR 1993 AND 2000 DEPTHS ARE ESTIMATED CONSIDERING THE DRAINAGE
*   PROJECTS UNDER CONSTRUCTION OR PLANNING.

	1980.DEPTH	(1988,1993,2000).DEPTH
01-UD	17.0	17
02-CBD	10.5	10
(03-RAY,05-MR)	15.0	15
04-UC	15.1	15
06-SAD	5.5	7
(07-FOR,22-USW,23-LSW,		
24-WAR,25-KAB)	8.0	8
22A-PHL	999.9	6
08-PAK	19.9	15
09-LD	17.9	17
10-LBD	18.1	18
11-JHA	13.6	14
12-GUG	14.8	14
13-UJ	10.0	10
14-LJ	8.1	9
(15-BAH,19-RAN,28-MUZ)	9.0	9
(16-MAI,17-SID)	19.0	19
18-HAV	10.1	10
20-PAN	8.4	9
21-ABB	8.2	9
26-THA	8.7	8
27-PAH	10.0	12
29-DGK	7.8	8
(31-P+D,32-BEG,36-DAD,39-ROH)		7
(34-NW,35-RIC,42-KAL,43-LCH,		
44-FUL,45-PIN)		6
33-GHO	5.5	7
37-KW	5.6	7
38-KE	5.4	7
41-NAR	4.9	7

2418 TABLE DEP2(IS,CNL,M) DEPTH TO WATER TABLE (FEET)

	JAN	FEB	MAR	APR	MAY	JUN	JUL	AUG	SEP	OCT	NOV	DEC
1980. 31-P+D	6.9	6.9	6.9	6.9	6.9	4.5	2.5	2.5	2.5	3.5	4.5	6.9

```
2421      1980. 32-BEG                              6.6  6.6  6.6  6.6  6.6  4.5  2.5  2.5  2.5  3.5  4.5  6.6
2422      1980.(34-NW,35-RIC)                       4.6  4.6  4.6  4.6  4.6  4.5  2.5  2.5  2.5  3.5  4.5  4.6
2423      1980. 36-DAD                              5.6  5.6  5.6  5.6  5.6  4.5  2.5  2.5  2.5  3.5  4.5  5.6
2424      1980. 39-ROH                              5.9  5.9  5.9  5.9  5.9  5.8  5.8  5.8  5.8  5.9  5.9  5.9
2425      1980.(42-KAL,43-LCH,44-FUL,45-PIN)        5.0  5.0  5.0  5.0  5.0  3.5  2.0  2.0  2.5  4.0  5.0  5.0
2426
2427      PARAMETER DEPTH(CNL,M)  DEPTH TO GROUNDWATER (FEET);
2428              LOOP(ISR,
2429                DEPTH(CNL,M) = DEP1(CNL,ISR,"DEPTH");  DEPTH(CNL,M)$(DEP2(ISR,CNL,M) NE 0 ) = DEP2(ISR,CNL,M) );
2430      OPTIONS DEPTH:1;    DISPLAY DEPTH ;
2431
2432  *   SUB-IRRIGATION AND EFFECTIVE RAIN CALCULATIONS
2433
2434      PARAMETERS
2435         EFR(CNL,M)           EFFECTIVE RAINFALL IN FEET
2436         EQEVAP(CNL,M)        EVAPORATION FROM THE EQUAIFER   (FEET)
2437         SUBIRR(CNL,M)        WATER SUPPLIED BY CAPILLARY ACTION FROM THE AQUIFER
2438      PARAMETER
2439      SUBIRRFAC(Z)  MAXIMUM SUB-IRRIGATION IN SALINE AREAS AS PROPORTION OF CROP REQ.(NET OF RAIN)
2440                    /(NWFP, PMW,PCW,PSW,PRW,SRWN,SCWS) .4,
2441                                                  SCWN .3,
2442                                                  SRWS .1/
2443                   ;
2444      SCALARS DRC  RUN-OFF PORTION OF RAINFALL /.15/
2445              THE1 PORTION OF EQUAIFER EVAPORATION USED BY CROPS /0.6/;
2446
2447      LOOP(CNL,
2448          EQEVAP(CNL,M) = MIN(1., 10.637/DEPTH(CNL,M)**2.558)*EVAP(CNL,M)  ;
2449          SUBIRR(CNL,M) = EQEVAP(CNL,M) * THE1
2450         );
2451      LOOP(ISR,
2452      EFR(CNL,M)    = (1.0 - DRC - (1-COMDEF(ISR,"FLDE",CNL)))* RAIN(CNL,M)/12.0 ) ;
2453
```

```
2455
2456    SET N   NODES OF THE INDUS RIVER SYSTEM/
2457            SULEM-B     SULEMANKI BARRAGE              - SUTLEJ RIVER
2458            ISLAM-B     ISLAM BARRAGE                  - SUTLEJ RIVER
2459            PANJNAD-B   PANJNAD BARRAGE                - PANJNAD RIVER
2460            RAVI-I      RAVI INFLOW AT MADHOPUR        - RAVI RIVER
2461            BALLOKI-B   BALLOKI BARRAGE                - RAVI RIVER
2462            SIDHNAI-B   SIDHANI BARRAGE                - RAVI RIVER
2463            MARALA-B    MARRALA BARRAGE                - CHENAB RIVER
2464            KHANKI-B    KHANKI BARRAGE                 - CHENAB RIVER
2465            QADIRA-B    QADIRABAD BARRAGE              - CHENAB RIVER
2466            TRIMMU-B    TRIMMU BARRAGE                 - CHENAB RIVER
2467            MANGLA-R    MANGLA RESERVOIR               - JEHLUM RIVER
2468            RASUL-B     RASUL BARRAGE                  - JHELUM RIVER
2469            TARBELA-R   TARBELA RESERVOIR              - INDUS RIVER
2470            AMANDA-H    AMANDARA HEAD WORKS            - SWAT RIVER
2471            MUNDA-H     MUNDA HEAD WORKS               - SWAT RIVER
2472            WARSAK-D    WARSAK RESERVOIR               - KABUL RIVER
2473            K-S-JCT     KABUL AND SWAT RIVER JCT       - KABUL RIVER
2474            KALABAGH-R  KALABAGH RESERVOIR             - INDUS RIVER
2475            CHASMA-R    CHASMA RESERVOIR               - INDUS RIVER
2476            TAUNSA-B    TAUNSA BARRAGE                 - INDUS RIVER
2477            GUDU-B      GUDU BARRAGE                   - INDUS RIVER
2478            SUKKUR-B    SUKKUR BARRAGE                 - INDUS RIVER
2479            NARA-JCT    NARA JUNCTION                  - NARA COMPLEX
2480            NARA-HEAD   HEAD WORKS FOR IRRI. DIVERSION - NARA COMPLEX
2481            CHOTIARI-R  CHOTIARI RESERVOIR             - NARA COMPLEX
2482            KOTRI-B     KOTRI BARRAGE                  - INDUS RIVER
2483            A-SEA       ARABIAN SEA
2484            A1          DIVERSION POINT FOR UJ UJ LINK AND R.P.C
2485            A2          DIV. FOR BRBD LINK UC LINK AND UC INT.
2486            A3          DIV. FOR MR CROSS LINK
2487            A4          MR CROSS TAIL AND BRBD LINK
2488            A5          DIV. FOR CBD AND UD CANALS
2489            A6          DIVERSION TO LCC FEEDER
2490            A7          DIV. FOR GUGERA AND JHANG CANALS
2491            A8          DIVESION POINT FOR MAILI+L PAKPATTAN CANALS
2492            A9          SMB LINK TO L-BAHAWL CANAL
2493            A10         DIVERSION POINT FOR LJ
2494            /
2495    I   SYSTEM INFLOWS /
2496            SWAT        SWAT RIVER AT CHAKDARA
2497            KABUL       KABUL RIVER AT WARSAK
2498            INDUS       INDUS RIVER AT TARBELA
2499            HARO        HARO RIVER AT GARIALL
2500            SOAN        SOAN RIVER AT DHOK PATHAN
2501            JEHLUM      JEHLUM RIVER AT MANGLS
2502            CHENAB      CHENAB RIVER AT MARRALA
2503            RAVI        RAVI RIVER BELOW MADHOPUR
2504            SUTLEJ      SUTLEJ RIVER BELOW FEROZPUR /
2505
2506    NC(N,CNL) NODE TO CANAL MAP/
2507            SULEM-B.    (06-SAD,07-FOR,08-PAK),    A8.         16-MAI
2508            A9.         15-BAH                ,   PANJNAD-B.  (20-PAN,21-ABB)
2509
2510            BALLOKI-B.  (10-LBD,09-LD)        ,   SIDHNAI-B.  17-SID
2511
2512            MARALA-B.   05-MR                 ,   A2.         (04-UC,03-RAY)
2513            A5.         (02-CBD,01-UD)        ,   A7.         (11-JHA,12-GUG)
2514            TRIMMU-B.   (19-RAN,18-HAV)
2515            A1.         13-UJ                 ,   A10.        14-LJ
2516            TARBELA-R.  22A-PHL
2517            AMANDA-H.   22-USW                ,   MUNDA-H.    23-LSW
```

```
2518              WARSAK-D.    24-WAR              ,     WARSAK-D.    25-KAB
2519              CHASMA-R.    (26-THA,27-PAH)
2520              TAUNSA-B.    (28-MUZ,29-DGK)     ,     GUDU-B.      (31-P+D,32-BEG,33-GHO)
2521              SUKKUR-B.    (34-NW,35-RIC,36-DAD,37-KW,38-KE,39-ROH)
2522              NARA-HEAD.   41-NAR
2523              KOTRI-B.     (42-KAL,43-LCH,44-FUL,45-PIN)/
2524                ;
2525
2526      ALIAS (N,N1);
2527      SET NN(N,N1) WATER FLOW SYSTEM NODE TO NODE /
2528      *-- SUTLEJ RAVI SYSTEM
2529              SULEM-B.         BALLOKI-B
2530              ISLAM-B.         SULEM-B
2531              A9.              (A8,        ISLAM-B)
2532              A8.              SIDHNAI-B
2533              PANJNAD-B.       (TRIMMU-B,SIDHNAI-B, ISLAM-B, TAUNSA-B)
2534      *-- RAVI CHENAB SYSTEM
2535              RAVI-I.          A3
2536              A3.              MARALA-B
2537              A4.              (A3,A2)
2538              A5.              A4
2539              A2.              MARALA-B
2540              BALLOKI-B.       (RAVI-I,    A2,        A6)
2541              A6.              QADIRA-B
2542              A7.              (A6,        KHANKI-B)
2543              SIDHNAI-B.       (TRIMMU-B, BALLOKI-B)
2544      *-- CHENAB JEHLUM SYSTEM
2545              KHANKI-B.        (A1,        MARALA-B)
2546              QADIRA-B.        (RASUL-B,   KHANKI-B)
2547              TRIMMU-B.        (QADIRA-B, RASUL-B, CHASMA-R)
2548              A1.              MANGLA-R
2549              RASUL-B.         (MANGLA-R, A10)
2550              A10.             (RASUL-B,  A1)
2551      *-- KABUL SWAT SYSTEM
2552              MUNDA-H.         AMANDA-H
2553              K-S-JCT.         (MUNDA-H,   WARSAK-D)
2554      *-- INDUS RIVER
2555              KALABAGH-R.      (K-S-JCT,   TARBELA-R)
2556              CHASMA-R.        KALABAGH-R
2557              TAUNSA-B.        CHASMA-R
2558              GUDU-B.          (PANJNAD-B,TAUNSA-B)
2559              SUKKUR-B.        GUDU-B
2560              NARA-JCT.        SUKKUR-B
2561              CHOTIARI-R.      NARA-JCT
2562              NARA-HEAD.       (NARA-JCT,  CHOTIARI-R)
2563              KOTRI-B.         SUKKUR-B
2564              A-SEA.           KOTRI-B /
2565
2566          NI(N,I) NODE TO RIM STATION INFLOW MAP /
2567              AMANDA-H.    SWAT      ,   WARSAK-D.    KABUL  ,
2568              TARBELA-R.   INDUS     ,   KALABAGH-R.  HARO
2569              KALABAGH-R.  SOAN      ,   MANGLA-R.    JEHLUM
2570              MARALA-B.    CHENAB    ,   RAVI-I.      RAVI
2571              SULEM-B.     SUTLEJ /
2572
2573      NB(N)   ; NB(N) = YES; NB("A-SEA")=NO;
2574
2575      *-   FLOW CAPACITIES
2576
2577      PARAMETER
2578        NCAP(N,N1) NODE TO NODE TRANSFER CAPACITY (MAF)/
2579              CHASMA-R.    TRIMMU-B  1.307,    TAUNSA-B.   PANJNAD-B   .724
2580              SUKKUR-B.    NARA-JCT  1.190,    NARA-JCT.   NARA-HEAD  1.190
2581              NARA-JCT.    CHOTIARI-R .416,    CHOTIARI-R.NARA-HEAD    .135
2582
```

```
2583          MANGLA-R.   A1         .685,      A1.        A10         .199
2584          RASUL-B.    A10        .328,      A1.        KHANKI-B    .422
2585          RASUL-B.    QADIRA-B  1.232,
2586
2587          MARALA-B.   A3         .985,      A3.        A4          .199
2588          A3.         RAVI-I     .938,      MARALA-B.  A2          .995
2589          A2.         BALLOKI-B  .698,      A2.        A4          .310
2590          A4.         A5         .310
2591
2592          KHANKI-B.   A7         .684,      QADIRA-B.  A6         1.146
2593          A6.         A7         .270,      A6.        BALLOKI-B   .808
2594
2595          BALLOKI-B.  SULEM-B   1.172,      TRIMMU-B.  SIDHNAI-B  1.03
2596          SIDHNAI-B.  A8         .662,      A8.        A9          .640
2597          ISLAM-B.    A9         .299 /
2598
2599      LLOSS(N,N1)  LINK CANAL LOSS FACTORS /
2600          CHASMA-R.   TRIMMU-B   .096,      TAUNSA-B.  PANJNAD-B   .0615
2601          SUKKUR-B.   NARA-JCT   .055,      NARA-JCT.  NARA-HEAD   .0001
2602          NARA-JCT.   CHOTIARI-R .001,      CHOTIARI-R. NARA-HEAD  .0001
2603
2604          A1.         A10        .004 ,     A1.        KHANKI-B    .099
2605          RASUL-B.    QADIRA-B   .039,      MARALA-B.  A3          .0705
2606          A3.         A4         .0075,     A2.        BALLOKI-B   .0705
2607          A2.         A4         .097 ,     A4.        A5          .007
2608
2609          KHANKI-B.   A7         .0435 ,    A6.        BALLOKI-B   .087
2610
2611          BALLOKI-B.  SULEM-B    .0525,     TRIMMU-B.  SIDHNAI-B   .075
2612          SIDHNAI-B.  A8         .0495 /
2613      LCEFF(N,N1)  LINK CANAL EFFICIENCY FROM HEAD TO TAIL
2614          ;
2615      LCEFF(N,N1)$LLOSS(N,N1) = 1 - LLOSS(N,N1) ;
2616
2617  * DISPLAY N,NC,NN, NI;
2618  * DISPLAY NCAP, LLOSS, LCEFF ;
```

```
INDUS BASIN MODEL REVISED (IBMR) FILENAME=WSISD41
FLOW ROUTING COEFFICINTS

2620   SET  CD /C,D/
2621   TABLE RIVERCD(N,CD)    COEFFICIENTS FOR RIVER ROUTING
2622                             C          D
2623       RASUL-B             .02732     .93766
2624       TRIMMU-B            .15348
2625       KHANKI-B                       2.2113
2626       BALLOKI-B
2627       SIDHNAI-B           .14063     1.0024
2628       ISLAM-B             .12294
2629       PANJNAD-B           .08077
2630       GUDU-B              .10947
2631       KALABAGH-R          .04856
2632       CHASMA-R            .04526
2633       TAUNSA-B            .07205
2634       SUKKUR-B            .02218
2635       KOTRI-B             .17054
2636
2637   TABLE RIVERB(N,N1)  COEFFICIENTS FOR RIVER ROUTING
2638
2639               MANGLA-R  RASUL-B  QADIRA-B  MARALA-B  RAVI-I  BALLOKI-B
2640   RASUL-B     1.01841
2641   TRIMMU-B              .82803   .9068
2642   KHANKI-B                                 .93361
2643   BALLOKI-B                                          1.2181
2644   SIDHNAI-B                                                   .70555
2645
2646   +           SULEM-B   TRIMMU-B  SIDHNAI-B  ISLAM-B  TAUNSA-B  PANJNAD-B
2647   ISLAM-B     .9086
2648   PANJNAD-B             .81033    1.13626    .80359
2649   GUDU-B                                              .94003    1.0
2650
2651   +           K-S-JCT   TARBELA-R  KALABAGH-R  CHASMA-R  GUDU-B  SUKKUR-B
2652
2653
2654   KALABAGH-R  1.04621   1.08686
2655   CHASMA-R                         .9849
2656   TAUNSA-B                                     .9497
2657   SUKKUR-B                                               .9908
2658   KOTRI-B                                                          .70625
2659
2660     ;
2661
2662    RIVERB(N,N1)$(RIVERB(N,N1) EQ 0 )    = 1;
2663    RIVERCD(N,"D")$(RIVERCD(N,"D") EQ 0) = 1;
2664
2665
```

INDUS BASIN MODEL REVISED (IBMR) FILENAME=WSISD41
RIM STATION AND TRIBUTORY INFLOWS

07/25/91 17:17:54
GAMS 2.21 IBM CMS

```
2667     SET S58/50, 80 /
2668
2669     TABLE INFL5080(S58,I,M1) SYSTEM INFLOWS MEASURED AT THE RIM STATIONS (MAF)
2670
2671  *    50 - 50 % PROBABILITY FLOWS AND 80 ARE 80 % PROBABILITY FLOWS.
2672  *
2673  * SOURCE:   WATER RESOUCE MANAGEMENT DIRECTORATE, WAPDA PUBLICATIONS AND
2674  *           RAP IRRIGATION AND DRAINAGE SUPPORTING REPORT
2675  *    HISTORIC DATA USED: INDUS   OCT 1936 - MARCH 1988
2676  *                        JEHLUM  APRIL 1922 - MARCH 1988
2677  *                        CHENAB  APRIL 1922 - MARCH 1988
2678  *           RAVI AND SUTLEJ APRIL 1966 - MARCH 1988
2679  *           ALL OTHERS      APRIL 1966 - MARCH 1976
2680
2681                  APR     MAY     JUN     JUL     AUG     SEP    KHARIF
2682     50.INDUS    1.931   3.984   7.191  16.148  17.437   6.196  52.887
2683     50.JEHLUM   3.826   4.355   2.993   3.429   2.236   0.846  17.685
2684     50.CHENAB   1.361   2.281   4.036   5.279   5.097   2.540  20.594
2685     50.RAVI     0.226   0.161   0.038   1.089   1.732   0.841   4.088
2686     50.SUTLEJ   0.031   0.045   0.052   0.483   1.003   1.198   2.810
2687
2688     50.HARO     0.016   0.013   0.014   0.048   0.204   0.073   0.369
2689     50.SOAN     0.012   0.007   0.090   0.225   0.426   0.068   0.829
2690     50.SWAT     0.483   0.636   1.049   0.863   0.642   0.346   4.020
2691     50.KABUL    1.512   1.883   3.467   3.490   2.619   0.975  13.945
2692
2693     80.INDUS    1.871   3.760   9.634  13.135  10.524   5.234  44.159
2694     80.JEHLUM   1.526   2.242   2.594   2.632   3.140   1.623  13.757
2695     80.CHENAB   1.120   2.064   3.133   4.515   3.537   2.844  17.213
2696     80.RAVI     0.049   0.020   0.011   0.696   1.346   0.313   2.434
2697     80.SUTLEJ   0.000   0.001   0.000   0.145   0.291   0.004   0.441
2698
2699     80.HARO     0.014   0.019   0.011   0.020   0.101   0.022   0.187
2700     80.SOAN     0.006   0.004   0.020   0.252   0.145   0.021   0.448
2701     80.SWAT     0.323   0.574   0.848   0.579   0.463   0.448   3.235
2702     80.KABUL    0.963   1.737   2.238   2.157   2.303   1.534  10.934
2703
2704   +              OCT     NOV     DEC     JAN     FEB     MAR    RABI
2705     50.INDUS    2.297   1.279   1.188   0.952   0.973   1.468   8.157
2706     50.JEHLUM   0.988   0.521   0.412   0.460   0.621   1.417   4.419
2707     50.CHENAB   0.728   0.487   0.407   0.480   0.711   1.079   3.891
2708     50.RAVI     0.188   0.134   0.131   0.090   0.026   0.455   1.023
2709     50.SUTLEJ   0.129   0.009   0.004   0.060   0.000   0.000   0.202
2710
2711     50.HARO     0.037   0.020   0.016   0.019   0.019   0.051   0.163
2712     50.SOAN     0.010   0.005   0.020   0.022   0.028   0.043   0.128
2713     50.SWAT     0.195   0.122   0.088   0.079   0.068   0.140   0.692
2714     50.KABUL    0.523   0.405   0.354   0.379   0.440   0.687   2.787
2715
2716     80.INDUS    2.133   1.281   1.060   0.898   0.835   1.025   7.232
2717     80.JEHLUM   0.405   0.270   0.309   0.306   0.526   1.514   3.331
2718     80.CHENAB   0.516   0.325   0.285   0.292   0.804   0.711   2.933
2719     80.RAVI     0.076   0.048   0.074   0.135   0.112   0.135   0.580
2720     80.SUTLEJ   0.004   0.002   0.004   0.005   0.002   0.000   0.017
2721
2722     80.HARO     0.024   0.016   0.018   0.017   0.018   0.018   0.112
2723     80.SOAN     0.012   0.010   0.009   0.010   0.019   0.031   0.091
2724     80.SWAT     0.103   0.070   0.061   0.064   0.085   0.212   0.597
2725     80.KABUL    0.273   0.420   0.354   0.331   0.429   0.359   2.167
2726
2727
2728     TABLE TRI(S58,N1,N,M1) TRIBUTARY INFLOWS (MAF)
2729  *    SOURCE: IRRIGATION DRAINAGE AND FLOOD MANAGEMENT SUPPORTING REPORT OF RAP. PLANNING DIVISION WAPDA.
```

INDUS BASIN MODEL REVISED (IBMR) FILENAME=WSISD41
RIM STATION AND TRIBUTORY INFLOWS

```
2730                          APR   MAY   JUN   JUL   AUG   SEP  KHARIF
2731  50.MARALA-B. KHANKI-B   0.004 0.006 0.023 0.120 0.240 0.056 0.449
2732  50.MANGLA-R. RASUL-B    0.120 0.180 0.150 0.170 0.320 0.210 1.150
2733  50.CHASMA-R. TAUNSA-B   0.051 0.013 0.021 0.048 0.021 0.012 0.166
2734  50.TARBELA-R.KALABAGH-R 0.041 0.087 0.182 0.350 0.602 0.180 1.442
2735  50.BALLOKI-B.SIDHNAI-B  0.000 0.001 0.009 0.096 0.206 0.041 0.353
2736  50.AMANDA-H. MUNDA-H    0.036 0.337 0.637 0.854 0.544 0.062 2.470
2737  50.MUNDA-H.  K-S-JCT    0.015 0.013 0.005 0.008 0.007 0.007 0.054
2738  50.WARSAK-D. K-S-JCT    0.126 0.102 0.059 0.070 0.056 0.056 0.469
2739
2740  80.MARALA-B. KHANKI-B   0.004 0.005 0.006 0.136 0.131 0.011 0.293
2741  80.MANGLA-R. RASUL-B    0.120 0.180 0.150 0.170 0.320 0.210 1.150
2742  80.CHASMA-R. TAUNSA-B   0.021 0.014 0.006 0.047 0.008 0.003 0.098
2743  80.TARBELA-R.KALABAGH-R 0.052 0.094 0.106 0.126 0.416 0.079 0.873
2744  80.BALLOKI-B.SIDHNAI-B  0.000 0.000 0.006 0.075 0.164 0.027 0.271
2745  80.AMANDA-H. MUNDA-H    0.121 0.410 0.161 0.132 0.000 0.019 0.844
2746  80.MUNDA-H.  K-S-JCT    0.006 0.009 0.003 0.005 0.005 0.004 0.031
2747  80.WARSAK-D. K-S-JCT    0.050 0.077 0.027 0.039 0.036 0.032 0.261
2748
2749  +                       OCT   NOV   DEC   JAN   FEB   MAR  RABI
2750  50.MARALA-B. KHANKI-B   0.017 0.007 0.008 0.007 0.006 0.007 0.052
2751  50.MANGLA-R. RASUL-B    0.060 0.000 0.000 0.000 0.000 0.000 0.060
2752  50.CHASMA-R. TAUNSA-B   0.004 0.005 0.010 0.006 0.015 0.026 0.065
2753  50.TARBELA-R.KALABAGH-R 0.100 0.064 0.038 0.055 0.052 0.130 0.439
2754  50.BALLOKI-B.SIDHNAI-B  0.002 0.003 0.008 0.005 0.007 0.006 0.030
2755  50.AMANDA-H. MUNDA-H    0.133 0.153 0.150 0.119 0.119 0.221 0.894
2756  50.MUNDA-H.  K-S-JCT    0.003 0.003 0.004 0.004 0.004 0.009 0.028
2757  50.WARSAK-D. K-S-JCT    0.024 0.027 0.035 0.035 0.038 0.070 0.229
2758
2759  80.MARALA-B. KHANKI-B   0.007 0.006 0.006 0.003 0.011 0.006 0.039
2760  80.MANGLA-R. RASUL-B    0.060 0.000 0.000 0.000 0.000 0.000 0.060
2761  80.CHASMA-R. TAUNSA-B   0.009 0.006 0.010 0.006 0.008 0.011 0.050
2762  80.TARBELA-R.KALABAGH-R 0.088 0.039 0.041 0.040 0.042 0.045 0.295
2763  80.BALLOKI-B.SIDHNAI-B  0.000 0.001 0.001 0.000 0.000 0.000 0.002
2764  80.AMANDA-H. MUNDA-H    0.090 0.075 0.045 0.047 0.019 0.121 0.397
2765  80.MUNDA-H.  K-S-JCT    0.005 0.004 0.003 0.003 0.000 0.005 0.022
2766  80.WARSAK-D. K-S-JCT    0.020 0.016 0.020 0.020 0.027 0.065 0.169
2767       ;
2768   PARAMETER INFLOW(I,M)  INFLOWS FOR THIS RUN (MAF)
2769             TRIB(N1,N,M) TRIBUTARY INFLOWS FOR THIS RUN (MAF) ;
2770   INFLOW(I,M) = INFL5080("50",I,M);  TRIB(N1,N,M) = TRI("50",N1,N,M);
2771
```

```
2773      TABLE   RRCAP(N,IS)    LIVE STORAGE CAPACITY OF RESERVOIRS (MAF)
2774
2775   *  SOURCE: WAPDA, WSIPS , JAN 1990.
2776   *  KALABAGH RETENTION LEVEL EL 915 FEET .
2777   *  TARBELA'S LIVE CAPACITY IN 2000 WITH KALABAGH=7.500 MAF, BASED ON
2778   *  THE TABRBELA-KALABAGH CONJUCTIVE OPERATION, WSIPS APP-E PP E-3.
2779   *
2780
2781                           1980         1988         2000
2782            MANGLA-R       5.251        4.880        4.62
2783            TARBELA-R      9.093        8.860        8.357
2784            KALABAGH-R                               6.10
2785            CHASMA-R       0.500         .44          .23
2786            CHOTIARI-R                               1.10
2787
2788      TABLE RULELO(N,M)  LOWER RULE CURVE
2789   *             END OF MONTH CONTENTS AS PERCENT OF LIVE CAPACITY
2790
2791               JAN  FEB  MAR  APR  MAY  JUN  JUL  AUG  SEP  OCT  NOV  DEC
2792   MANGLA-R     16    6    0    7   21   48   74   95   83   55   35   24
2793   TARBELA-R    43   30   14    4    0   10   46  100  100   64   59   53
2794   KALABAGH-R   55   49   34   28    0    0   24   70  100   96   78   61
2795   CHASMA-R      0    0    0    0    0    0    0    0    0    0    0    0
2796   CHOTIARI-R    0    0    0    0    0    0    0    0    0    0    0    0
2797
2798      TABLE RULEUP(N,M) UPPER RULE CURVE
2799   *             END OF MONTH CONTENTS AS PERCENT OF LIVE CAPACITY
2800   *  KALABAGH SLUICING .
2801
2802               JAN  FEB  MAR  APR  MAY  JUN  JUL  AUG  SEP  OCT  NOV  DEC
2803   MANGLA-R     39   28   19   31   42   59   87  100  100   90   68   49
2804   TARBELA-R    69   47   37   26   13   48   73  100  100  100   87   77
2805   KALABAGH-R  100  100  100  100    0    0  100  100  100  100  100  100
2806   CHASMA-R    100  100  100  100  100  100  100  100  100  100  100  100
2807   CHOTIARI-R  100  100  100  100  100  100  100  100  100  100  100  100
2808
2809      TABLE  REVAPL(N,M)  EVAPORATION LOSSES FROM RESERVOIRS (KAF)
2810
2811               JAN  FEB  MAR  APR  MAY  JUN  JUL  AUG  SEP  OCT  NOV  DEC
2812   MANGLA-R      1    5    4    5   12   17  -1.9 -2.5    4   22   10    5
2813   TARBELA-R     1    5    7    9   16   28   10    1   12   29   12    5
2814   KALABAGH-R    1    4    6    9   12    5    9    1    9   24    9    5
2815   CHASMA-R      3    3    1              6    3    7   15    6    3
2816   ;
2817   * KALABAGH IS NOT INCLUDED FOR YEAR 2000 RUNS, BOTH WITH OR WITHOUT PLAN
2818   * CASES. CHOTIARI IS INCLUDED IN WITH CASE (PART OF LBOD)
2819      PARAMETER RCAP(N)  LIVE CAPACITY OF RESRVOIRS (MAF);
2820         LOOP(ISR,
2821             RCAP(N)         = RRCAP(N,ISR) );
2822
2823         RCAP("KALABAGH-R") = 0;
2824         RCAP("CHOTIARI-R")$INCP("WITHOUT") = 0 ;
2825      DISPLAY RCAP;
2826
2827   SET  POW /R-ELE  RESERVOIR ELEVATION (FEET FROM SPD)
2828             P-CAP  INSTALLED CAPACITY OF THE POWER HOUSE AT R-ELE (MW)
2829             G-CAP  GENERATION CAPABILITY (KWH PER AF) AT R-ELE
2830             R-CAP  GROSS RESERVOIR CAPACITY (MAF) AT R-ELE/
2831        PN(N) NODES WITH POWER HOUSE /MANGLA-R,TARBELA-R,KALABAGH-R/
2832        V   /1*27/;
2833
2834      TABLE POWERCHAR(N,*,V)     POWER GENERATION CHRACTERSITICS OF HRDRO STATIONS
2835
```

INDUS BASIN MODEL REVISED (IBMR) FILENAME=WSISD41
RESERVOIR CHARACTERISTICS

```
2836    * TOTAL NUMBER OF UNITS AT MANGLA=8, TARBELA=10, KALABAGH CAPACITY=2400 MW.
2837
2838                         1      2      3      4      5      6      7      8      9     10     11     12     13
2839    TARBELA-R.R-ELE   1300   1310   1320   1330   1340   1350   1360   1370   1380   1390   1400   1410   1420
2840    TARBELA-R.P-CAP    480    552    630    700    780    832    894    942    992   1052   1114   1184   1266
2841    TARBELA-R.G-CAP    136    166    183    192    202    209    216    224    233    237    247    257    263
2842    TARBELA-R.R-CAP  1.313  1.439  1.571  1.707  1.848  1.995  2.236  2.486  2.745  3.014  3.297  3.631  3.977
2843
2844    MANGLA-R .R-ELE   1040   1050   1055   1060   1065   1070   1075   1080   1085   1090   1095   1100   1105
2845    MANGLA-R .P-CAP    384    424    440    464    488    504    528    544    568    584    608    632    648
2846    MANGLA-R .G-CAP    141    155    164    173    183    192    198    205    209    212    220    228    233
2847    MANGLA-R .R-CAP  0.417  0.509  0.555  0.601  0.648  0.696  0.744  0.792  0.883  0.973  1.064  1.155  1.271
2848
2849    KALABAGH-R.R-ELE   710    750    825    830    835    840    845    850    855    860    865    870    875
2850    KALABAGH-R.P-CAP     0      0   1207   1294   1365   1466   1550   1630   1700   1787   1850   1924   1990
2851    KALABAGH-R.G-CAP     0      0    110    115    120    126    133    137    144    149    155    160    165
2852    KALABAGH-R.R-CAP  .025   .250  1.762  1.950  2.250  2.360  2.650  2.850  3.200  3.360  3.650  4.000  4.400
2853
2854            +           14     15     16     17     18     19     20     21     22     23     24     25     26     27
2855    TARBELA-R.R-ELE   1430   1440   1450   1460   1470   1480   1490   1500   1510   1520   1530   1540   1550   1550
2856    TARBELA-R.P-CAP   1308   1366   1416   1476   1530   1596   1678   1730   1792   1848   1908   1936   1952   1952
2857    TARBELA-R.G-CAP    269    281    288    295    303    310    318    336    346    356    356    368    376    376
2858    TARBELA-R.R-CAP  4.334  4.705  5.095  5.520  5.960  6.412  6.882  7.369  7.865  8.381  8.917  9.478 10.063 20.000
2859
2860    MANGLA-R .R-ELE   1110   1115   1120   1130   1135   1140   1145   1155   1160   1165   1170   1180   1202   1202
2861    MANGLA-R .P-CAP    672    696    712    760    776    800    816    856    880    896    920    960    960    960
2862    MANGLA-R .G-CAP    237    247    252    263    263    269    275    281    288    295    303    318    356    356
2863    MANGLA-R .R-CAP  1.386  1.501  1.617  1.933  2.091  2.249  2.453  2.862  3.066  3.316  3.566  4.067  5.364 20.000
2864
2865    KALABAGH-R.R-ELE   880    885    890    895    900    905    910    915    915    915    915    915    915    925
2866    KALABAGH-R.P-CAP  2051   2100   2159   2200   2255   2300   2340   2370   2370   2370   2370   2370   2370   2436
2867    KALABAGH-R.G-CAP   170    175    180    185    190    195    199    204    204    204    204    204    204    212
2868    KALABAGH-R.R-CAP  4.700  5.100  5.475  5.750  6.350  6.750  7.350  7.862  7.862  7.862  7.862  7.862  7.862  9.375
2869           ;
2870    * SEGMENT 27 WOULD BE USED IF THE DAM IS RAISED TO CREATE ADDITIONAL
2871    * STORAGE ON TOP OF THE EXISTING.
2872
2873    * ADJUST THE  CAPACITY CURVE FOR SEDIMENT/DEAD STORAGE
2874    POWERCHAR(PN,"R-CAP",V) = MAX(0, POWERCHAR(PN,"R-CAP",V)-
2875                          (POWERCHAR(PN,"R-CAP","26") -RCAP(PN)));
2876    POWERCHAR(PN,"R-CAP","1")= 0;
2877
        *-   REPORT ON THE INFLOWS ETC.

            PARAMETER REP7
                      REP8
                         ;
        REP7(I,M) = INFLOW(I,M);
        REP7(I,SEA1)=SUM(M$SEA1M(SEA1,M), INFLOW(I,M));
        REP7("TOTAL",M1)       = SUM(I, REP7(I,M1));
        DISPLAY " SYSTEM INFLOWS AT RIM STATIONS (MAF) ", REP7;

        *
        REP8(N,N1,M)=TRIB(N,N1,M);
        REP8(N,N1,SEA1) = SUM(M$SEA1M(SEA1,M), TRIB(N,N1,M));
        DISPLAY " TRIBUTORY INFLOW IN (MAF)", REP8;
        *-  -  -  -
2895
```

```
2897    SET P3 /FINANCIAL,ECONOMIC, EXPORT,IMPORT/
2898
2899    TABLE PRICES(PS,CQ,P3)  1988  PRICES
2900
2901               FINANCIAL  ECONOMIC   EXPORT  IMPORT
2902  87-88.BASMATI     6.        6.       4.8
2903  87-88.IRRI        2.9       2.4      2.3
2904  87-88.COTTON      4.5       6.       4.5
2905  87-88.GRAM        3.9       3.8
2906  87-88.MAIZE       2.5       2.3
2907  87-88.MUS+RAP     3.75      3.4
2908  87-88.SC-GUR      3.0       3.0
2909  87-88.SC-MILL      .3        .4                .5
2910  87-88.WHEAT       2.0       2.2      2.0
2911  87-88.ORCHARD     4.8       4.3
2912  87-88.POTATOES    2.6       2.6
2913  87-88.ONIONS      2.9       2.9      2.8
2914  87-88.CHILLI     10.       10.
2915  87-88.COW-MILK    5.4       2.46
2916  87-88.BUFF-MILK   5.4       3.24
2917  87-88.MEAT       13.8      14.4
2918
2919
2920    TABLE FINSDWTPR(C,PS,*)  PRICES OF SEED(RS PER KG) AND WATER(RS PER ACRE )
2921
2922           87-88.SEED  87-88.WATER  87-88.MISCC
2923  BASMATI     5.          32.           75.
2924  IRRI        2.5         32.           75.
2925  COTTON      5.5         34.          500.
2926  RAB-FOD    40.          11.
2927  GRAM        5.7         16.
2928  MAIZE       4.          14.
2929  MUS+RAP     4.3         22.
2930  KHA-FOD     2.5         14.
2931  SC-GUR       .2         64.
2932  SC-MILL      .2         64.
2933  WHEAT       2.9         22.
2934  ORCHARD                 84.
2935  POTATOES    2.6         16.         1650.
2936  ONIONS      2.9         16.         1020.
2937  CHILLI     10.          42.          525.
2938
2939
2940    TABLE ECNSDWTPR(C,PS,*)  PRICES OF SEED(RS PER KG) AND WATER(RS PER ACRE )
2941
2942           87-88.SEED  87-88.WATER  87-88.MISCC
2943  BASMATI     7.          16.           75.
2944  IRRI        3.7         16.           75.
2945  COTTON      6.7         14.4         500.
2946  RAB-FOD    36.           6.4
2947  GRAM        5.1          8.
2948  MAIZE       7.           9.6
2949  MUS+RAP     5.9         12.
2950  KHA-FOD     2.25         6.4
2951  SC-GUR       .43        33.
2952  SC-MILL      .43        33.
2953  WHEAT       3.6         10.
2954  ORCHARD                 64.
2955  POTATOES    2.6         12.         1650.
2956  ONIONS      2.9         12.         1020.
2957  CHILLI     10           32.          525.
2958
2959  *       WAGES, FERTILIZER AND OTHER INPUT PRICES
```

INDUS BASIN MODEL REVISED (IBMR) FILENAME=WSISD41
PRICES

```
2960   SET P1 / NITROGEN, PHOSPHATE , PROTEIN, TWINVT, TRINVT,TWOPC,TROPC/
2961       P11 /FINANCIAL, ECONOMIC /
2962
2963    TABLE PRI1(PS,P11,P1)  FERTILIZER TUBEWELL TRACTOR AND PROTEIN PRICES
2964
2965  *      FERTILIZER AND PROTEIN PRICES ARE IN RS/KG, TWINVT AND TRINVT ARE
2966  *      ANNUALIZED COST FOR A TUBEWELL AND TRACTOR (RUPEES)
2967  *      TWOPC AND TROPC ARE COST OF TUBEWELL WATER(RS/ACRE FOOT) AND
2968  *      COST OF TRACTOR (RS/TRACTOR HOUR)
2969  *
2970                 NITROGEN PHOSPHATE  PROTEIN  TWINVT  TRINVT TWOPC  TROPC
2971    87-88.FINANCIAL  5.8     7.0       9.     10000   25000   225    45.
2972    87-88.ECONOMIC   8.3    10.9       9.     10000   25000   170    50.
2973
2974        TABLE WAGEPS(PS,P11,M)    WAGE RATES RS PER MAN HOUR
2975
2976               JAN  FEB  MAR APR MAY  JUN  JUL  AUG  SEP  OCT NOV DEC
2977   87-88.FINANCIAL 3.   3.   3.  6.  6.   3.   3.   3.   3.   6.  6.  3.
2978   87-88.ECONOMIC  2.75 2.75 2.75 5.  5.  2.75 2.75 2.75 2.75 5.  5.  2.75
2979
2980
2981  * MISCELLANEOUS PARAMETRES
2982     SCALARS   LSTD     STANDARD LABOR LIMIT ( HOURS PER MONTH )            /  200    /
2983               TRCAP    TRACTOR CAPACITY IN TRACTOR HOURS PER MONTH         /  250    /
2984               TWCAP    NAMEPLATE CAPACITY OF THE PRIVATE TUBEWELL( AF PER MONTH)  /  59.5041/
2985               NTWUCAP  EFFECTIVE CAPACITY OF NEW TUBEWELLS(AF PER MONTH)
2986               TWEFAC   FACTOR TO CONVERT WC LOSSES TO FROM PRIVATE TUBEWELL LOSSES /  0.5   /
2987               LABFAC   FACTOR TO CONVERT WAGE TO THE RESERVATION WAGE      /  0.5    /
2988                  ;
2989     PARAMETER TWUTIL(*) EFFECTIVE CAPACITY OF TUBEWELLS(PROPORTION OF NAME PLATE CAPACITY)
2990              / EXISTING   .60
2991                NEW        .75 /;
```

```
2993    TABLE TOTPROD(Z,CQ ) TOTAL PRODUCTION  1988 (000'S TONS)
2994
2995  *   SC-MILL IS TOTAL CANE PRODUCTION (INCLUDING THAT USED FOR GUR)
2996  *   SC-GUR IS REFINED PORTION OF TOTAL CANE - SAME AS CONSUMPTION
2997
2998           WHEAT  BASMATI   IRRI  MAIZE  SC-MILL SC-GUR  COTTON  MUS+RAP  GRAM  CHILLI POTATOES ONIONS
2999
3000   NWFP    161.8     0.0    0.0   42.1   2473.7   49.5     0.0    14.8   22.4    4.8    29.7   25.2
3001   PCW    4636.8   182.8  175.7   23.0  10885.2  217.7  3318.4   103.8  132.4   38.7   199.5  191.5
3002   PMW     747.1     0.0    0.0    0.0   1364.8   27.3   153.6     7.5   50.6    7.1    36.8   35.1
3003   PRW    1720.1   636.0  279.6    5.6   1065.7   21.3    46.9    16.6   43.0   15.2    78.5   75.0
3004   PSW    2327.3   141.2    0.0   27.2   7148.6  143.0   397.0    14.6   36.4   21.6   110.8  106.2
3005   SCWN    830.1     0.0   18.1    0.0   3030.7   60.6   485.5    17.4   16.7    5.9    33.7   27.8
3006   SCWS    847.7     0.0  157.3    0.0   4947.9   99.0   437.4     6.8    6.2    7.0    39.9   32.4
3007   SRWN    414.2     0.0  920.3    0.0    289.5    5.8     6.1    27.8   23.1    7.2    41.2   33.6
3008   SRWS    112.7     0.0  257.3    0.0   1832.0   36.6    18.9     3.4   10.1    2.2    12.3   10.1
3009
3010    TABLE FARMCONS(Z,CQ ) ON-FARM CONSUMPTION 1988 (000'S TONS)
3011
3012          WHEAT  BASMATI   IRRI  MAIZE  SC-GUR  MUS+RAP  GRAM  CHILLI POTATOES ONIONS
3013
3014   NWFP   110.0                   18.9    49.5    6.7   10.1    2.2    13.4   11.3
3015   PCW   3013.9    27.4   21.1    10.3   217.7   46.7   59.6   17.4    89.8   86.2
3016   PMW    485.6                           27.3    3.4   22.8    3.2    16.6   15.8
3017   PRW   1118.1    95.4   33.6            21.3    7.5   19.4    6.8    35.3   33.7
3018   PSW   1512.7    21.2           12.3   143.0    6.6   16.4    9.7    49.9   47.8
3019   SCWN   697.3            11.8           60.6    7.8    7.5    2.7    15.2   12.5
3020   SCWS   712.1           102.2           99.0    3.1    2.8    3.1    18.0   14.6
3021   SRWN   347.9           171.1            5.8   12.5   10.4    2.9    16.8   13.5
3022   SRWS    94.6            27.1           36.6    1.5    4.5    1.3     7.4    6.1
3023
3024     TABLE DEMAND(*,CQ ) MARKET DEMAND BY ZONE (000 TONS OR MILLION LITERS)
3025
3026  *   BASMATI & IRRI ADJUSTED FOR EXPECTED EXPORTS
3027  *   SC-MILL ADJUSTED FOR EXPECTED IMPORTS OF REFINED SUGAR
3028
3029         WHEAT BASMATI   IRRI MAIZE SC-MILL COTTON MUS+RAP  GRAM CHILLI POTATOES ONIONS COW-MILK BUFF-MILK
3030
3031   NWFP   51.8                  23.1  2337.6    0.0    8.1   12.3   2.6   16.3   13.9    167      163
3032   PCW  1622.9   155.4  49.3        10286.5  3318.4   57.1   72.8  21.3  109.7  105.3    190      267
3033   PMW   261.5                       1289.8   153.6    4.1   27.8   3.9   20.2   19.3    709     2298
3034   PRW   602.0   290.6  32.2         1007.1           9.1   23.7   8.3   43.2   41.2    260     1591
3035   PSW   814.5   120.2        15.0   6755.5   397.0    8.0   20.0  11.9   61.0   58.4    180     1142
3036   SCWN  132.8           6.3         2864.0   485.5    9.6    9.2   3.3   18.5   15.3    201      374
3037   SCWS   66.3          55.0         4675.8   437.4    3.8    3.4   3.8   22.0   17.8    110      216
3038   SRWN   18.03        171.2          273.5           15.3   12.7   3.9   22.7   18.5    187      469
3039   SRWS   28.0          27.1         1731.3            1.8    5.5   1.2    6.7    5.5    206      366
3040
3041    ;
3042
3043  * ADUJUST THE DEMANDS FOR COW AND BUFFALO POPULATION IN THE IRRIGATED AREA
3044  *&&Z    ZONE3XXXXX(Z,CQ, "DEMAND") = DEMAND(Z,CQ ) ;
3045    SCALAR
3046     COWF ADJUSTMENT FACTOR FOR COWS POPULATION IN THE IRRIGATED AREAS/.5/
3047     BUFF ADJUSTMENT FACTOR FOR BUFFLOES POP.   IN THE IRRIGATED AREAS/.8/
3048      ;
3049
3050
3051     DEMAND(Z,"COW-MILK")  = DEMAND(Z,"COW-MILK" )*COWF;
3052     DEMAND(Z,"BUFF-MILK") = DEMAND(Z,"BUFF-MILK")*BUFF;
3053
3054
3055     PARAMETER ELAST(CQ ) ELASTICITY OF DEMAND FOR CROP AND LIVESTOCK COMODITIES
```

INDUS BASIN MODEL REVISED (IBMR) FILENAME=WSISD41
DEMAND AND CONSUMPTION DATA

```
3056          /
3057    (BASMATI,IRRI)           -.4
3058    COTTON                   -.9
3059    GRAM                     -.22
3060    MAIZE                    -.4
3061    MUS+RAP                  -.35
3062    SC-MILL                  -.7
3063    WHEAT                    -.23
3064    (POTATOES,ONIONS,CHILLI) -.7
3065    (COW-MILK,BUFF-MILK)     -.44
3066          /
3067
3068        GROWTHRD(CQ )  GROWTH RATE OF REFERENCE DEMAND (PERCENT)/
3069              BASMATI   4.5,   IRRI      4.0,   COTTON    5.0,
3070              GRAM      4.7,   WHEAT     4.1,   MUS+RAP   3.4,
3071              SC-MILL   5.5,
3072              COW-MILK  6.3,   BUFF-MILK 6.3,   MEAT      6.6 /
3073         ;
3074
3075   TABLE CONSRATIO(Z,G) PROPORTION OF CONSUMPTION BY GROWUNDWATER TYPE
3076
3077                  FRESH
3078      NWFP        1
3079      PMW          .85
3080      PCW          .9
3081      PSW          .83
3082      PRW         1
3083      SCWN         .55
3084      SRWN         .33
3085      SCWS         .34
3086      SRWS        0
3087         ;
3088   PARAMETER NATEXP(CQ) NATIONAL EXPORTS (000 TONS)
3089             /BASMATI  250, IRRI 1100, COTTON 700, ONIONS 10 /
3090             EXPLIMIT(Z,CQ) EXPORT LIMITS BY ZONE
3091   SCALAR     EXPLIMITGR    GROWTH OF EXPORT LIMITS (PERCENT) /5 /
3092         ;
3093   TABLE EXPPV(PV,CQ) PROVINCIAL EXPORTS AS PROPORTION OF NATIONAL
3094
3095                 BASMATI   IRRI   COTTON   ONIONS
3096      NWFP                                  0.14
3097      PUNJAB     1.00      0.29   0.84      0.33
3098      SIND                 0.71   0.16      0.53
3099
3100   TABLE EXPZO(Z,CQ) ZONAL EXPORTS AS PROPORTION OF PROVINCIAL
3101
3102                 BASMATI   IRRI   COTTON   ONIONS
3103      NWFP                                  1
3104      PCW                  0.33   1.00      0.52
3105      PRW        1.00      0.67              0.22
3106      PMW                                    0.09
3107      PSW                                    0.17
3108
3109      SCWN                         0.74     0.41
3110      SRWN                 0.74              0.30
3111      SCWS                         0.26     0.22
3112      SRWS                 0.26              0.08
3113         ;
3114
3115   GROWTHRD(CQ)$(GROWTHRD(CQ) EQ 0) = 3.0 ;
3116   CONSRATIO(Z,"SALINE") = 1 - CONSRATIO(Z,"FRESH") ;
3117   EXPLIMIT(Z,CQ) = NATEXP(CQ)*
3118                     SUM(PV$PVZ(PV,Z), EXPPV(PV,CQ))*EXPZO(Z,CQ) ;
3119   EXPLIMIT(Z,CQ) = SUM(IS$ISR(IS),(1+EXPLIMITGR/100)**
3120                     (ORD(IS)+1979-BASEYEAR)*EXPLIMIT(Z,CQ)) ;
```

INDUS BASIN MODEL REVISED (IBMR) FILENAME=WSISD41
DEMAND AND CONSUMPTION DATA

```
3121
3122   *&&Z     ZONE3XXXXX(Z,CQ,"TOTPROD")  = TOTPROD(Z,CQ)  ;
3123   *&&Z     ZONE3XXXXX(Z,CQ,"FARMCONS") = FARMCONS(Z,CQ) ;
3124   *&&Z     ZONE3XXXXX(Z,CQ,"ELAST")    = ELAST(CQ)      ;
3125   *&&Z     ZONE3XXXXX(Z,CQ,"GROWTHRD") = GROWTHRD(CQ)   ;
3126   *&&Z     ZONE3XXXXX(Z,CQ,"EXPLIMIT") = EXPLIMIT(Z,CQ) ;
3127
```

INDUS BASIN MODEL REVISED (IBMR) FILENAME=WSISD41   07/25/91 17:17:54
DATA TRANSFORMATION TO ZONES   GAMS 2.21 IBM CMS

```
3129      SET SR1(DC) /CCA, CCAP /
3130      ALIAS (G,G1)
3131      PARAMETER
3132          ZWT(Z,CNL,SA)        WEIGHTING FACTOR TO MAP RAIN EVAP AND EFFICIENCIES TO ZONES
3133          EQEVAPZ(Z,M)         EVAPORATION FROM THE EQUAIFER BY ACZ (FEET)
3134          SUBIRRZ(Z,M)         SUBIRRIGATION BY ACZ                 (FEET)
3135          EFRZ(Z,M)            EFFECTIVE RAIN  BY ACZ               (FEET)
3136          RESOURCE(Z,G,R1)     ENDOWMENTS BY ACZ AND GROUNDWATER QUALITY
3137          CNEFF(CNL)           CANAL EFFICIENCY FROM CANAL HEAD TO THE WATERCOURSE HEAD
3138          WCEFF(CNL,M)         WATERCOURSE COMMAND DELIVERY EFFICIENCY
3139          TWEFF(CNL,M)         DELIVERY EFFICIENCY FROM PRIVATE TUBEWELL TO THE ROOT ZONE
3140          CNEFFZ(Z)            WEIGHTED CANAL DELIVERY EFFICIENCY FROM CANAL HEAD TO WATERCOURSE HEAD
3141          TWEFFZ(Z,M)          WEIGHTED PRIVATE TUBEWELL DELIVERY EFFICIENCY BY ZONE
3142          WCEFFZ(Z,M)          WEIGHTED WATER COURSE COMMAND DELIVERY EFFICIENCY BY ZONE
3143          FLEFFZ(Z)            WEIGHTED FIELD EFFICIENCY BY ZONE
3144          CANALWZ(Z,G,M)       CANAL WATER AVAILABLILITY AT THE CANAL HEAD(MAF)
3145          CANALWRTZ(Z,G,M)     CANAL WATER AVAILABLILITY AT THE ROOT ZONE(MAF)
3146          GWTSA(CNL,SA,M)      GOVERNMENT TUBEWELL PUMPAGE BY CANAL AND SUBAREA(KAF)
3147          GWT1(Z,G,M)          PUBLIC TUBEWELL PUMPAGE AT THE ROOT ZONE (KAF)
3148          RATIOFS(Z,G)         FRESH AND SALINE CCA AS A PROPORTION OFF TOTAL
3149          ;
3150      LOOP(ISR,
3151      RESOURCE(Z,G,SR1)          = SUM((CNL,SA)$(ZSA(Z,CNL,SA)$GWFG(CNL,SA,G)), COMDEF(ISR,SR1,CNL) *SUBDEF(SA,CNL) );
3152      ZWT(Z,CNL,SA)$ZSA(Z,CNL,SA) = COMDEF(ISR,"CCA",CNL)*SUBDEF(SA,CNL)/SUM(G,RESOURCE(Z,G,"CCA") );
3153      CNEFF(CNL)       = COMDEF(ISR,"CEFF",CNL)  ;
3154      WCEFF(CNL,M)     = SUM( WCE$WCEM(WCE,M), COMDEF(ISR,WCE,CNL) );
3155      TWEFF(CNL,M)$COMDEF(ISR,"FLDE",CNL)
3156             =(1- (1-WCEFF(CNL,M)/COMDEF(ISR,"FLDE",CNL))*TWEFAC)*COMDEF(ISR,"FLDE",CNL);
3157      CNEFFZ(Z)        = SUM( (CNL,SA),    COMDEF(ISR,"CEFF",CNL)*ZWT(Z,CNL,SA)   )          ;
3158      FLEFFZ(Z)        = SUM( (CNL,SA),    COMDEF(ISR,"FLDE",CNL)*ZWT(Z,CNL,SA)   )          ;
3159      CANALWRTZ(Z,G,M) = SUM( (CNL,SA)$(ZSA(Z,CNL,SA)$GWFG(CNL,SA,G)), DIVPOST(CNL,M)*SUBDEF(SA,CNL)*
3160                                                                  COMDEF(ISR,"CEFF",CNL)*WCEFF(CNL,M) );
3161      GWTSA(CNL,SA,M)$CAREA(CNL,"FRESH")=SUBDEF(SA,CNL)$GWF(CNL,SA)*COMDEF(ISR,"CCA",CNL)/CAREA(CNL,"FRESH")*GWT(CNL,M)
3162      );
3163
3164      RATIOFS(Z,G)     = RESOURCE(Z,G,"CCA")/ SUM(G1, RESOURCE(Z,G1,"CCA") ) ;
3165      CANALWZ(Z,G,M)   = SUM( (CNL,SA)$(ZSA(Z,CNL,SA)$GWFG(CNL,SA,G)), DIVPOST(CNL,M)*SUBDEF(SA,CNL) );
3166      EQEVAPZ(Z,M)     = SUM( (CNL,SA),       EQEVAP(CNL,M)*ZWT(Z,CNL,SA)   )         ;
3167      SUBIRRZ(Z,M)     = SUM( (CNL,SA),       SUBIRR(CNL,M)*ZWT(Z,CNL,SA)   )         ;
3168      EFRZ(Z,M)        = SUM( (CNL,SA),       EFR(CNL,M)*ZWT(Z,CNL,SA)      )         ;
3169      TWEFFZ(Z,M)      = SUM( (CNL,SA),       TWEFF(CNL,M)*ZWT(Z,CNL,SA)    )         ;
3170      WCEFFZ(Z,M)      = SUM( (CNL,SA),       WCEFF(CNL,M)*ZWT(Z,CNL,SA)    )         ;
3171      GWT1(Z,"FRESH",M) = SUM( (CNL,SA)$ZSA(Z,CNL,SA), GWTSA(CNL,SA,M)*WCEFF(CNL,M) ) ;
3172
3173
3174  * EQEVAPZ, AND SUBIRRZ SHOULD BE REPLACED BY EQEVAPZ1 AND SUBIRRZ1 ;
3175
3176      TABLE DEPTH1(Z,G,M) DEPTH FOR YEAR 2000 FROM GW MODEL (FEET)
3177
3178              JAN  FEB  MAR  APR  MAY  JUN  JUL  AUG  SEP  OCT  NOV  DEC
3179
3180  NWFP.FRESH  5.   5.   5.   4.9  4.9  5.   5.1  5.2  5.4  5.5  5.2  4.9
3181  PMW .FRESH 13.8 14.  14.4 14.5 14.3 14.1 13.7 13.5 13.5 13.7 13.7 13.7
3182  PMW .SALINE 5.1  4.9  4.9  5.   5.3  5.5  5.5  5.6  5.7  5.6  5.4  5.1
3183  PCW .FRESH 30.9 31.5 32.1 31.9 31.5 31.2 30.7 30.4 30.9 31.4 31.8 31.9
3184  PCW .SALINE 4.8  4.8  4.9  5.1  5.2  5.3  5.2  5.1  5.1  5.   4.9  4.8
3185  PSW .FRESH 29.9 30.5 31.2 31.  30.8 30.6 30.2 30.  30.3 31.  31.1 31.1
3186  PSW .SALINE 4.9  4.9  4.9  5.   5.2  5.3  5.3  5.4  5.3  5.   5.1  5.
3187  PRW .FRESH 70.9 71.6 72.7 72.8 72.6 72.8 72.9 73.2 74.1 74.8 75.  75.2
3188  SCWN.FRESH  5.6  5.9  6.2  6.3  6.2  6.   5.6  5.3  5.4  5.6  5.5  5.4
3189  SCWN.SALINE 4.6  4.6  4.7  4.9  5.   5.1  4.8  4.6  4.7  4.7  4.6  4.5
3190  SRWN.FRESH  6.3  6.5  6.7  6.9  7.1  6.7  6.1  5.8  5.8  6.1  6.2  6.3
3191  SRWN.SALINE 5.5  5.5  5.7  5.9  6.1  5.9  5.5  5.3  5.3  5.3  5.4  5.4
```

INDUS BASIN MODEL REVISED (IBMR) FILENAME=WSISD41
DATA TRANSFORMATION TO ZONES

07/25/91 17:17:54
GAMS 2.21 IBM CMS

```
3192      SCWS.FRESH  45.9 46.2 46.8 46.6 46.5 47.  47.3 47.7 48.3 48.2 48.5 48.5
3193      SCWS.SALINE  4.9  4.8  4.9  5.   5.2  5.3  5.1  5.   5.   5.   4.9  4.8
3194      SRWS.SALINE  5.3  5.4  5.7  5.9  6.   5.7  5.3  5.1  5.   5.   5.1  5.3
3195    ;
3196  * COMPUTE EVAPZ1, SUBIRR AND EQEVAP
3197     PARAMETER EVAPZ1(Z,G,* ) WEIGHTED EVAPORATION BY ZONE AND GW TYPE
3198               EQEVAPZ1(Z,G,* ) EQ. EVAP BY ZONE ANF GW TYPE
3199               SUBIRRZ1(Z,G,* ) SUBIRRIGATION BY ZONE AND GW TYPE
3200     ;
3201     LOOP(ISR,
3202     EVAPZ1(Z,G,M)$RESOURCE(Z,G,"CCA") =
3203             SUM( (CNL,SA)$(ZSA(Z,CNL,SA)$GWFG(CNL,SA,G)),
3204                EVAP(CNL,M)*COMDEF(ISR,"CCA",CNL)*SUBDEF(SA,CNL))/
3205                   RESOURCE(Z,G,"CCA")       );
3206     EVAPZ1(Z,G,SEA1)= SUM(M$SEA1M(SEA1,M), EVAPZ1(Z,G,M));
3207     EQEVAPZ1(Z,G,M) = MIN(1., 10.637/DEPTH1(Z,G,M)**2.558)*EVAPZ1(Z,G,M);
3208     EQEVAPZ1(Z,G,SEA1) = SUM(M$SEA1M(SEA1,M), EQEVAPZ1(Z,G,M));
3209     SUBIRRZ1(Z,G,M) = EQEVAPZ1(Z,G,M) * THE1;
3210     SUBIRRZ1(Z,G,SEA1) = SUM(M$SEA1M(SEA1,M) ,SUBIRRZ1(Z,G,M));
3211   DISPLAY DEPTH1, EVAPZ1, EQEVAPZ1, SUBIRRZ1;
3212   DISPLAY   WCEFF, TWEFF, GWTSA, GWT1 ;
       *- REPORT WATER COMPONENTS
          SET CRT /CANAL, CANAL-RT,GOVT-TW, GWT-RT /
              RAS /RAIN, SUBIRR, TOTAL /
       PARAMETER REP5(Z,*,*)
                 REP51
                 REP6
                 ;
          REP5(Z,"RAIN",M)      = EFRZ(Z,M)*12   ;
          REP5(Z,"SUBIRR",M)    = SUBIRRZ(Z,M)*12 ; REP5(Z,"TOTAL",M) = SUM(RAS, REP5(Z,RAS,M) ) ;
          REP5(Z,RAS,SEA1)      = SUM(M$SEA1M(SEA1,M), REP5(Z,RAS,M)) ;

          REP51(Z,"CNLDELEFF",M)= CNEFFZ(Z)*WCEFFZ(Z,M) ;
          REP51(Z,"WCDELEFF",M) = WCEFFZ(Z,M) ;
          REP51(Z,"TWDELEFF",M) = TWEFFZ(Z,M) ;

          REP6(Z,G,"CANAL",M)           = CANALWZ(Z,G,M)              ;
          REP6(Z,"TOTAL","CANAL",M)     = SUM(G,CANALWZ(Z,G,M) )      ;
          REP6(Z,G,"CANAL-RT",M)        = CANALWRTZ(Z,G,M)            ;
          REP6(Z,"TOTAL","CANAL-RT",M)  = SUM(G, CANALWRTZ(Z,G,M))    ;
          REP6(Z,"FRESH","GOVT-TW",M)   = SUM( (CNL,SA)$ZSA(Z,CNL,SA), GWTSA(CNL,SA,M) )/1000;
          REP6(Z,"TOTAL","GOVT-TW",M)   = REP6(Z,"FRESH","GOVT-TW",M);
          REP6(Z,G,"GWT-RT",M)          = GWT1(Z,G,M)/1000;
          REP6(Z,"TOTAL","GWT-RT",M)    = GWT1(Z,"FRESH",M)/1000;
          REP6(Z,T1,CRT,SEA1)           = SUM(M$SEA1M(SEA1,M), REP6(Z,T1,CRT,M));
          DISPLAY ZWT  ;
          DISPLAY "RAIN       EFFECTIVE RAIN(INCHES) "
                  "SUBIRR     WATER AVAILABLE TO THE CROPS BY CAPPILARY ACTION FROM THE GW(INCHES) "
                  "TOTAL      TOTAL WATER AVAILABLE FROM RAIN AND SUBIRR(INCHES) ", REP5 ;
          DISPLAY "CNLDELEFF  WEIGHTED CANAL DELIVERY EFFICIENCY FROM CANAL HEAD TO THE ROOT ZONE "
                  "WCDELEFF   WEIGHTED WATERCOURSE COMMAND  EFF. FROM CANAL HEAD TO THE ROOT ZONE "
                  "TWDELEFF   WEIGHTED PERIVATE TUBEWELL EFF. FROM CANAL HEAD TO THE ROOT ZONE ", REP51 ;
          DISPLAY "CANAL      CANAL WATER AVAILABLE AT THE CANAL HEAD  (MAF)     "
                  "CANNAL-RT  CANAL WATER AVAILABLE AT THE ROOT ZONE (MAF)       "
                  "GOVT-TW    GOVERNMENT TUBEWELL PUMPAGE AT THE WATERCOURSE HEAD(MAF)"
                  "GWT-RT     GOVERNMENT TUBEWELL PUMPAGE AT THE ROOT ZONE(MAF)", REP6 ;
          DISPLAY EQEVAPZ ;
3251   *--
3252
3253   SET FTT(R1)/FARMPOP,FARMHH,TRACTORS,TUBEWELLS,BULLOCKS,COWS,BUFFALOS/;
3254
3255         TABLE RES88(R1 ,Z) AVAILABLE RESOURCES 1988
3256
3257  * FARMPOP -- THOSAND WORKERS, FARMHH -- THOUSAND HOUSEHOLDS
3258  * BULLOCKS, COWS AND BUFFALOS ARE IN THOUSANDS
```

```
3259  *    TUBEWELLS ARE 1 CFS TUBEWELLS
3260
3261                NWFP     PMW      PCW      PSW      PRW     SCWN     SRWN     SCWS     SRWS
3262
3263  FARMPOP        371     478     2589     1441     1016      484      589      567      176
3264  FARMHH         174     226     1123      586      513      211      253      275       88
3265  TRACTORS      7465   12044    86817    45777    44564     8177     4154     8839     2702
3266  TUBEWELLS     2638   12265   109658    37952    49897     9573     4863    10348     3163
3267  BULLOCKS       151     332     1978      952      368      399      486      444      143
3268  COWS           505     736     2549     1055      590      442      570      630      256
3269  BUFFALOS       319     487     4146     2911     2279      584      587      557      249
3270
3271
3272     TABLE CROPAREA(Z,C)  CROPPED AREA 1988 (000'S ACRES)
3273
3274  *     SC-MILL AREA INCLUDES GUR AREA
3275
3276        WHEAT BASMATI IRRI MAIZE SC-MILL COTTON MUS+RAP GRAM CHILLI POTATOES ONIONS ORCHARD RAB-FOD KHA-FOD
3277  NWFP    224      0    0    78     143                52   132     9       8        6      27       47     14
3278  PCW    6324    396  277    41     722   3820        275   746    55      51       40     300     1174   1332
3279  PMW    1168     25   29           100    270         27   389    14      13       10      93      158    328
3280  PRW    2141   1189  272     9      79     78         42   233    21      19       15      56      506    374
3281  PSW    2688    391   42    51     495    614         41   218    33      30       24     242      675    698
3282  SCWN   1004      0   19           169    743         63    58    10       9        7      49      131     72
3283  SCWS    921      0  196           238    758         25    22    12      11        8     103      137     59
3284  SRWN    590      0  951            24     13        124   100    15      14       11       7      117      4
3285  SRWS    101      0  325           111     41         13    38     4       4        3      41       30      7
3286
3287     TABLE GROWTHRES(R1 ,Z) GROWTH RATE OF FARM POPULATION TRACTORS AND TUBWELLS (PERCENT)
3288
3289                 (NWFP,PMW,PCW,PSW,PRW)    (SCWN,SCWS,SRWN,SRWS)
3290  FARMPOP
3291  FARMHH
3292  TRACTORS              3                         3
3293  TUBEWELLS             1
3294  BULLOCKS              2.35                      2.35
3295  COWS                  1.21                      1.21
3296  BUFFALOS              1.21                      1.21
3297
3298  PARAMETER
3299    ORCHAREA(Z)    AREA UNDER ORCHARDS BY ZONE (THOUSAND ACRES )
3300    ORCHGROWTH(Z) GROWTH RATE OF ORCHARD AREA
3301                   /(NWFP,PMW,PCW,PSW,PRW) 6.43
3302                    (SCWN,SCWS,SRWN,SRWS)  2.591/
3303                   ;
3304  DEMAND(Z,CQ) = DEMAND(Z,CQ)* SUM(IS$ISR(IS),
3305                    (1+GROWTHRD(CQ)/100)**(ORD(IS)+1979-BASEYEAR) );
3306  ORCHAREA(Z)  = CROPAREA(Z,"ORCHARD")* SUM(IS$ISR(IS),
3307                    (1+ORCHGROWTH(Z)/100)**(ORD(IS)+1979-BASEYEAR) );
3308  RESOURCE(Z,G,FTT) = RES88(FTT,Z)*RATIOFS(Z,G)*SUM(IS$ISR(IS),
3309                    (1+GROWTHRES(FTT,Z)/100)**(ORD(IS)+1979-BASEYEAR) );
3310
3311  NTWUCAP  = TWUTIL("NEW")*TWCAP ;
3312  RESOURCE(Z,"FRESH","TUBEWELLS")$RESOURCE(Z,"FRESH","CCA") =
3313                             SUM(G, RESOURCE(Z,G,"TUBEWELLS"));
3314  RESOURCE(Z,"SALINE","TUBEWELLS")= 0 ;
3315  RESOURCE(Z,"FRESH","TWC")$RESOURCE(Z,"FRESH","CCA") =
3316                        RESOURCE(Z,"FRESH","TUBEWELLS")*TWCAP*
3317                        TWUTIL("EXISTING") /1000  ;
3318
3319  PARAMETER  SCMILLCAP(Z)  SUGARCANE MILL CAPACITY (THOUSAND TONNS PER YEAR )
3320            /NWFP  6937,
3321             PMW   5750,   PCW  8550,   PSW  12000,   PRW  3600
3322             SCWN  6000,   SCWS 8370,   SRWN 3800,    SRWS 3730 /;
3323  OPTION TOTPROD:1, FARMCONS:1, DEMAND:1, CONSRATIO:2 ;
```

INDUS BASIN MODEL REVISED (IBMR) FILENAME=WSISD41
DATA TRANSFORMATION TO ZONES

```
3324     DISPLAY RESOURCE, TOTPROD, FARMCONS,CONSRATIO, DEMAND,EXPLIMIT ;
3325     SET CNL1(CNL)       CANALS EXCLUDING NWFP CANALS          ;
3326         CNL1(CNL) = YES;   CNL1(CNL)$PVCNL("NWFP",CNL) = NO ;
3327
3328     PARAMETER
3329        POSTT     AVERAGE CANAL DIVERSIONS BY SEASON
3330        PROTARB   SHARES AS PROPORTION OF TOTAL HISTORIC DIVERSIONS (EXCLUDING NWFP)
3331          ;
3332       POSTT(CNL,SEA) = SUM(M$SEAM(SEA,M),    DIVPOST(CNL,M));
3333       POSTT(PV2,SEA) = SUM(CNL$PVCNL(PV2,CNL), POSTT(CNL,SEA));
3334       PROTARB(CNL1,SEA) = 0.999 * POSTT(CNL1,SEA)/ ( POSTT("PUNJAB",SEA) + POSTT("SIND",SEA)) ;
3335       PROTARB(PV2,SEA)  = 0.999 * POSTT(PV2,SEA)  / ( POSTT("PUNJAB",SEA) + POSTT("SIND",SEA)) ;
3336       OPTION POSTT:5, PROTARB:5 ;
3337     DISPLAY CNL1, POSTT, PROTARB ;
3338     PARAMETER DIVNWFP(M) MONTHLY TOTAL DIVERSION TO NWFP ZONE (MAF);
3339
3340          TABLE MAXNWFP(*,M) MAXIMUM POSSIBLE DIVERSIONS IN NWFP (MAF)
3341
3342     * THESE DIVERSION WERE DETRMINED BY RUNNING THE ZONAL MODEL WITHOUT
3343     * HISTORIC DIVERSIONS AND CANAL CAPACITY BOUNDS.
3344
3345             JAN  FEB  MAR  APR  MAY  JUN  JUL  AUG  SEP  OCT  NOV  DEC
3346
3347     1988    .158 .187 .270 .184 .232 .325 .333 .333 .332 .200 .063 .133
3348     2000    .132 .199 .231 .170 .195 .268 .293 .276 .333 .333 .235 .158
3349     2000PLAN .143 .219 .252 .196 .223 .310 .336 .307 .439 .450 .264 .176
3350     ;
3351       DIVNWFP(M) = SUM(ISR, MAXNWFP(ISR,M));
3352       DIVNWFP(M)$INCP("WITH") = MAXNWFP("2000PLAN",M);
3353     DISPLAY DIVNWFP;
```

**** FILE SUMMARY FOR USER KDAM

INPUT	D41	GAMS	A
OUTPUT	D41	LISTING	A
SAVE	D41	WORK*	A

COMPILATION TIME   =   5.100 SECONDS   VER: IBM-TB-003

Appendix A.2:

Model Specification

INDUS BASIN MODEL REVISED(IBMR) FILENAME=WSISM41
MODEL SETUP

This file contains the equation and model specification. Set PSR and
PSR1 contains the prices scenarios for the model.
This file is excuted after the data setup (WSISD*) setup as GAMS RESTART.

```
3363  SCALARS  CNLLF  CANAL LOSSES RECOVERY IN SALINE AREAS /0/
3364           WCLF   WATER COURSE COMMAND LOSSES RECOVER IN SALINE AREAS/0/
3365  SETS
3366  PSR(PS)  PRICE SCENARIO FOR THE MODEL(FINANCIAL PRICES) /87-88 /
3367  PSR1(PS) PRICE SCENARIO FOR REPORT(ECONOMIC PRICES)    /87-88 /
3368  Z1(Z)    ZONE SELECTION FOR THIS RUN
3369           / NWFP,PMW,PCW,PSW,PRW,SCWN,SCWS,SRWN,SRWS /
3370  *
3371  *  TOTAL COMODITIES ARE 18,
3372  *       WITH ENDOGENOUS PRICES= 13, FIXED PRICES=2, FODDER=2
3373  *       CONSUMPTION ONLY = 1
3374
3375  CN(CQ)     COMODITIES  ENDOGENOUS PRICES
3376             /BASMATI,   IRRI,      COTTON,
3377              GRAM,      MAIZE,     MUS+RAP,
3378              SC-MILL,   WHEAT,     POTATOES,
3379              ONIONS,    CHILLI,
3380              COW-MILK,  BUFF-MILK /
3381  CCN(CQ)    CROP COMODITIES WITH ENDOGENOUS PRICES
3382             /BASMATI,   IRRI,      COTTON,
3383              GRAM,      MAIZE,     MUS+RAP,
3384              SC-MILL,   WHEAT,     POTATOES,
3385              ONIONS,    CHILLI /
3386  QN(CQ)     LIVESTOCK COMODITIES ENDOGENOUS PRICES
3387             /COW-MILK, BUFF-MILK/
3388  NCN(CQ)    CROPS WITH FIXED PRICES EXCLUDING FODDER
3389             /ORCHARD, MEAT /
3390  CE(CQ)     EXPORTABLE COMODITIES
3391             / BASMATI, IRRI, COTTON,ONIONS,WHEAT /
3392  CM(CQ)     COMODITIES WHICH COULD BE IMPORTED/SC-MILL/
3393
3394  EX(Z,G)            TO CHECK FRESH OR SALINE AREA WITHIN A ZONE
3395  TECHC(Z,CQ)        COMODITIES BY ZONES
3396                ;
3397  EX(Z1,G)$RESOURCE(Z1,G,"CCA") = YES ;
3398  DISPLAY  CQ,CN, CCN, QN, NCN,CE,CM,EX;
3399  DISPLAY  FERT;
3400
3401        TABLE TEC(C,T,S,W,Z)  CROP TECHNOLOGY DISABLED FOR 1988 RUN
3402                                                          SCWN  SRWN  SCWS  SRWS
3403  WHEAT. (BULLOCK,SEMI-MECH).LA-PLANT.(HEAVY,JANUARY)       1     1     1     1
3404  WHEAT. (BULLOCK,SEMI-MECH).(QK-HARV,STANDARD). HEAVY            1     1     1
3405  WHEAT. (BULLOCK,SEMI-MECH).(QK-HARV,STANDARD).JANUARY           1           1
3406  WHEAT. (BULLOCK,SEMI-MECH).(QK-HARV,STANDARD,LA-PLANT).LIGHT                1
3407        ;
3408  SCALARS BIG BIG NUMBER USED FOR ARTIFICAL PRODUCTION /4000/
3409          PAWAT  BIG NUMBER FOR ARTIFICIAL WATER        /999999/
3410          PAFOD  BIG NUMBER FOR ARTIFICIAL FODDER       /1000/
3411  PARAMETER
3412     RVAL(N)  VALUE OF WATER STORED IN THE RESERVOIRS
3413               /TARBELA-R 1, KALABAGH-R .8, MANGLA-R .6
3414                CHASMA-R .2, CHOTIARI-R .2, A-SEA   .1 /
3415     FSALEP(CQ)  FINANCIAL SALE PRICE FOR CROP AND LIVESTOCK COMODITIES (RS PER KG OR PER LITER)
3416     PP          FINANCIAL PURCHASE PRICE OF PROTEIN (RS PER KGS)
3417     MISC(*)     FINANCIAL MISCELLENIOUS PRICES
3418     SEEDP(C)    FINANCIAL SEED PRICE           (RS PER KGS)
3419     WAGE(M)     FINANCIAL WAGE RATES           (RS PER MAN-HOUR)
3420     MISCCT(C)   FINANCIAL WATER CHARGES AND MISCILLENIOUS COSTS(RS PER ACRE)
3421
3422     ESALEP(CQ)  ECONOMIC SALE PRICE FOR CROP AND LIVESTOCK COMODITIES  (RS PER KG OR PER LITER)
3423     EPP         ECONOMIC PRICE OF PROTEIN CONCENTRATE    (RS PER KGS)
3424     EMISC(*)    ECONOMIC MISCELLENIOUS PRICES
3425     ESEEDP(C)   ECONOMIC SEED PRICE
3426     EWAGE(M)    ECONOMIC WAGE RATE       (RS PER MAN-HOUR)
3427     EMISCCT(C)  ECONOMIC WATER CHARGES AND MISCILLENIOUS COSTS(RS PER ACRE)
```

```
3428
3429      IMPORTP(CQ)  IMPORT PRICES FOR THE  SCENARIO
3430      EXPORTP(CQ)  EXPORT PRICES FOR THE  SCENARIO
3431
3432      WNR(C,Z,T,S,W,M) WATER REQUIREMENTS NET OF RAIN
3433               ;
3434
3435      LOOP (PSR,
3436         FSALEP(CQ)    = PRICES(PSR,CQ,"FINANCIAL")        ;
3437         PP            = PRI1(PSR,"FINANCIAL","PROTEIN");
3438
3439         MISC(P1)      = PRI1(PSR,"FINANCIAL",P1)     ;
3440         SEEDP(C)      = FINSDWTPR(C,PSR,"SEED")      ;
3441         WAGE(M)       = WAGEPS(PSR,"FINANCIAL",M)    ;
3442         MISCC(C)      = FINSDWTPR(C,PSR,"WATER")+FINSDWTPR(C,PSR,"MISCC");
3443         IMPORTP(CQ)   = PRICES(PSR,CQ,"IMPORT")      ;
3444         EXPORTP(CQ)   = PRICES(PSR,CQ,"EXPORT")
3445      );
3446      LOOP (PSR1,
3447         ESALEP(CQ)    = PRICES(PSR1,CQ ,"ECONOMIC")       ;
3448         EPP           = PRI1(PSR1,"ECONOMIC","PROTEIN") ;
3449
3450         EMISC(P1)     = PRI1(PSR1,"ECONOMIC",P1)          ;
3451         ESEEDP(C)     = ECNSDWTPR(C,PSR1,"SEED")          ;
3452         EWAGE(M)      = WAGEPS(PSR1,"ECONOMIC",M)         ;
3453         EMISCCT(C)    = ECNSDWTPR(C,PSR1,"WATER")+ECNSDWTPR(C,PSR1,"MISCC")
3454      );
    * Economic prices are set to finacial
         ESALEP(cq) =  FSALEP(cq);
         EMISC(P1)  =  MISC(P1)   ;
         ESEEDP(C)  =  SEEDP(C)   ;
         EPP        =  PP         ;
         EWAGE(M)   =  WAGE(M)    ;
3463
3464      WNR(C,Z1,T,S,W,M)   = MAX(0, (WATER(C,Z1,T,S,W,M)-EFRZ(Z1,M)) ) ;
3465 *- SET COTON IN PRW TO NO
3466      TEC("COTTON",T,S,W,"PRW") = 1   ;
3467      TEC("MAIZE",T,S,W,"PRW") = 1    ;
3468
3469      TECH(Z1,C,T,S,W)$TEC(C,T,S,W,Z1) = NO ;
3470      TECHC(Z1,C)$SUM( (T,S,W), TECH(Z1,C,T,S,W) ) = YES   ;
3471      TECHC(Z1,CF)=NO;  TECHC(Z1,Q) = YES ;
3472 *  DISPLAY TECH;
3473      DISPLAY   TECHC ;
3474
3475   SCALARS
3476      TOLCNL   ALLOWED DEVIATION FROM PROPORTIONAL ALLOCATION BY CANAL    /0.0 /
3477      TOLPR    ALLOWED DEVIATION FROM PROPORTIONAL ALLOCATION BY PROVINCE / 0.0 /
3478      TOLNWFP  NWFP DIVERSION TOLERANCE / 0 /
3479              ;
3480 *-
3481    PARAMETER BETA(CQ,Z1)   GRADIENT COMODITIES DEMAND CURVE
3482              ALPHA(CQ,Z1) DEMAND CURVE INTECEPT
3483              ;
3484    SCALAR    BETAF        BETA FACTOR /.5 / ;
3485    BETA(CN,Z1 )$DEMAND(Z1 ,CN) = FSALEP(CN) / DEMAND(Z1 ,CN) / ELAST(CN) ;
3486    ALPHA(CN,Z1 )                = FSALEP(CN) - BETA(CN,Z1 )*DEMAND(Z1 ,CN)   ;
3487
3488 *-- LINEARIZATION OF THE DEMAND FUNCTION.
3489
3490    SET   P    GRID POINTS FOR LINEARIZATION / 1*20 /
3491    PARAMETER
3492        PMAX(CQ,Z1)      MAXIMUM PRICE FOR SEGMENTS
3493        PMIN(CQ,Z1)      MINIMUM PRICE FOR SEGMENTS
3494        QMAX(CQ,Z1)      MAX NATIONAL CONSUMPTION
```

```
3495        QMIN(CQ,Z1)      MIN NATIONAL CONSUMPTION
3496        INCR(CQ,Z1)      INCREMENT
3497        WS(CQ,Z1,P)      WELFARE SEGMENTS    (MILLION RUPEES)
3498        RS(CQ,Z1,P)      REVENUE DEFINITION (MILLION RUPEES)
3499        QS(CQ,Z1,P)      QUANTITY DEFINITION(THOUSAND TONS OR MILLION LITERS)
3500        ENDPR(CQ,Z1,P)   PRICE (RUPEES PER KGS OR LITER )
3501              ;
3502  PMIN(CN,Z1) = 0.5*FSALEP(CN) ;
3503  PMAX(CN,Z1) = MIN(ALPHA(CN,Z1), 2*FSALEP(CN) ) ;
3504 * PMAX(CE,Z1) = MIN(ALPHA(CE,Z1 ),2*FSALEP(CE) ) ;
3505 * PMIN(CE,Z1) =      FSALEP(CE)  ;
3506  QMIN(CN,Z1)$BETA(CN,Z1 ) = (PMAX(CN,Z1 )-ALPHA(CN,Z1 ))/BETA(CN,Z1 );
3507  QMAX(CN,Z1)$BETA(CN,Z1 ) = (PMIN(CN,Z1 )-ALPHA(CN,Z1 ))/BETA(CN,Z1 );
3508  INCR(CN,Z1) = (QMAX(CN,Z1 )-QMIN(CN,Z1 ))/(CARD(P)-1);
3509
3510  QS(CN,Z1,P) = QMIN(CN,Z1) + INCR(CN,Z1)*(ORD(P)-1);
3511  WS(CN,Z1,P) = ALPHA(CN,Z1)*QS(CN,Z1,P) + BETAF*BETA(CN,Z1)*SQR(QS(CN,Z1,P)) ;
3512  RS(CN,Z1,P) = ALPHA(CN,Z1)*QS(CN,Z1,P) + BETA(CN,Z1)*SQR(QS(CN,Z1,P));
3513  ENDPR(CN,Z1,P)= ALPHA(CN,Z1)          + BETA(CN,Z1)*QS(CN,Z1,P);
3514 *DISPLAY PMAX, PMIN, QMAX, QMIN,INCR, QS, WS, RS, ENDPR ;
3515 *--
3516 *DISPLAY ALPHA , BETA ;
```

INDUS BASIN MODEL REVISED(IBMR) FILENAME=WSISM41
EQUATIONS AND VARIABLES

```
3518    VARIABLES   CPS    CONSUMER PLUS PRODUCERS SURPLUS (MILLION RUPEES)
3519    POSITIVE VARIABLES
3520       ACOST(Z,G)           FARM COST IN              (MILLION RUPEES)
3521       PPC(Z,G,SEA)         PURCHASES OF PROTEIN CONCENTRATES          (THOUSAND METRIC TONS)
3522       X(Z,G,C,T,S,W)       CROPPED AREA BY TECHNOLOGY                 (THOUSAND ACRES)
3523       ANIMAL(Z,G,A)        PRODUCTION OF LIVESTOCK TYPE A             (THOUSANDS)
3524       PRODT(Z,G,CQ)        PRODUCTION (CROP COMMODITIES 000 METRIC TONS LIVESTOCK COMM MILL. KGS OR LITERS)
3525       PRODA(Z,G,CQ)        ARTIFICIAL SUPPLY
3526       IMPORT(Z ,CQ )       IMPORT OF COMODITIES (CROP COMM. 000 M. TONS LIVESTOCK MILL. KGS OR LITERS)
3527       EXPORT(Z ,CQ )       EXPORT OF COMODITIES                       (000 METRIC TONNS)
3528       CONSUMP(Z,G,CQ )     ON FARM CONSUMPTION                        (000 METRIC TONNS)
3529       FAMILYL(Z,G,M)       FAMILY LABOR USED                          (MILLION MAN HOURS)
3530       HIREDL(Z,G,M)        HIRED LABOR USED                           (MILLION MAN HOURS)
3531       ITW(Z)               INVESTMENT IN INCREASED PRIVATE TUBEWELL CAPACITY  (KAF PER MONTH)
3532       TW(Z,M)              PRIVATE TUBEWELL WATER USED  BY MONTH M    (KAF)
3533       ITR(Z,G)             INVESTMENT IN INCREASED TRACTOR CAPACITY   (000 TRACTOR-HRS PER MONTH)
3534       TS(Z,G,M)            PRIVATE TRACTOR SERVICES USE BY MONTH      (THOUSAND HRS)
3535       F(N,N1,M)            FLOW TO NODE N FROM NODE N1                (MAF)
3536       RCONT(N,M)           END OF THE MONTH RESRVOIR CONTENTS         (MAF)
3537       CANALDIV(CNL,M)      CANAL DIVERSION AT THE CANAL HEAD          (MAF)
3538       CNLDIVSEA(CNL,SEA)   CANAL DIVERSION BY SEASON                  (MAF)
3539       PRSEA(PV2,SEA)       CANAL DIVERSION BY PROVINCE (SIND AND PUNJAB)  (MAF)
3540       TCDIVSEA(SEA)        TOTAL CANAL DIVERSION IN SIND AND PUNJAB BY SEASON (MAF)
3541       WDIVRZ(Z1,G,M)       SURFACE WATER DIVERSION AT THE ROOT ZONE   (KAF)
3542       SLKLAND(Z,G,M)       SLACK LAND                                 (THOUSAND ACRES)
3543       SLKWATER(Z,G,M)      SLACK WATER AT THE ROOT ZONE               (KAF)
3544       ARTFOD(Z1,G,SEA)     ARTIFICIAL FODDER SUPPLY  EQUAIVALENT OF RAB-FOD  (000 TONNS)
3545       ARTWATER(Z,G,M)      WATER FROM IMAGINARY SOURCE AT THE ROOT ZONE  (KAF)
3546       ARTWATERND(N,M)      WATER FROM IMAGINARY SOURCE AT NODES       (MAF)
3547       NAT(CQ,Z,P)          PROVINCIAL DEMAND LINEARIZED
3548       NATN(CQ,Z)           PROVINCIAL DEMAND NON-LINEAR
3549
3550    EQUATIONS
3551       OBJZ    OBJECTIVE FUNCTION FOR THE ZONE MODEL LINEAR VERSION      (MILLION RUPEES)
3552       OBJZN   OBJECTIVE FUNCTION FOR THE ZONE MODEL NON-LINEAR VERSION  (MILLION RUPEES)
3553       OBJN    OBJECTIVE FUNCTION FOR THE INDUS MODEL LINEAR VERSION     (MILLION RUPEES)
3554       OBJNN   OBJECTIVE FUNCTION FOR THE INDUS MODEL NON-LINEAR VERSION (MILLION RUPEES)
3555       COST(Z,G)            ANNUAL FARM COST                           (MILLION RUPEES)
3556       CONV(CQ,Z)           CONVEX COMBINATION FOR AGGREGATE CONSUMPTION
3557       DEMNAT(CQ,Z)         PROVINCIAL DEMAND BALANCE LINEAR   (000 TONS OR MILLION LITERS)
3558       DEMNATN(CQ,Z)        ZONAL DEMAND BALANCE NON-LINEAR    (000 TONS OR MILLION LITERS)
3559       CCOMBAL(Z,G,C)       COMMODITY BALANCES FOR CROPS       (000 TONS)
3560       QCOMBAL(Z,G,Q)       LIVESTOCK COMODITY BALANCES        (000 TONS OR M LITERS)
3561       CONSBAL(Z,G,CQ)      CONSUMPTION BALANCE                (000 TONS OR M LITERS)
3562       LABORC(Z,G,M)        MONTHLY LABOR CONSTRAINT           (MILLION MAN HOURS)
3563       FODDER(Z,G,SEA)      SEASONAL MAINTENANCE OF FODDER SUPPLIES   (000 METRIC TONS)
3564       PROTEIN(Z,G,SEA)     PROTEIN REQUIREMENTS OF LIVESTOCK BY SEASON (000 METRIC TONS)
3565       GRNFDR(Z,G,SEA)      GREEN FODDER REQUIREMENTS                  (000 METRIC TONS)
3566       BDRAFT(Z,G,M)        BULLOCK DRAFT POWER CONSTRAINT             (MILLION BULLOCK HOURS)
3567       BREPCO(Z,G)          BULLOCK REPRODUCTION CONSTRAINT
3568       BULLOCKC(Z1)         BULLOCK POPULATION CONSTRAINT              (000 BULLOCKS)
3569       TDRAFT(Z,G,M)        TRACTOR DRAFT POWER BALANCE                (000 TRACTOR HOURS)
3570       TRCAPC(Z,M)          TRACTOR CAPACITY CONSTRAINT                (000 TRACTOR HOURS)
3571       TWCAPC(Z,M)          TUBEWELL CAPACITY CONSTRAINT               (KAF)
3572       LANDC(Z,G,M)         LAND CONSTRAINT                            (000 ACRES)
3573       ORCHAREAC(Z)         ORCHARD AREA CONSTRAINT                    (000 ACRES)
3574       SCMILLC(Z)           SUGAR CANE TO MILL CONSTRAINT              (000 ACRES)
3575       WATERBALN(Z,G,M)     WATER BALANCE AT THE ROOT ZONE             (KAF)
3576       WATALCZ(Z,G,M)       SURFACE WATER BY ZONE                      (KAF)
3577       SUBIRRC(Z,G,M)       SUBIRRIGATION CONSTRAINT                   (KAF)
3578       NBAL(N,M)            WATER BALANCE AT A NODE                    (MAF)
3579       WATALCSEA(CNL,SEA)   WATER ALLOCATIONS BY SEASON                (MAF)
3580       DIVSEA(SEA)          TOTAL CANAL DIVERSIONS IN SIND AND PUNJAB  (MAF)
3581       DIVCNLSEA(CNL,SEA)   CANAL DIVERSION BY SEASON                  (MAF)
3582       WATALCPRO(PV,SEA)    WATER ALLOCATION BY PROVINCE               (MAF)
```

```
INDUS BASIN MODEL REVISED(IBMR) FILENAME=WSISM41                           07/25/91 17:18:07
EQUATIONS AND VARIABLES                                                   GAMS 2.21 IBM CMS


3583       PRSEAW(PV,SEA)        DIVERSIONS BY PROVINCE AND SEASON        (MAF)
3584       NWFPALC(M)            WATER ALLOCATIONS TO THE NWFP ACZ        (MAF)
3585              ;
3586
3587  OBJZ..
3588    CPS  =E=
3589     SUM(Z1, SUM(G$EX(Z1,G), SUM(NCN, FSALEP(NCN)*PRODT(Z1,G,NCN)  )
3590                            - ACOST(Z1,G) - SUM(SEA, ARTFOD(Z1,G,SEA))*PAFOD
3591                            - SUM(M, ARTWATER(Z1,G,M))*PAWAT
3592                            - SUM(CQ$TECHC(Z1,CQ), PRODA(Z1,G,CQ )*BIG)) )
3593    - SUM(Z1,  SUM(CM$TECHC(Z1,CM),      IMPORT(Z1,CM)*IMPORTP(CM))      )
3594    + SUM(Z1,  SUM(CE$TECHC(Z1,CE),      EXPORT(Z1,CE)*EXPORTP(CE))      )
3595    + SUM(Z1,  SUM( (CN,P)$TECHC(Z1,CN), NAT(CN,Z1,P)*WS(CN,Z1 ,P))      ) ;
3596
3597  OBJZN..
3598    CPS  =E=
3599     SUM(Z1, SUM(G$EX(Z1,G), SUM(NCN, FSALEP(NCN)*PRODT(Z1,G,NCN)  )
3600                            - ACOST(Z1,G) - SUM(SEA, ARTFOD(Z1,G,SEA))*PAFOD
3601                            - SUM(M, ARTWATER(Z1,G,M))*PAWAT
3602                            - SUM(CQ$TECHC(Z1,CQ), PRODA(Z1,G,CQ )*BIG)) )
3603    - SUM(Z1,  SUM(CM$TECHC(Z1,CM),      IMPORT(Z1,CM)*IMPORTP(CM))      )
3604    + SUM(Z1,  SUM(CE$TECHC(Z1,CE),      EXPORT(Z1,CE)*EXPORTP(CE))      )
3605    + SUM(Z1,  SUM(CN$TECHC(Z1,CN),  ALPHA(CN,Z1)*NATN(CN,Z1)
3606                          + BETAF*BETA(CN,Z1)*SQR(NATN(CN,Z1))  )   ) ;
3607
3608  OBJN..
3609    CPS  =E=
3610     SUM(Z1, SUM(G$EX(Z1,G), SUM(NCN, FSALEP(NCN)*PRODT(Z1,G,NCN)  )
3611                            - ACOST(Z1,G) - SUM(SEA, ARTFOD(Z1,G,SEA))*PAFOD
3612                            - SUM(M, ARTWATER(Z1,G,M))*PAWAT
3613                            - SUM(CQ$TECHC(Z1,CQ), PRODA(Z1,G,CQ )*BIG)) )
3614    - SUM(Z1,  SUM(CM$TECHC(Z1,CM),      IMPORT(Z1,CM) *IMPORTP(CM))     )
3615    + SUM(Z1,  SUM(CE$TECHC(Z1,CE),      EXPORT(Z1,CE) *EXPORTP(CE))     )
3616    + SUM(Z1,  SUM( (CN,P)$TECHC(Z1,CN), NAT(CN,Z1,P)*WS(CN,Z1,P))       )
3617    + SUM((N,M), -ARTWATERND(N,M)*PAWAT + RVAL(N)*RCONT(N,M)$RCAP(N)
3618                          + RVAL("A-SEA")*F("A-SEA","KOTRI-B",M) ) ;
3619
3620  OBJNN..
3621    CPS  =E=
3622     SUM(Z1, SUM(G$EX(Z1,G), SUM(NCN, FSALEP(NCN)*PRODT(Z1,G,NCN)  )
3623                            - ACOST(Z1,G) - SUM(SEA, ARTFOD(Z1,G,SEA))*PAFOD
3624                            - SUM(M, ARTWATER(Z1,G,M))*PAWAT
3625                            - SUM(CQ$TECHC(Z1,CQ), PRODA(Z1,G,CQ )*BIG)) )
3626    - SUM(Z1,  SUM(CM$TECHC(Z1,CM),      IMPORT(Z1,CM)*IMPORTP(CM))      )
3627    + SUM(Z1,  SUM(CE$TECHC(Z1,CE),      EXPORT(Z1,CE)*EXPORTP(CE))      )
3628    + SUM(Z1,  SUM(CN$TECHC(Z1,CN),  ALPHA(CN,Z1)*NATN(CN,Z1)
3629                          + BETAF*BETA(CN,Z1)*SQR(NATN(CN,Z1))  )        )
3630    + SUM((N,M), -ARTWATERND(N,M)*PAWAT + RVAL(N)*RCONT(N,M)$RCAP(N)
3631                          + RVAL("A-SEA")*F("A-SEA","KOTRI-B",M) ) ;
3632
3633
3634     COST(Z1,G)$EX(Z1,G)..    ACOST(Z1,G) =E= (SUM((C,T,S,W)$TECH(Z1,C,T,S,W), (SUM(P2, FERT(P2,C,Z1)*MISC(P2))+
3635                                             MISCCT(C)+SEEDP(C)*SYLDS(C,Z1,T,S,W,"SEED") )*X(Z1,G,C,T,S,W) )+
3636                                        SUM(M, MISC("TWOPC")*TW(Z1,M)$GF(G) + MISC("TROPC")*TS(Z1,G,M) ) +
3637                                        MISC("TWINVT")*ITW(Z1)$GF(G)  + MISC("TRINVT")*ITR(Z1,G) +
3638                                        SUM(A, IOLIVE(A,Z1,"FIX-COST")*ANIMAL(Z1,G,A))  )/1000+
3639                                        SUM(SEA, PP*PPC(Z1,G,SEA)) +
3640                                        SUM(M, (FAMILYL(Z1,G,M)*LABFAC + HIREDL(Z1,G,M))*WAGE(M) ) ;
3641     CONV(CN,Z1)$TECHC(Z1,CN)..    SUM(P, NAT(CN,Z1,P)) =L= 1;
3642
3643     DEMNAT(CQ,Z1 )$TECHC(Z1,CQ)..
3644        SUM(G$EX(Z1,G),
3645              PRODT(Z1,G,CQ) - CONSUMP(Z1,G,CQ)$CC(CQ)+PRODA(Z1,G,CQ) )-
3646        EXPORT(Z1,CQ)$CE(CQ) + IMPORT(Z1,CQ)$CM(CQ)
3647                    =G= SUM(P, NAT(CQ,Z1,P)*QS(CQ,Z1 ,P))$CN(CQ);
```

```
     DEMNATN(CQ,Z1 )$TECHC(Z1,CQ)..
        SUM(G$EX(Z1,G),
               PRODT(Z1,G,CQ) - CONSUMP(Z1,G,CQ)$CC(CQ)+PRODA(Z1,G,CQ) )-
        EXPORT(Z1,CQ)$CE(CQ) + IMPORT(Z1,CQ)$CM(CQ)
                              =G= NATN(CQ,Z1)$CN(CQ) ;

     CCOMBAL(Z1,G,C)$(CNF(C)$EX(Z1,G)).. SUM((T,S,W)$TECH(Z1,C,T,S,W), YIELD(C,T,S,W,Z1)*X(Z1,G,C,T,S,W))=E= PRODT(Z1,G,C);

     QCOMBAL(Z1,G,Q)$EX(Z1,G)..          SUM(A, IOLIVE(A,Z1,Q)*ANIMAL(Z1,G,A) ) /1000            =E= PRODT(Z1,G,Q);

     CONSBAL(Z1,G,CC)$(EX(Z1,G)$TECHC(Z1,CC) )..
               PRODT(Z1,G,CC) + PRODA(Z1,G,CC) =G= CONSUMP(Z1,G,CC) ;

     LABORC(Z1,G,M)$EX(Z1,G)..   (SUM( (C,T,S,W)$TECH(Z1,C,T,S,W), LABOR(C,Z1,T,S,W,M)*X(Z1,G,C,T,S,W) ) +
                                   SUM( A,  IOLIVE(A,Z1,"LABOR")*ANIMAL(Z1,G,A)))/1000 =L= FAMILYL(Z1,G,M)+HIREDL(Z1,G,M);

     FODDER(Z1,G,SEA)$EX(Z1,G)..  SUM(A, IOLIVE(A,Z1,"TDN")*ANIMAL(Z1,G,A) ) =L= SUM( (C,T,S,W)$TECH(Z1,C,T,S,W), (
              YIELD(C,T,S,W,Z1)*SYLDS(C,Z1,T,S,W,"STRAW-YLD")*SCONV("TDN",SEA,C) + WEEDY(Z1,SEA,C)*
              SCONV("TDN","RABI","RAB-FOD")    ) * X(Z1,G,C,T,S,W)  ) +
              SUM(M$SEAM(SEA,M), SLKLAND(Z1,G,M)*GRAZ(Z1,SEA)*SCONV("TDN","RABI","RAB-FOD") +
              ARTFOD(Z1,G,SEA)*SCONV("TDN","RABI","RAB-FOD");

     PROTEIN(Z1,G,SEA)$EX(Z1,G).. SUM(A, IOLIVE(A,Z1,"DP")*ANIMAL(Z1,G,A) ) =L= PPC(Z1,G,SEA)+
              SUM( (C,T,S,W)$TECH(Z1,C,T,S,W), (YIELD(C,T,S,W,Z1)*SYLDS(C,Z1,T,S,W,"STRAW-YLD")*
              SCONV("DP",SEA,C) + WEEDY(Z1,SEA,C)*SCONV("DP","RABI","RAB-FOD") )*X(Z1,G,C,T,S,W) )+
              SUM(M$SEAM(SEA,M), SLKLAND(Z1,G,M)*GRAZ(Z1,SEA)*SCONV("DP","RABI","RAB-FOD") +
              ARTFOD(Z1,G,SEA)*SCONV("DP","RABI","RAB-FOD");

     GRNFDR(Z1,G,SEA)$EX(Z1,G)..  GR*SUM(A, IOLIVE(A,Z1,"TDN")*ANIMAL(Z1,G,A) ) =L=  SUM( (CF,T,S,W)$TECH(Z1,CF,T,S,W),
              YIELD(CF,T,S,W,Z1)*SCONV("TDN",SEA,CF)*X(Z1,G,CF,T,S,W)  ) +
              SUM( (C,T,S,W)$TECH(Z1,C,T,S,W),
              WEEDY(Z1,SEA,C)*SCONV("TDN","RABI","RAB-FOD")*X(Z1,G,C,T,S,W)  ) +
              ARTFOD(Z1,G,SEA)*SCONV("TDN","RABI","RAB-FOD");

     BDRAFT(Z1,G,M)$EX(Z1,G)..   SUM((C,T,S,W)$TECH(Z1,C,T,S,W), BULLOCK(C,Z1,T,S,W,M)*X(Z1,G,C,T,S,W) )/1000 =L=
                                                                BP(M)*ANIMAL(Z1,G,"BULLOCK")/1000 ;

     BREPCO(Z1,G)$EX(Z1,G)..     ANIMAL(Z1,G,"BULLOCK")   =L=   REPCO*ANIMAL(Z1,G,"COW")   ;

     BULLOCKC(Z1)..              SUM(G$EX(Z1,G), ANIMAL(Z1,G,"BULLOCK") ) =L= RES88("BULLOCKS",Z1)  ;

     TDRAFT(Z1,G,M)$EX(Z1,G)..   SUM((C,T,S,W)$TECH(Z1,C,T,S,W), TRACTOR(C,Z1,T,S,W,M)*X(Z1,G,C,T,S,W) ) =E= TS(Z1,G,M) ;

     TRCAPC(Z1,M)..   SUM(G$EX(Z1,G), TS(Z1,G,M) ) =L= SUM(G$EX(Z1,G),
                       (RESOURCE(Z1,G,"TRACTORS")/1000 +ITR(Z1,G))*TRCAP )  ;

     TWCAPC(Z1,M)$EX(Z1,"FRESH")..

         TW(Z1,M) =L=  RESOURCE(Z1,"FRESH","TWC") + NTWUCAP*ITW(Z1) ;

     LANDC(Z1,G,M)$EX(Z1,G)..    SUM((C,T,S,W)$TECH(Z1,C,T,S,W), LAND(C,Z1,T,S,W,M)*X(Z1,G,C,T,S,W)) + SLKLAND(Z1,G,M)=E=
                                                                                    RESOURCE(Z1,G,"CCA")*1000 ;
     ORCHAREAC(Z1)..             SUM( (G,T,S,W)$EX(Z1,G), X(Z1,G,"ORCHARD",T,S,W)$TECH(Z1,"ORCHARD",T,S,W))=L=ORCHAREA(Z1);

     SCMILLC(Z1)..   SUM(G$EX(Z1,G),  PRODT(Z1,G,"SC-MILL") ) =L= SCMILLCAP(Z1)   ;


     WATERBALN(Z1,G,M)$EX(Z1,G)..
        SUM((C,T,S,W)$TECH(Z1,C,T,S,W),
          MAX( (WNR(C,Z1,T,S,W,M)-SUBIRRZ1(Z1,G,M)*LAND(C,Z1,T,S,W,M)) , 0.0)*
            X(Z1,G,C,T,S,W))+ SLKWATER(Z1,G,M) =E=
            TWEFFZ(Z1,M)*TW(Z1,M)$GF(G)  + GWT1(Z1,G,M) +
            ARTWATER(Z1,G,M) + WDIVRZ(Z1,G,M) ;
```

```
INDUS BASIN MODEL REVISED(IBMR) FILENAME=WSISM41                    07/25/91 17:18:07
EQUATIONS AND VARIABLES                                             GAMS 2.21 IBM CMS

3713    WATALCZ(Z1,G,M)$EX(Z1,G)..   WDIVRZ(Z1,G,M) =E= SUM( (CNL,SA)$(ZSA(Z1,CNL,SA)$GWFG(CNL,SA,G) ),
3714                                                 (CNEFF(CNL)+(1-CNEFF(CNL))*CNLLF$GS(G))
3715                                                 *(WCEFF(CNL,M)+(1-WCEFF(CNL,M))*WCLF$GS(G))
3716                                                 *CANALDIV(CNL,M)*SUBDEF(SA,CNL)*1000 ) ;
3717
3718    DIVCNLSEA(CNL,SEA)..
3719       CNLDIVSEA(CNL,SEA) =E= SUM(M$SEAM(SEA,M), CANALDIV(CNL,M));
3720    PRSEAW(PV2,SEA)..
3721       PRSEA(PV2,SEA)     =E= SUM(CNL1$PVCNL(PV2,CNL1),
3722                                    CNLDIVSEA(CNL1,SEA)) ;
3723    DIVSEA(SEA)..
3724       TCDIVSEA(SEA)      =E= SUM(PV2,  PRSEA(PV2,SEA) )                 ;
3725    WATALCSEA(CNL1,SEA)..
3726       PROTARB(CNL1,SEA)*(1-TOLCNL)*TCDIVSEA(SEA)    =L=
3727                                    CNLDIVSEA(CNL1,SEA)                  ;
3728    WATALCPRO(PV2,SEA)..
3729       PROTARB(PV2,SEA)*(1-TOLPR)*TCDIVSEA(SEA)      =L=
3730                                    PRSEA(PV2,SEA)                       ;
3731
3732    NWFPALC(M)..  SUM(CNL$PVCNL("NWFP",CNL), CANALDIV(CNL,M))=G=
3733                                      (1-TOLNWFP)*DIVNWFP(M) ;
3734
3735   SUBIRRC(Z1,G,M)$(EX(Z1,G)$GS(G))..  WDIVRZ(Z1,G,M) =G=
3736                   (1-SUBIRRFAC(Z1) )*SUM((C,T,S,W)$TECH(Z1,C,T,S,W), WNR(C,Z1,T,S,W,M)*X(Z1,G,C,T,S,W)) ;
3737
3738   NBAL(N,M)$NB(N)..
3739     SUM(I$NI(N,I), INFLOW(I,M)) +
3740     SUM(N1, RIVERCD(N,"D")*TRIB(N1,N,M)+RIVERCD(N,"C")*TRIB(N1,N,M--1) )+
3741     SUM(N1$NN(N,N1),
3742         F(N,N1,M)*LCEFF(N1,N)$LCEFF(N1,N) +
3743         (RIVERB(N,N1)*F(N,N1,M)+
3744                 RIVERCD(N,"C")*F(N,N1,M--1)  )$(LCEFF(N1,N) EQ 0) )-
3745     SUM(N1$NN(N1,N),
3746         F(N1,N,M)) + ( RCONT(N,M--1) - RCONT(N,M) - REVAPL(N,M)/1000)$RCAP(N) -
3747     SUM(CNL$NC(N,CNL), CANALDIV(CNL,M)) + ARTWATERND(N,M) =E= 0 ;
3748
3749  *- BOUNDS FOR THE NETWORK LINK CANAL CAPACITIES.
3750     ARTWATERND.LO(N,M) = 0   ;
3751     F.UP(N,N1,M)        = INF ;
3752     F.UP(N,N1,M)$( NCAP(N1,N) NE 0 ) = NCAP(N1,N);
3753
3754  *-  BOUNDS ON FAMILY LABOR
3755     FAMILYL.UP(Z1,G,M) =  RESOURCE(Z1,G,"FARMPOP")*LSTD/1000 ;
3756
3757  * CONSUMPTION BOUNDS
3758     CONSUMP.FX(Z1,G,CQ) = FARMCONS(Z1,CQ)*CONSRATIO(Z1,G) ;
3759     EXPORT.UP(Z1,CE)    = EXPLIMIT(Z1,CE) ;
3760     ITR.FX(Z1,G)=0; ITW.FX(Z1)  =0;
3761
3762  *-
3763  MODEL WSISZ AGROCLIMATIC ZONES MODEL LINEAR OBJECTIVE /
3764         OBJZ,     COST,     CONV,     DEMNAT,   CCOMBAL,  QCOMBAL,
3765         CONSBAL,  LABORC,   FODDER,   PROTEIN,  GRNFDR,   BDRAFT,
3766         BREPCO,   TDRAFT,   TRCAPC,   TWCAPC,   LANDC,    ORCHAREAC,
3767         SCMILLC,  WATERBALN, WATALCZ,  SUBIRRC                    /
3768
3769  MODEL WSISZN AGROCLIMATIC ZONES MODEL NON-LINEAR OBJECTIVE /
3770         OBJZN,    COST,               DEMNATN,  CCOMBAL,  QCOMBAL,
3771         CONSBAL,  LABORC,   FODDER,   PROTEIN,  GRNFDR,   BDRAFT,
3772         BREPCO,   TDRAFT,   TRCAPC,   TWCAPC,   LANDC,    ORCHAREAC,
3773         SCMILLC,  WATERBALN, WATALCZ,  SUBIRRC                    /
3774  MODEL WSISN  IBMR MODEL WITH WATER NETWORK  LINEAR /
3775         OBJN,     COST,     CONV,     DEMNAT,   CCOMBAL,  QCOMBAL,
3776         CONSBAL,  LABORC,   FODDER,   PROTEIN,  GRNFDR,   BDRAFT,
3777         BREPCO,   TDRAFT,   TRCAPC,   TWCAPC,   LANDC,    ORCHAREAC,
```

```
3778                     WATERBALN,  WATALCZ,    SUBIRRC,
3779         DIVCNLSEA, PRSEAW,    DIVSEA,     WATALCSEA, WATALCPRO, NWFPALC,
3780         NBAL /
3781
3782   MODEL WSISNN IBMR MODEL WITH WATER NETWORK  NON-LINEAR /
3783         OBJNN,    COST,                  DEMNATN,   CCOMBAL,   QCOMBAL,
3784         CONSBAL,  LABORC,    FODDER,     PROTEIN,   GRNFDR,    BDRAFT,
3785         BREPCO,   TDRAFT,    TRCAPC,     TWCAPC,    LANDC,     ORCHAREAC,
3786                   WATERBALN, WATALCZ,    SUBIRRC,
3787         DIVCNLSEA, PRSEAW,   DIVSEA,     WATALCSEA, WATALCPRO, NWFPALC,
3788         NBAL /
3789
3790         ;
3791     OPTION ITERLIM = 25000;  OPTION RESLIM = 900 ;
3792     OPTION LIMROW=0; OPTION LIMCOL=0 ;
3793     OPTION LP=MPSX ;
3794
3795   *- RIVER FLOW TESTS
3796     TRIB("CHASMA-R","TAUNSA-B",M)   = 0 ;
3797     TRIB("TARBELA-R","KALABAGH-R",M) = 0 ;
3798     INFLOW("HARO",M) = 0;   INFLOW("SOAN",M) = 0 ;
3799
3800   *
3801   * YEAR 2000 RUN
3802   * RESTART FROM WSISD41
3803   * DIVERSIONS PROPORTIONAL TO POST TARBELA AVERAGE (SHARES BY SEASON)
3804   * INVESTMENTS ARE ALLOWED
3805
3806   * NOTE: INFLOWS FROM RAVI AND SUTLEJ RIVER ARE SET TO ZERO HERE.
3807     INFLOW("RAVI",M) = 0 ;   INFLOW("SUTLEJ",M) = 0 ;
3808
3809   *-
3810   * IRRIGATION CANAL CAPICITY BOUNDS
3811     CANALDIV.UP(CNL,M) = SUM(ISR, COMDEF(ISR,"CCAP",CNL) ) ;
3812
3813   * RESERVOIR OPERATING RULE BOUNDS
3814   *
3815     RCAP("KALABAGH-R") = 0 ;
3816     RCONT.LO(N,M) = RULELO(N,M)*RCAP(N)/100;
3817     RCONT.UP(N,M) = RULEUP(N,M)*RCAP(N)/100 ;
3818     ITR.UP(Z1,G)= INF ;
3819     ITW.UP(Z1)  = INF ;
3820     EXPLIMIT(Z1,"WHEAT") = INF ;
3821     EXPORT.UP(Z1,CE) = EXPLIMIT(Z1,CE) ;
3822
3823     CANALDIV.UP(CNL,M) = SUM(ISR, COMDEF(ISR,"CCAP",CNL) ) ;
3824
3825     TOLNWFP= 0 ;
3826     TOLCNL = 0 ;
3827     TOLPR  = 0 ;
3828
3829     SOLVE WSISN  MAXIMIZING CPS USING  LP ;
3830
3831   * END OF PROGRAM
```

**** FILE SUMMARY FOR USER KDAM

```
RESTART   D41    WORK*     A
INPUT     M41    GAMS      A
OUTPUT    M41    LISTING   A
SAVE      M41    WORK*     A

COMPILATION TIME    =      1.360 SECONDS      VER: IBM-TB-003
```

Appendix A.3:

Report Program

```
3834  * $OFFSYMLIST  OFFSYMXREF
3835  ** CONTINUED FROM WSISM* GAMS **
3836
3837  SETS
3838     C2 /RABI, KHARIF, ANNUAL /
3839     C3
3840     SEAC(SEA,C) SEASON TO CROP MAP /
3841         RABI.  (MUS+RAP, RAB-FOD, GRAM,  WHEAT, POTATOES,
3842                ONIONS,SC-MILL, SC-GUR,ORCHARD)
3843         KHARIF.(BASMATI, IRRI,MAIZE, KHA-FOD, COTTON,
3844                SC-MILL, SC-GUR,ORCHARD) /
3845     RR1     /GPV,VAL-ADDED, FARM-INC,FL-COST,HL-COST,SEED-COST,
3846             FERT-COST,MISCC-COST,TW-OPC,TR-OPC,TW-INVT,TR-INVT
3847             ANIML-COST,PROT-COST, TOTAL-COST/
3848     R9      /SEEP-RAIN, SEEP-PTW, SEEP-GTW, SEEP-CANAL,
3849             SEEP-WCFLD,SEEP-LINK,SEEP-RIVER,P-TUBEWELL,G-TUBEWELL
3850             TOT-INF, TOT-OUTF, INF-OUTF, P-EVAP-GW, EVAP-GW,BALANCE /
3851     R2(RR1) /FL-COST,HL-COST,SEED-COST,FERT-COST,MISCC-COST,
3852             TW-OPC,TR-OPC,TW-INVT,TR-INVT,ANIML-COST,PROT-COST,
3853             TOTAL-COST/
3854     R17(RR1) /HL-COST,SEED-COST,FERT-COST,TW-OPC,TR-OPC,ANIML-COST,
3855              PROT-COST  /
3856     R18(RR1) /HL-COST,SEED-COST,FERT-COST,MISCC-COST,TW-OPC,TR-OPC,
3857              ANIML-COST,PROT-COST  /
3858     R3      /WATER-REQ, RAIN, SUBIRR, CANAL, P-TUBEWELL, G-TUBEWELL,
3859             ARTWATER, SLKWATER, TOT-SUPPLY /
3860     R4(R3)  /CANAL, RAIN, SUBIRR, P-TUBEWELL, G-TUBEWELL, ARTWATER /
3861     R6(R9)  /SEEP-RAIN, SEEP-PTW, SEEP-GTW, SEEP-CANAL, SEEP-WCFLD,
3862             SEEP-LINK,SEEP-RIVER /
3863     R7(R9)  /P-TUBEWELL, G-TUBEWELL   /
3864     R8(RR1) /GPV, VAL-ADDED, FARM-INC, FL-COST, HL-COST, TOTAL-COST /
3865     R10     /M-NWFP,A-NWFP,M-PMW,A-PMW,M-PCW,A-PCW,M-PSW,A-PSW,M-PRW,A-PRW
3866             M-SCWN,A-SCWN,M-SRWN,A-SRWN,M-SCWS,A-SCWS,M-SRWS,A-SRWS /
3867     R11(R10) /M-NWFP,M-PMW,M-PCW,M-PSW,M-PRW,M-SCWN,M-SCWS,M-SRWN,M-SRWS /
3868     R12(R10) /A-NWFP,A-PMW,A-PCW,A-PSW,A-PRW,A-SCWN,A-SCWS,A-SRWN,A-SRWS /
3869     R11Z(R11,Z) /M-NWFP.NWFP, M-PMW.PMW,   M-PCW.PCW,   M-PSW.PSW,   M-PRW.PRW
3870             M-SCWN.SCWN, M-SCWS.SCWS, M-SRWN.SRWN, M-SRWS.SRWS /
3871     R12Z(R12,Z) /A-NWFP.NWFP, A-PMW.PMW,   A-PCW.PCW,   A-PSW.PSW,   A-PRW.PRW
3872             A-SCWN.SCWN, A-SCWS.SCWS, A-SRWN.SRWN, A-SRWS.SRWS /
3873     R13         /M-PUNJAB, A-PUNJAB/
3874     R14         /M-SIND, A-SIND/
3875     R15         /M-PAKISTAN, A-PAKISTAN /
3876     R16         /IMP-COST, AFOD-COST, ARTWATER /
3877     INT         /R-INT, K-INT, A-INT /
3878     INTC2(INT,C2) /R-INT.RABI, K-INT.KHARIF, A-INT.ANNUAL /
3879     R22(R10)    /M-SCWN,M-SCWS,M-SRWN,M-SRWS /
3880     R23(R10)    /M-NWFP,M-PMW,M-PCW,M-PSW,M-PRW /
3881     R22A(R10)   /A-SCWN,A-SCWS,A-SRWN,A-SRWS /
3882     R23A(R10)   /A-NWFP,A-PMW,A-PCW,A-PSW,A-PRW /
3883     R24
3884     R25      /NWFP, MARALA, MANGLA, U-INDUS, L-INDUS/
3885     R26(R25,CNL) /MARALA. (01-UD, 02-CBD,03-RAY,04-UC ,05-MR)
3886                   MANGLA. (06-SAD,07-FOR,08-PAK,09-LD ,10-LBD
3887                          11-JHA,12-GUG,13-UJ ,14-LJ ,15-BAH)
3888                   U-INDUS.(16-MAI,17-SID,18-HAV,19-RAN,20-PAN
3889                          21-ABB,26-THA,27-PAH,28-MUZ,29-DGK)/
3890              ;
3891     C3(CQ)  = YES;  C3(C2) = YES ;
3892     R24(R10) = YES; R24(R13)=YES; R24(R14)= YES; R24(R15)=YES;
3893     R26("NWFP",CNL)   = PVCNL("NWFP",CNL) ;
3894     R26("L-INDUS",CNL) = PVCNL("SIND",CNL) ;
3895
3896     SCALAR
```

```
3897        SPFC PORTION OF CANAL LOSSES TO GROUNDWATER        / 0.7 /
3898        SPFWF PORTION OF WATERCOURSE AND FIELD LOSSES TO GW / 0.8 /
3899        SPFRV PORTION OF RIVER SEE
3900            ;
3901
3902        TABLE LNKSP(N1,N,Z) PROPORTION OF SEE
3903
3904                              PMW   PCW   PSW   PRW
3905
3906    CHASMA-R.     TRIMMU-B    .6
3907    TAUNSA-B.     PANJNAD-B          1
3908
3909    A1.           KHANKI-B                .6
3910    RASUL-B.      QADIRA-B                 1
3911
3912    MARALA-B.     A3                             .2
3913    A3.           A4                              1
3914    A2.           BALLOKI-B                       1
3915    A2.           A4                              1
3916    A4.           A5                 1
3917    KHANKI-B.     A7                              1
3918    QADIRA-B.     A6                              1
3919    A6.           BALLOKI-B                       1
3920    TRIMMU-B.     SIDHNAI-B               1
3921
3922    BALLOKI-B.    SULEM-B            1
3923    SIDHNAI-B.    A8                 1
3924    A8.           A9                 1
3925
3926       TABLE RIVSP(N1,N,Z) PROPORTION OF RIVER SEE
3927
3928                         NWFP  PMW   PCW   PSW   PRW   SCWN  SRWN  SCWS
3929
3930    SUKKUR-B.KOTRI-B                                    .3    .2    .1
3931    GUDU-B.SUKKUR-B                                     .3    .3
3932    CHASMA-R.TAUNSA-B     .4
3933    TAUNSA-B.GUDU-B             .3
3934    PANJNAD-B.GUDU-B            .2
3935    TRIMMU-B.PANJNAD-B          .3    .3
3936    SIDHNAI-B.PANJNAD-B         .3    .4
3937    ISLAM-B.PANJNAD-B           1
3938    SULEM-B.ISLAM-B             1
3939    BALLOKI-B.SIDHNAI-B         .5    .5
3940    RAVI-I.BALLOKI-B                        1
3941    QADIRA-B.TRIMMU-B                 .5
3942    KHANKI-B.QADIRA-B                 .4
3943    RASUL-B.TRIMMU-B     .05          .5
3944    WARSAK-D.K-S-JCT      .4
3945
3946    PARAMETER
3947        CA                TOTAL CROPPED AREA (FRESH + SALINE THOUSAND ACRES)
3948        CAG               CROPPED AREA BY GROUND WATER QUALITY(000 ACRES)
3949        CAT               CROPPED AREA BY TECHNOLOGY (THOUSAND ACRES)
3950        CAC
3951        CADIFF            CROPPED AREA DIFFERENCE MODEL - ACTUAL
3952        WYC               WEIGTHED YIELD           (METRIC TONS)
3953        REP45
3954        REP46             SLACK LAND BY MONTH       (THOUSAND ACRES)
3955        REP47             LAND USED  BY MONTH       (THOUSAND ACRES)
3956        REP48             FERTLIZER NUTRIENTS USED (THOUSAND TONNS)
3957        REP49A            REPORT ON PRODUCTION(THOUSAND TONNS MILK IN MILLION LITERS)
3958        REP49B            PRODUCTION COMPARISON (UNITS AS IN REP49A)
3959        REP49C            CONSUMPTION           (UNITS AS IN REP49A)
3960        REP49D            IMPORTS AND EXPORTS   (UNITS AS IN REP49A)
3961        REP50(*,RR1,*)    REPORT ON TOTAL INCOME AND COST (FINANCIAL PRICES)  (MILLION RUPEES)
```

INDUS BASIN MODEL REVISED (IBMR) FILENAME=WSISR41
REPORT ON THE SOLUTION

```
3962        REP52(*,R8,*)         REPORT ON INCOME AND COST PER ACRE OF CCA            (RUPEES)
3963        REP54(*,R8,*)         REPORT ON INCOME AND COST PER FARM HOUSEHOLD         (RUPEES)
3964        REP56(*,RR1,*)        REPORT ON INCOME AND COST BASED ON INTERNATIONAL PRICES (MILLION RUPEES)
3965        REP57(*,R8,*)         REPORT ON INCOME AND COST PER ACRE OF CCA BASED ON INTERNATIONAL PRICES (RUPEES)
3966        REP58(*,R8,*)         REPORT ON INCOME AND COST PER HOUSEHOLD BASED ON INTERERNATIONAL PRICES (RUPEES)
3967        REP59                 REPORT ON LABOR UTILIZATION                          (MILLION MAN HOURS )
3968        REP60(*,*,R3,*)       REPORT ON WATER BALANCE AT THE ROOT ZONE             (MAF)
3969        REP70(*,*,R3,*)       REPORT ON WATER BALANCE AT THE WATERCOURSE HEAD      (MAF)
3970        REP72                 SURFACE WATER DIVERSIONS AT THE CANAL HEAD BY CANAL  (MAF)
3971        REP72A                SURFACE WATER DIVERSIONS AT THE CANAL HEAD BY REGION (MAF)
3972        REP73                 POST TARBELA DIVERSIONS  AT THE CANAL HEAD BY CANAL  (MAF)
3973        REP74                 DIFFERENCE OF CANAL DIVERSIONS BY THE MODEL AND POST TARBELA (MAF)
3974        REP75                 SURFACE WATER DIVERSION AT THE CANAL HEAD BY ZONE    (MAF)
3975        REP77                 SURFACE WATER FLOW  TO NODE N FROM NODE N1           (MAF)
3976        REP80(*,*,R9)         REPORT ON GROUNDWATER BALANCE                        (MAF)
3977        REP82                 REPORT ON GROUNDWATER BALANCE                        (MAF)
3978        REP85                 REPORT ON GROUNDWATER BALANCE PER ACRE OFF CCA       (AF )
3979        REP90
3980        REP95
3981        CHILI(*,T1,SEA)
3982        REP77A                REPORT ON RIVER LOSSES AND GAINS
3983        LINKLOSS              REPORT ON LOSSES IN THE LINK CANALS
3984        LOSSGAIN              LOSSES AND GAINS BY RIVER REACH
3985        ENDPRICE
3986             ;
3987
3988     DISPLAY PSR, Z1, ISR ;
3989     PARAMETER CNEFFG(CNL,G)     CANAL EFFICIENCY BY GW TYPE
3990               WCEFFG(CNL,G,M)   WC COMMAND EFFIENCY BY GW TYPE
3991               WCEFFZG(Z  ,G,M)  WC COMMAND EFFIENCY BY ZONE AND GW TYPE;
3992
3993      CNEFFG(CNL,G)   = CNEFF(CNL)+(1-CNEFF(CNL))*CNLLF$GS(G) ;
3994      WCEFFG(CNL,G,M) = WCEFF(CNL,M)+(1-WCEFF(CNL,M))*WCLF$GS(G);
3995      WCEFFZG(Z1,G,M) = SUM( (CNL,SA), WCEFFG(CNL,G,M)*ZWT(Z1,CNL,SA) );
3996  DISPLAY CNEFFG, WCEFFG, WCEFFZG;
3997  *-
3998    REP45(G,R1,Z)       = RESOURCE(Z,G,R1)           ;
3999    REP45("TOTAL",R1,Z) = SUM(G, REP45(G,R1,Z) )  ;
4000    REP45(T1,R1,PV)     = SUM(Z$PVZ(PV,Z), REP45(T1,R1,Z) ) ;
4001
4002    REP46(Z1,G,M) = SLKLAND.L(Z1,G,M) ; REP46(Z1,"TOTAL",M) = SUM(G, REP46(Z1,G,M) ) ;
4003    REP46(PV,T1,M) = SUM(Z1$PVZ(PV,Z1), REP46(Z1,T1,M) ) ;
4004    REP47(Z1,T1,M) = REP45(T1,"CCA",Z1)*1000 - REP46(Z1,T1,M) ;
4005    REP47(PV,T1,M) = SUM(Z1$PVZ(PV,Z1), REP47(Z1,T1,M) ) ;
4006
4007    REP48(G,P2,C,Z1)       = SUM( (T,S,W), X.L(Z1,G,C,T,S,W)*FERT(P2,C,Z1)) /1000 ;
4008    REP48("TOTAL",P2,C,Z1) = SUM(G, REP48(G,P2,C,Z1) );
4009    REP48(T1,P2,C,PV)      = SUM(Z1$PVZ(PV,Z1), REP48(T1,P2,C,Z1));
4010    REP48(T1,P2,"TOTAL",Z1)= SUM(C,            REP48(T1,P2,C,Z1));
4011    REP48(T1,P2,"TOTAL",PV)= SUM(Z1$PVZ(PV,Z1), REP48(T1,P2,"TOTAL",Z1));
4012
4013    CAT(G,T,C,Z1)        = SUM( (S,W),  X.L(Z1,G,C,T,S,W) ) ;
4014    CAT("TOTAL",T,C,Z1)  = SUM(G, CAT(G,T,C,Z1) ) ;
4015    CAT(T1,T,C,PV)       = SUM(Z1$PVZ(PV,Z1), CAT(T1,T,C,Z1) ) ;
4016
4017    CAC(T1,C,Z1)    = SUM(T,  CAT(T1,T,C,Z1) ) ;
4018    CAC(T1,C,PV)    = SUM(T,  CAT(T1,T,C,PV) ) ;
4019  *- REPORT ON CROPPED AREA
4020    CAG(G,C,R11)           =  SUM(Z1$R11Z(R11,Z1), CAC(G,C,Z1) );
4021    CAG("TOTAL",C,R11)     =  SUM(G, CAG(G,C,R11) )   ;
4022    CHILI(R22,T1,"RABI")   =  CAG(T1,"CHILLI",R22)    ;
4023    CHILI(R23,T1,"KHARIF") =  CAG(T1,"CHILLI",R23)    ;
4024    CAG(T1,SEA,R11)      = SUM(C$SEAC(SEA,C), CAG(T1,C,R11)) +CHILI(R11,T1,SEA);
4025    CAG(T1,"ANNUAL",R11) = SUM(SEA, CAG(T1,SEA,R11) ) ;
4026
```

INDUS BASIN MODEL REVISED (IBMR) FILENAME=WSISR41
REPORT ON THE SOLUTION

```
4027  *- TOTAL CROPPED AREA REPORT AND COMPARISION WITH ACTUAL
4028     CA(C3 ,R11)  = CAG("TOTAL",C3,R11) ;
4029
4030
4031     CA(C  ,R12)  = SUM(Z$R12Z(R12,Z), CROPAREA(Z,C ) );
4032     CA(SEA,R12)  = SUM(C$SEAC(SEA,C),  CA(C,R12) )           ;
4033     CA("RABI",R22A )  = CA("RABI",R22A)   + CA("CHILLI",R22A);
4034     CA("KHARIF",R23A) = CA("KHARIF",R23A) + CA("CHILLI",R23A);
4035     CA("ANNUAL",R12) = SUM(SEA, CA(SEA,R12));
4036
4037     CA(C3,"M-PUNJAB")= SUM(R11$(SUM(Z$PVZ("PUNJAB",Z), R11Z(R11,Z)) ),  CA(C3,R11) ) ;
4038     CA(C3,"A-PUNJAB")= SUM(R12$(SUM(Z$PVZ("PUNJAB",Z), R12Z(R12,Z)) ),  CA(C3,R12) ) ;
4039     CA(C3,"M-SIND")  = SUM(R11$(SUM(Z$PVZ("SIND",Z),   R11Z(R11,Z)) ),  CA(C3,R11) ) ;
4040     CA(C3,"A-SIND")  = SUM(R12$(SUM(Z$PVZ("SIND",Z),   R12Z(R12,Z)) ),  CA(C3,R12) ) ;
4041     CA(C3,"M-PAKISTAN") = SUM(R11, CA(C3,R11) ) ;
4042     CA(C3,"A-PAKISTAN") = SUM(R12, CA(C3,R12) ) ;
4043
4044     CA(INT,R11)$(SUM(Z$R11Z(R11,Z), REP45("TOTAL","CCA",Z))) = SUM(C2$INTC2(INT,C2), CA(C2,R11)  /
                                           SUM(Z$R11Z(R11,Z),
4045                                                           REP45("TOTAL","CCA",Z) )/1000*100 );
4046     CA(INT,R12)$(SUM(Z$R12Z(R12,Z), REP45("TOTAL","CCA",Z))) = SUM(C2$INTC2(INT,C2), CA(C2,R12)  /
                                           SUM(Z$R12Z(R12,Z),
4047                                                                    REP45("TOTAL","CCA",Z) )/1000*100 );
4048     CA(INT,R13) = SUM(C2$INTC2(INT,C2), CA(C2,R13)/REP45("TOTAL","CCA","PUNJAB")/1000*100);
4049     CA(INT,R14) = SUM(C2$INTC2(INT,C2), CA(C2,R14)/REP45("TOTAL","CCA","SIND")/1000*100  );
4050     CA(INT,R15) = SUM(C2$INTC2(INT,C2), CA(C2,R15)/REP45("TOTAL","CCA","PAKISTAN")/1000*100);
4051
4052     CA("SC-MILL",R24) = CA("SC-MILL",R24) + CA("SC-GUR",R24) ;
4053     CA("SC-GUR", R24) = 0 ;
4054
4055
4056  *- CROPPED AREA DIFFERENCE
4057     CADIFF(C3,Z) = SUM(R11$R11Z(R11,Z), CA(C3,R11) ) -
4058                    SUM(R12$R12Z(R12,Z), CA(C3,R12) ) ;
4059     CADIFF(C3,"PUNJAB")     = CA(C3,"M-PUNJAB")    - CA(C3,"A-PUNJAB") ;
4060     CADIFF(C3,"SIND")       = CA(C3,"M-SIND")      - CA(C3,"A-SIND") ;
4061     CADIFF(C3,"PAKISTAN")   = CA(C3,"M-PAKISTAN")  - CA(C3,"A-PAKISTAN") ;
4062
4063  *-  REPORT ON WEIGHTED YIELD
4064     WYC(C,Z1)$CAC("TOTAL",C,Z1)=SUM((T,S,W), YIELD(C,T,S,W,Z1)*SUM(G,X.L(Z1,G,C,T,S,W))/CAC("TOTAL",C,Z1));
4065     WYC(C,PV)$CAC("TOTAL",C,PV) = SUM(Z1$PVZ(PV,Z1), WYC(C,Z1)*CAC("TOTAL",C,Z1))/CAC("TOTAL",C,PV);
4066  *- REPORT ON INCOME AND COST
4067     REP49A(G,Z1,CQ)           = PRODT.L(Z1,G,CQ) ;
4068     REP49A(G,Z1,"SC-MILL") = REP49A(G,Z1,"SC-MILL") +
4069                              REP49A(G,Z1,"SC-GUR" )*12.0 ;
4070     REP49A("TOTAL",Z1,CQ) = SUM(G, REP49A(G,Z1,CQ));
4071     REP49A(T1,PV,CQ)       = SUM(Z1$PVZ(PV,Z1), REP49A(T1,Z1,CQ));
4072
4073     REP49B(CQ ,R11)  = SUM(Z1$R11Z(R11,Z1), REP49A("TOTAL",Z1,CQ)    );
4074     REP49B(CQ ,"M-PUNJAB")  =     REP49A("TOTAL","PUNJAB",CQ)     ;
4075     REP49B(CQ ,"M-SIND"  )  =     REP49A("TOTAL","SIND",CQ)       ;
4076     REP49B(CQ ,"M-PAKISTAN")=     REP49A("TOTAL","PAKISTAN",CQ)   ;
4077
4078     REP49B(CQ ,R12)  = SUM(Z1$R12Z(R12,Z1), TOTPROD(Z1,CQ ) );
4079     REP49B(CQ ,"A-PUNJAB")  = SUM(Z1$PVZ("PUNJAB",Z1)  ,TOTPROD(Z1,CQ )) ;
4080     REP49B(CQ ,"A-SIND"  )  = SUM(Z1$PVZ("SIND",Z1)    ,TOTPROD(Z1,CQ )) ;
4081     REP49B(CQ ,"A-PAKISTAN")= SUM(Z1$PVZ("PAKISTAN",Z1),TOTPROD(Z1,CQ )) ;
4082     REP49B("TOTAL",R24) = SUM(CQ , REP49B(CQ ,R24)) ;
4083  *- CONSUMPTION IMPORT EXPORT REPORT
4084     REP49C(Z1,CQ ) = SUM(G, CONSUMP.L(Z1,G,CQ )) ;
4085     REP49C(PV,CQ ) = SUM(Z1$PVZ(PV,Z1), REP49C(Z1,CQ ) );
4086
4087     REP49D("IMPORTS",Z1,CQ ) = IMPORT.L(Z1,CQ ) ;
4088     REP49D("IMPORTS",PV,CQ)  = SUM(Z1$PVZ(PV,Z1), REP49D("IMPORTS",Z1,CQ));
4089     REP49D("EXPORTS",Z1,CQ ) = EXPORT.L(Z1,CQ ) ;
```

INDUS BASIN MODEL REVISED (IBMR) FILENAME=WSISR41
REPORT ON THE SOLUTION

```
4090    REP49D("EXPORTS",PV,CQ) = SUM(Z1$PVZ(PV,Z1), REP49D("EXPORTS",Z1,CQ));
4091
4092    ENDPRICE(CN,Z1)         = SUM(P, NAT.L(CN,Z1,P)*ENDPR(CN,Z1,P)) ;
4093  * COST COMPONENTS
4094    REP50(G,"FL-COST",Z1)   = SUM(M,    FAMILYL.L(Z1,G,M)*WAGE(M) )*LABFAC *1000 ;
4095    REP50(G,"HL-COST",Z1)   = SUM(M,    HIREDL.L(Z1,G,M) *WAGE(M))*1000 ;
4096    REP50(G,"SEED-COST",Z1) = SUM( (C,T,S,W),
4097                              X.L(Z1,G,C,T,S,W)*SYLDS(C,Z1,T,S,W,"SEED")*SEEDP(C) );
4098    REP50(G,"FERT-COST",Z1) = SUM( (C,T,S,W,P2),
4099                              X.L(Z1,G,C,T,S,W)*FERT(P2,C,Z1)*MISC(P2) )     ;
4100    REP50(G,"MISCC-COST",Z1)= SUM(C,  CAC(G,C,Z1)*MISCCT(C)    )   ;
4101    REP50("FRESH","TW-OPC",Z1) = SUM(M, TW.L(Z1,M)*MISC("TWOPC") ) ;
4102    REP50("FRESH","TW-INVT",Z1)=         ITW.L(Z1)*MISC("TWINVT")       ;
4103    REP50(G,"TR-OPC",Z1)    = SUM(M, TS.L(Z1,G,M)*MISC("TROPC") )   ;
4104    REP50(G,"TR-INVT",Z1)   =         ITR.L(Z1,G)*MISC("TRINVT")    ;
4105    REP50(G,"ANIML-COST",Z1)= SUM(A, IOLIVE(A,Z1,"FIX-COST")*ANIMAL.L(Z1,G,A) ) ;
4106    REP50(G,"PROT-COST",Z1) = SUM(SEA, PP*PPC.L(Z1,G,SEA))*1000       ;
4107    REP50(G,"TOTAL-COST",Z1)= SUM(R2, REP50(G,R2,Z1) ) ;
4108  * CONVERT ALL COSTS TO MILLIONS OF RUPEES
4109    REP50(G,R2,Z1)          = REP50(G,R2,Z1)/1000        ;
4110    REP50(G,"GPV",Z1)       =
4111      SUM(CQ , FSALEP(CQ )* PRODT.L(Z1,G,CQ) ) ;
4112    REP50(G,"FARM-INC",Z1)  = REP50(G,"GPV",Z1) - SUM(R18, REP50(G,R18,Z1) ) ;
4113    REP50("TOTAL",RR1,Z1)   = SUM(G, REP50(G,RR1,Z1) ) ;
4114    REP50(T1,RR1,PV)        = SUM(Z1$PVZ(PV,Z1), REP50(T1,RR1,Z1));
4115 *- REPORT ON PER ACRE AND PER FARM HOUSEHOLD (RUPEES)
4116    REP52(T1,R8,Z1)$REP45(T1,"CCA",Z1) = REP50(T1,R8,Z1)       / REP45(T1,"CCA",Z1)       ;
4117    REP52(T1,R8,PV)$REP45(T1,"CCA",PV) = REP50(T1,R8,PV) / REP45(T1,"CCA",PV)        ;
4118    REP54(T1,R8,Z1)$REP45(T1,"CCA",Z1) = REP50(T1,R8,Z1)*1000/ REP45(T1,"FARMHH",Z1)       ;
4119    REP54(T1,R8,PV)$REP45(T1,"FARMHH",PV) = REP50(T1,R8,PV) *1000/ REP45(T1,"FARMHH",PV)    ;
4120 *- REPORT ON INCOME AT COST USING INTRNATIONAL PRICES
4121  * COST COMPONENTS
4122    REP56(G,"FL-COST",Z1)   =  SUM(M,  FAMILYL.L(Z1,G,M)*EWAGE(M) ) *LABFAC ;
4123    REP56(G,"HL-COST",Z1)   =  SUM(M,  HIREDL.L(Z1,G,M) *EWAGE(M) )  ;
4124    REP56(G,"SEED-COST",Z1) = SUM( (C,T,S,W),
4125          X.L(Z1,G,C,T,S,W)*SYLDS(C,Z1,T,S,W,"SEED")*ESEEDP(C) )/1000;
4126    REP56(G,"FERT-COST",Z1) = SUM( (C,T,S,W,P2),
4127          X.L(Z1,G,C,T,S,W)*FERT(P2,C,Z1)*EMISC(P2) )/1000 ;
4128    REP50(G,"MISCC-COST",Z1)= SUM(C,  CAC(G,C,Z1)*EMISCCT(C)     )   ;
4129    REP56(G,"TW-OPC",Z1)    = REP50(G,"TW-OPC",Z1)*EMISC("TWOPC")/MISC("TWOPC")       ;
4130    REP56(G,"TW-INVT",Z1)   = REP50(G,"TW-INVT",Z1)*EMISC("TWINVT")/MISC("TWINVT")    ;
4131    REP56(G,"TR-OPC",Z1)    = REP50(G,"TR-OPC",Z1)*EMISC("TROPC")/MISC("TROPC")       ;
4132    REP56(G,"TR-INVT",Z1)   = REP50(G,"TR-INVT",Z1)*EMISC("TRINVT")/MISC("TRINVT")    ;
4133    REP56(G,"ANIML-COST",Z1)= REP50(G,"ANIML-COST",Z1) ;
4134    REP56(G,"PROT-COST",Z1) = REP50(G,"PROT-COST",Z1)*EPP/PP ;
4135    REP56(G,"TOTAL-COST",Z1)= SUM(R2, REP56(G,R2,Z1) ) ;
4136    REP56(G,"GPV",Z1)       =
4137      SUM(CQ , ESALEP(CQ )* PRODT.L(Z1,G,CQ) ) ;
4138    REP56(G,"VAL-ADDED",Z1) = REP56(G,"GPV",Z1)- SUM(R17, REP56(G,R17,Z1)) ;
4139    REP56("TOTAL",RR1,Z1)   = SUM(G, REP56(G,RR1,Z1) ) ;
4140    REP56(T1,RR1,PV)        = SUM(Z1$PVZ(PV,Z1), REP56(T1,RR1,Z1));
4141 *- REPORT ON RS PER ACRE AND PER HOUSEHOLD  WITH INTERNATIONAL PRICES
4142    REP57(T1,R8,Z1)$REP45(T1,"CCA",Z1) = REP56(T1,R8,Z1)       / REP45(T1,"CCA",Z1)       ;
4143    REP57(T1,R8,PV)$REP45(T1,"CCA",PV) = REP56(T1,R8,PV)       / REP45(T1,"CCA",PV)        ;
4144    REP58(T1,R8,Z1)$REP45(T1,"CCA",Z1) = REP56(T1,R8,Z1)*1000/ REP45(T1,"FARMHH",Z1)        ;
4145    REP58(T1,R8,PV)$REP45(T1,"FARMHH",PV) = REP56(T1,R8,PV)     *1000/ REP45(T1,"FARMHH",PV)      ;
4146
4147 *- REPORT ON LABOR USAGE
4148    REP59(G,M,"FAMILY-LAB",Z1) = FAMILYL.L(Z1,G,M) ;   REP59(G,M,"HIRED-LAB",Z1) = HIREDL.L(Z1,G,M) ;
4149    REP59("TOTAL",M,"FAMILY-LAB",Z1)      = SUM(G, REP59(G,M,"FAMILY-LAB",Z1) )    ;
4150    REP59(T1,SEA,"FAMILY-LAB",Z1)         = SUM(M$SEAM(SEA,M), REP59(T1,M,"FAMILY-LAB",Z1) ) ;
4151    REP59(T1,"ANNUAL","FAMILY-LAB",Z1)    = SUM(M, REP59(T1,M,"FAMILY-LAB",Z1) )  ;
4152    REP59(T1,M1,"FAMILY-LAB",PV)          = SUM(Z1$PVZ(PV,Z1), REP59(T1,M1,"FAMILY-LAB",Z1) ) ;
4153
4154    REP59("TOTAL",M,"HIRED-LAB",Z1)       = SUM(G, REP59(G,M,"HIRED-LAB",Z1) )                        ;
```

INDUS BASIN MODEL REVISED (IBMR) FILENAME=WSISR41
REPORT ON THE SOLUTION

```
4155    REP59(T1,SEA,"HIRED-LAB",Z1)       = SUM(M$SEAM(SEA,M), REP59(T1,M,"HIRED-LAB",Z1) ) ;
4156    REP59(T1,"ANNUAL","HIRED-LAB",Z1)  = SUM(M, REP59(T1,M,"HIRED-LAB",Z1) )             ;
4157    REP59(T1,M1,"HIRED-LAB",PV)        = SUM(Z1$PVZ(PV,Z1), REP59(T1,M1,"HIRED-LAB",Z1) ) ;
4158
4159  *- REPORT ON WATER BALANCE AT THE ROOT ZONE
4160    REP60(Z1,G,"WATER-REQ",M) = SUM( (C,T,S,W), X.L(Z1,G,C,T,S,W)*WATER(C,Z1,T,S,W,M) );
4161    REP60(Z1,G,"RAIN",M) = REP60(Z1,G,"WATER-REQ",M)-SUM( (C,T,S,W), X.L(Z1,G,C,T,S,W)*WNR(C,Z1,T,S,W,M));
4162    REP60(Z1,G,"SUBIRR",M)= SUM( (C,T,S,W), MIN( SUBIRRZ1(Z1,G,M)*LAND(C,Z1,T,S,W,M),  WNR(C,Z1,T,S,W,M))*
4163                                     X.L(Z1,G,C,T,S,W));
4164    REP60(Z1,G,"CANAL",M)     = SUM( (CNL,SA)$(ZSA(Z1,CNL,SA)$GWFG(CNL,SA,G) ),
4165                                        CNEFFG(CNL,G)*WCEFFG(CNL,G,M)*
4166                                        CANALDIV.L(CNL,M)*SUBDEF(SA,CNL)*1000);
4167    REP60(Z1,"FRESH","P-TUBEWELL",M) = TW.L(Z1,M)*TWEFFZ(Z1,M) ;
4168    REP60(Z1,G,"G-TUBEWELL",M) = GWT1(Z1,G,M)               ;
4169    REP60(Z1,G,"ARTWATER",M)   = ARTWATER.L(Z1,G,M)         ;
4170    REP60(Z1,G,"SLKWATER",M)   = SLKWATER.L(Z1,G,M)         ;
4171    REP60(Z1,G,"TOT-SUPPLY",M) = SUM(R4, REP60(Z1,G,R4,M) ) ;
4172    REP60(Z1,G,R3,SEA)         = SUM(M$SEAM(SEA,M), REP60(Z1,G,R3,M) )  ;
4173    REP60(Z1,G,R3,"ANNUAL")    = SUM(SEA,           REP60(Z1,G,R3,SEA)) ;
4174    REP60(Z1,G,R3,M1)          = REP60(Z1,G,R3,M1)/1000                 ;
4175    REP60(Z1,"TOTAL",R3,M1)    = SUM(G, REP60(Z1,G,R3,M1) )             ;
4176    REP60(PV,T1,R3,M1)         = SUM(Z1$PVZ(PV,Z1), REP60(Z1,T1,R3,M1) ) ;
4177  *- REPORT ON WATER BALANCE AT THE WATERCOURSE ROOT ZONE (MAF)
4178    REP70(Z1,G ,R3,M)         = REP60(Z1,G ,R3,M)/WCEFFZG(Z1,G,M)       ;
4179    REP70(Z1,"TOTAL",R3,M)    = SUM(G, REP70(Z1,G,R3,M))                ;
4180    REP70(Z1,T1,R3,SEA)       = SUM(M$SEAM(SEA,M), REP70(Z1,T1,R3,M) )  ;
4181    REP70(Z1,T1,R3,"ANNUAL")  = SUM(SEA,           REP70(Z1,T1,R3,SEA) ) ;
4182    REP70(PV,T1,R3,M1)        = SUM(Z1$PVZ(PV,Z1), REP70(Z1,T1,R3,M1) ) ;
4183  *- REPORT ON THE CANAL DIVERSIONS (MAF)
4184    REP72(CNL,M)             = CANALDIV.L(CNL,M)                     ;
4185    REP72(CNL,SEA1)          = SUM(M$SEA1M(SEA1,M), CANALDIV.L(CNL,M) ) ;
4186    REP72(PV,M1)             = SUM(CNL$PVCNL(PV,CNL), REP72(CNL,M1) ) ;
4187    REP72("PAKISTAN",M1)     = SUM(CNL, REP72(CNL,M1) ) ;
4188    REP72A(M1,R25) = SUM(CNL$R26(R25,CNL), REP72(CNL,M1));
4189
4190    REP73(CNL,M)             = DIVPOST(CNL,M);
4191    REP73(CNL,SEA1)          = SUM(M$SEA1M(SEA1,M), REP73(CNL,M) ) ;
4192    REP73(PV,M1)             = SUM(CNL$PVCNL(PV,CNL), REP73(CNL,M1) );
4193    REP73("PAKISTAN",M1)     = SUM(CNL,   REP73(CNL,M1) );
4194    REP74(CNL,C2)            = REP72(CNL,C2) - REP73(CNL,C2) ;
4195    REP74(PV,C2)             = REP72(PV,C2)  - REP73(PV,C2) ;
4196    REP75(G,Z1,M1)           = SUM((CNL,SA)$(ZSA(Z1,CNL,SA)$GWFG(CNL,SA,G)),REP72(CNL,M1)*SUBDEF(SA,CNL) );
4197    REP75("TOTAL",Z1,M1)     = SUM(G,  REP75(G,Z1,M1) )             ;
4198    REP75(T1,PV,M1)          = SUM(Z1$PVZ(PV,Z1),  REP75(T1,Z1,M1) )       ;
4199  *- REPORT ON RIVER FLOWS
4200
4201    REP77(N,N1,M)         = F.L(N,N1,M) ;
4202    REP77(N,N1,SEA1)      = SUM(M$SEA1M(SEA1,M), REP77(N,N1,M) ) ;
4203
4204
4205    LINKLOSS(N1,N,M)  =  LLOSS(N1,N)*F.L(N,N1,M) ;
4206    LINKLOSS(N1,N,SEA1) =  SUM(M$SEA1M(SEA1,M), LINKLOSS(N1,N,M));
4207    LINKLOSS("TOTAL","TOTAL",M1) = SUM( (N,N1), LINKLOSS(N1,N,M1)) ;
4208
4209    LOSSGAIN(N,N1,M)$(LLOSS(N1,N) EQ 0) = F.L(N,N1,M) +TRIB(N1,N,M) -
4210              RIVERCD(N,"D")*TRIB(N1,N,M) -
4211              RIVERB(N,N1)*F.L(N,N1,M) - RIVERCD(N,"C")*F.L(N,N1,M--1)
4212              - TRIB(N1,N,M--1)*RIVERCD(N,"C") ;
4213    LOSSGAIN(N,N1,SEA1) = SUM(M$SEA1M(SEA1,M), LOSSGAIN(N,N1,M)) ;
4214
4215    REP77A(N,M1) = SUM(N1, LOSSGAIN(N,N1,M1)) ;
4216    REP77A(N,M1) = ROUND(REP77A(N,M1),5) ;
4217    REP77A("TOTAL",M1) = SUM(N, REP77A(N,M1));
4218
4219  *-  COMPUTE POWER GENRATION FROM HYDRO PLANTS
```

```
4220  *
4221   SET RPG/AVE-RCONT,R-ELE,R-OUTFLOW, P-CAP,G-CAP,ENERGY-G, ENERGY-S /
4222       RPG1(RPG)/R-OUTFLOW, ENERGY-G, ENERGY-S /
4223
4224
4225  PARAMETER  REP79(PN,* ,RPG)
4226             TEST(V) ;
4227
4228   REP79(PN,M,"AVE-RCONT") = (RCONT.L(PN,M) + RCONT.L(PN,M--1) )/2. ;
4229   REP79(PN,M,"R-OUTFLOW") = SUM(N, F.L(N,PN,M)) ;
4230
4231  LOOP( (PN,M),
4232
4233   TEST(V) = 0 ;
4234   TEST(V) = (REP79(PN,M,"AVE-RCONT") GE POWERCHAR(PN,"R-CAP",V)) AND
4235             (REP79(PN,M,"AVE-RCONT") LT POWERCHAR(PN,"R-CAP",V+1)) ;
4236   REP79(PN,M,"R-ELE") = SUM(V$TEST(V),
4237                        (POWERCHAR(PN,"R-ELE",V)+(REP79(PN,M,"AVE-RCONT")-POWERCHAR(PN,"R-CAP",V))*
4238                         (POWERCHAR(PN,"R-ELE",V+1)-POWERCHAR(PN,"R-ELE",V))/
4239                         (POWERCHAR(PN,"R-CAP",V+1)-POWERCHAR(PN,"R-CAP",V)) ));
4240   REP79(PN,M,"P-CAP") = SUM(V$TEST(V),
4241                        (POWERCHAR(PN,"P-CAP",V)+(REP79(PN,M,"AVE-RCONT")-POWERCHAR(PN,"R-CAP",V))*
4242                         (POWERCHAR(PN,"P-CAP",V+1)-POWERCHAR(PN,"P-CAP",V))/
4243                         (POWERCHAR(PN,"R-CAP",V+1)-POWERCHAR(PN,"R-CAP",V)) ));
4244   REP79(PN,M,"G-CAP") = SUM(V$TEST(V),
4245                        (POWERCHAR(PN,"G-CAP",V)+(REP79(PN,M,"AVE-RCONT")-POWERCHAR(PN,"R-CAP",V))*
4246                         (POWERCHAR(PN,"G-CAP",V+1)-POWERCHAR(PN,"G-CAP",V))/
4247                         (POWERCHAR(PN,"R-CAP",V+1)-POWERCHAR(PN,"R-CAP",V)) ))
4248      ) ;
4249
4250  REP79(PN,M,"ENERGY-G") = REP79(PN,M,"G-CAP")*REP79(PN,M,"R-OUTFLOW")/1000 ;
4251  REP79(PN,M,"ENERGY-S") = MAX(0, REP79(PN,M,"ENERGY-G") - REP79(PN,M,"P-CAP")*24*365/12*10**(-6)) ;
4252  REP79(PN,M,"ENERGY-G") = REP79(PN,M,"ENERGY-G")-REP79(PN,M,"ENERGY-S");
4253  REP79(PN,SEA1,RPG1)    = SUM(M$SEA1M(SEA1,M), REP79(PN,M,RPG1) ) ;
4254
4255  *- REPORT ON GROUNDWATER INFLOWS AND OUTFLOWS (MAF)
4256   REP80(Z1,G,"SEEP-RAIN")  = SUM( (CNL,SA,M), ZWT(Z1,CNL,SA)*RAIN(CNL,M))
4257                             /12*RESOURCE(Z1,G,"CCA")*(1-DRC)*
                                                        (1-FLEFFZ(Z1))*SPFWF ;
4258   REP80(Z1,"FRESH","SEEP-PTW")  = SUM(M,
4259                             TW.L(Z1,M)*(1-TWEFFZ(Z1,M) ))*SPFWF/1000 ;
4260   REP80(Z1,G,"SEEP-GTW")   = SUM(M, REP70(Z1,G,"G-TUBEWELL",M)*(1-WCEFFZ(Z1,M)) )*SPFWF   ;
4261   REP80(Z1,G,"SEEP-CANAL") = SUM( (CNL,SA)$(ZSA(Z1,CNL,SA)$GWFG(CNL,SA,G) ), SUM(M, CANALDIV.L(CNL,M)62
                               (1-CNEFFG(CNL,G))*SUBDEF(SA,CNL) )*SPFC ;
4263  REP80(Z1,G,"SEEP-WCFLD")=SUM((CNL,SA)$(ZSA(Z1,CNL,SA)$GWFG(CNL,SA,G)), SUM(M,
CANALDIV.L(CNL,M)*CNEFFG(CNL,G)*(1-WCEFFG(CNL,G,M)) )*SUBDEF(SA,CNL)  ) *SPFWF ;
4265
4266   REP80(Z1,G,"SEEP-LINK")  = SUM( (N,N1),
4267                              LINKLOSS(N1,N,"ANNUAL")*LNKSP(N1,N,Z1) )
4268                                            *RATIOFS(Z1,G)*SPFC ;
4269   REP80(Z1,G,"SEEP-RIVER") = SUM( (N,N1),
4270                              LOSSGAIN(N,N1,"ANNUAL")*RIVSP(N1,N,Z1))
4271                                            *SPFRV*RATIOFS(Z1,G) ;
4272   REP80(Z1,"FRESH","P-TUBWELL") = SUM(M,  TW.L(Z1,M) )/1000 ;
4273   REP80(Z1,G,"G-TUBWELL")$GF(G) = SUM( (CNL,SA,M)$ZSA(Z1,CNL,SA), GWTSA(CNL,SA,M) )/1000 ;
4274   REP80(Z1,G,"TOT-INF")    = SUM(R6, REP80(Z1,G,R6) )                  ;
4275   REP80(Z1,G,"TOT-OUTF")   = SUM(R7, REP80(Z1,G,R7) )                  ;
4276   REP80(Z1,G,"INF-OUTF")   = REP80(Z1,G,"TOT-INF") - REP80(Z1,G,"TOT-OUTF");
4277   REP80(Z1,G,"P-EVAP-GW")  = SUM(M, EQEVAPZ1(Z1,G,M)) *RESOURCE(Z1,G,"CCA")  ;
4278   REP80(Z1,G,"EVAP-GW")    = MIN( REP80(Z1,G,"P-EVAP-GW"), REP80(Z1,G,"INF-OUTF") ) ;
4279   REP80(Z1,G,"EVAP-GW")$(REP80(Z1,G,"EVAP-GW") LT 0) =   REP80(Z1,G,"P-EVAP-GW")  ;
4280   REP80(Z1,G,"BALANCE")    = REP80(Z1,G,"INF-OUTF") -REP80(Z1,G,"EVAP-GW")       ;
4281   REP80(Z1,"TOTAL",R9)     = SUM(G,  REP80(Z1,G,R9)  )                 ;
4282  *-
4283   REP82(T1,R9,Z1)          = REP80(Z1,T1,R9)       ;
```

```
4284     REP82(T1,R9,PV)           = SUM(Z1$PVZ(PV,Z1) , REP82(T1,R9,Z1) )  ;
4285  *-
4286     REP85(T1,R9,Z1)$REP45(T1,"CCA",Z1) = REP82(T1,R9,Z1) / REP45(T1,"CCA",Z1) ;
4287     REP85(T1,R9,PV)$REP45(T1,"CCA",PV) = REP82(T1,R9,PV)/ REP45(T1,"CCA",PV) ;
4288  *-
4289   OPTION
4290      REP45:3:2:1 ,    REP46:1:2:1,   REP47:1:2:1,   CAG:1:2:1,
4291      CA:1:1:1    ,    REP48:1:2:1,   REP49A:1:2:1,  REP49B:1:1:1,
4292      CADIFF:1:1:1,    WYC:3:1:1  ,   REP50:1:2:1,
4293      REP54:1:2:1,     REP56:1:2:1,   REP58:1:2:1,   REP59:1:2:2
4294      REP60:3:2:1 ,    REP70:3:2:1,   REP72:3:1:1,   REP73:3:1:1,
4295      REP74:3:1:1 ,    REP75:3:2:1,   REP77:3:2:1,   REP77A:3:1:1,
4296      REP79:3:1:1 ,    LINKLOSS:3:2:1, REP82:3:2:1,   REP85:3:2:1;
4297
4298     OPTION EJECT ;
4299     DISPLAY "CCA     - CULTURABLE COMMANDED AREA     (MILLION ACRES)"
4300             "CCAP    - CANAL CAPACITY AT THE CANAL HEAD (MAF/MONTH)"
4301             "POP     - FARM POPULATION               (THOUSANDS) "
4302             "TRACTORS - NUMBER OF EXISTING TRACTORS           "
4303             "TWC     - EXISTNG PRIVATE TUBEWELL CAPACITY (KAF/MONTH)", REP45 ;
4304     DISPLAY CAT ,REP46,REP47 ;
4305     OPTION EJECT ;
4306     DISPLAY "CROPPED AREA(000 ACRES) ", CAG;
4307     DISPLAY "CROPPED AREA COMPARISON( AREA 000 ACRES, INTENSITIES %"
4308             "SC-MILL IS SC-MILL + SC-GUR AREAS                   ", CA;
4309     DISPLAY CADIFF ;
4310     OPTION EJECT ;
4311     DISPLAY REP48 ;
4312     OPTION EJECT ;
4313     DISPLAY "WEIGHTED CROP YIELD (METRIC TONS) ", WYC ;
4314  *-
4315     OPTION EJECT ;
4316     DISPLAY "SC-MILL IS TOTAL CANE PRODUCSTION = SC-MILL + SC-GUR*12"
4317             REP49A, REP49B, REP49C, REP49D ;
4318     OPTION EJECT ;
4319     DISPLAY "GPV       - GROSS PRODUCTION VALUE                    (MILLION RUPEES) "
4320             "VAL-ADDED - VALUE ADDED                                (MILLION RUPEES) "
4321             "FARM-INC  - FARM INCOME                                (MILLION RUPEES) "
4322             "FL-COST   - COST OF FAMILY LABOR                       (MILLION RUPEES)"
4323             "HL-COST   - COST OF HIRED LABOR                        (MILLION RUPEES)"
4324             "SEED-COST - COST OF SEED USED                          (MILLION RUPEES)"
4325             "FERT-COST - COST OF FERTILIZER(NITROGEN AND PHOSPHOROUS) USED (MILLION RUPEES) "
4326             "MISCC-COST- COST OF CANAL WATER   (MILLION RUPEES) "
4327             "TW-OPC    - OPERATING COST OF PRIVATE TUBEWELLS        (MILLION RUPEES) "
4328             "TW-INVT   - INVESTMENT IN PRIVATE TUBEWELLS            (MILLION RUPEES) "
4329             "TR-OPC    - OPERATING COST OF TRACTORS                 (MILLION RUPEES) "
4330             "TR-INVT   - INVESTMENT IN PRIVATE TUBEWELLS            (MILLION RUPEES) "
4331             "ANIML-COST- FIXED COST OF ANIMALS                      (MILLION RUPEES) "
4332             "PROT-COST - COST OF PRTEIN CONCENTRATE PURCHASED FOR ANIMALS (MILLION RUPEES)"
4333             "            EQUATION TO AVOID INFEASABILITY  "
4334             "TOTAL-COST  COST OFF ALL ACTIVITIES                   (MILLION RUPEES) ", REP50 ;
4335
4336  * DISPLAY "GPV        - GROSS PRODUCTION VALUE PER ACRE OF CCA    (RUPEES) "
4337  *         "FARM-INC   - NET FARM INCOME PER ACRE OF CCA           (RUPEES) "
4338  *         "FL-COST    - COST OF FAMILY LABOR PER ACRE OF CCA      (RUPEES) "
4339  *         "HL-COST    - COST OF HIRED  LABOR PER ACRE OF CCA      (RUPEES) "
4340  *         "TOTAL-COST   COST OFF ALL ACTIVITIES PER ACRE OF CCA   (RUPEES) ", REP52 ;
4341     OPTION EJECT ;
4342     DISPLAY "GPV        - GROSS PRODUCTION VALUE PER FARM HOUSEHOLD(RUPEES) "
4343             "FARM-INC   - NET FARM INCOME PER FARM HOUSEHOLD       (RUPEES) "
4344             "FL-COST    - COST OF FAMILY LABOR PER HOUSEHOLD       (RUPEES) "
4345             "HL-COST    - COST OF HIRED  LABOR PER HOUSEHOLD       (RUPEES) "
4346             "TOTAL-COST - TOTAL COST PER FARM HOUSEHOLD            (RUPEES) ", REP54 ;
4347  * OPTION EJECT ;
4348     DISPLAY REP56, REP57, REP58            ;
```

INDUS BASIN MODEL REVISED (IBMR) FILENAME=WSISR41
REPORT ON THE SOLUTION

```
4349      OPTION EJECT ;
4350      DISPLAY REP59 ;
4351      OPTION EJECT ;
4352      DISPLAY "WATER-REQ  - WATER REQUIREMNETS AT THE ROOT ZONE (MAF) "
4353              "SUBI+RAIN  - WATER REQUIREMNETS NET OF SUBIRR AND EFFECTIVE RAIN AT THE "
4354              "                                        THE ROOT ZONE       (MAF)"
4355              "CANAL      - CANAL WATER USED AT THE ROOT ZONE               (MAF)"
4356              "P-TUBEWELL - PRIVATE TUBEWELL PUM
4357              "G-TUBEWELL - GOVERNMENT TUBEWELL PUM
4358              "ARTWATER   - WATER FROM IMMAGINARY AT THE ROOT ZONE          (MAF)"
4359              "SLKWATER   - SLACK WATER AT THE ROOT ZONE                    (MAF)"
4360              "TOT-SUPPLY - TOTAL SUPPLY OF WATER AT THE ROOT ZONE          (MAF)", REP60 ;
4361      OPTION EJECT ;
4362      DISPLAY " ALL FIGURES ARE REPRESENTED AT THE WATER COURSE HEAD USING "
4363              " WEIGHTED DELIVERY EFFICIENCIES (MAF)         ", REP70 ;
4364      OPTION EJECT ;
4365      DISPLAY REP72, REP72A ;
4366      OPTION EJECT ;
4367      DISPLAY REP73, REP74;
4368      DISPLAY "REPORTS REP77, LINKLOSS, REP77A, RCONT.L AND REP79"
4369              " ARE NOT APPLICABLE IN CASE OF  ZONAL MODELS "
4370              REP77, LINKLOSS, REP77A;
4371      OPTION EJECT ;
4372      DISPLAY RCONT.L ;
4373      DISPLAY "AVE-RCONT AVERAGE RESERVOIR CONTENTS DURING THE MONTH (MAF)"
4374              "R-ELE     RESERVOIR ELEVETAION (FEET FROM SPD)"
4375              "R-OUTFLOW RESERVOIR OUTFLOW (MAF) "
4376              "P-CAP     POWER CAPACITY AT R-ELE MW"
4377              "G-CAP     GENERATION CAPABILITY KWH/AF"
4378              "ENERGY-G  ENERGY GENERATION BILLION KILOWATT HOURS(BKWH)"
4379              "ENERGY-S  ENERGY SPILLED (BKWH)", REP79 ;
4380      OPTION EJECT ;
4381      DISPLAY REP75  ;
4382      OPTION EJECT ;
4383      DISPLAY "SEEP-RAIN   - SEE
4384              "SEEP-PTW    - SEE
4385              "SEEP-GTW    - SEE
4386              "SEEP-CANAL  - CANAL WATER SEE
4387              "SEEP-WCFLD  - CANAL WATER SEE
4388              "              TO THE GROUDWATER             (MAF) "
4389              "SEEP-LINK   - SEE
4390              "SEEP-RIVER  - SEE
4391              "P-TUBWELL   - PRIVATE TUBEWELL PUM
4392              "G-TUBWELL   - GOVERNEMENT TUBEWELL PUM
4393              "TOT-INF     - TOTAL INFLOW TO THE GROUNDWATER FROM ALL SOURCES (RAIN,"
4394              "              PRIVATE TUBEWELLS,GOVERNMENT TUBEWELL AND CANALWATER     "
4395              "              SEE
4396              "TOT-OUTF    - TOTAL OUFLOW FROM THE GROUNDWATER (PRIVATE AND GOVT. TUBEWELL"
4397              "              PUM
4398              "INF-OUTF    - DIFFERECE IN THE INFLOW AND OUTFLOW (MAF) "
4399              "P-EVAP-GW   - POTENTIAL EVAP. FROM GW FOR THE GIVEN GW LEVEL (MAF)"
4400              "EVAP-GW     - EVAPORATION FROM GOUNDWATER   (MAF) "
4401              "BALANCE     - BALANCE AFTER THE EVAPORATION FROM GROUNDWATER(MAF)", REP82, REP85 ;
4402
4403      REP90(G,Q,Z1)            = PRODT.L(Z1,G,Q)                  ;
4404      REP90("TOTAL",Q,Z1)      = SUM(T1,        REP90(T1,Q,Z1))   ;
4405      REP90(T1,Q,PV)           = SUM(Z1$PVZ(PV,Z1), REP90(T1,Q,Z1) ) ;
4406
4407      REP95(G,A,R11)           = SUM(Z1$R11Z(R11,Z1), ANIMAL.L(Z1,G,A) ) ;
4408      REP95("TOTAL","BULLOCK",R12) = SUM(Z$R12Z(R12,Z), RES88("BULLOCKS",Z))        ;
4409      REP95("TOTAL","BUFFALO",R12) = SUM(Z$R12Z(R12,Z), RES88("BUFFALOS",Z))*BUFF ;
4410      REP95("TOTAL","COW",R12)     = SUM(Z$R12Z(R12,Z), RES88("COWS",Z))*COWF;
4411
4412      REP95("TOTAL",A,R11) = SUM(G, REP95(G,A,R11) )  ;
4413      REP95(T1,A,"M-PUNJAB")= SUM(R11$(SUM(Z$PVZ("PUNJAB",Z), R11Z(R11,Z)) ),   REP95(T1,A,R11) ) ;
```

```
4414    REP95(T1,A,"A-PUNJAB")= SUM(R12$(SUM(Z$PVZ("PUNJAB",Z), R12Z(R12,Z)) ), REP95(T1,A,R12) ) ;
4415    REP95(T1,A,"M-SIND")  = SUM(R11$(SUM(Z$PVZ("SIND",Z), R11Z(R11,Z)) ), REP95(T1,A,R11) ) ;
4416    REP95(T1,A,"A-SIND")  = SUM(R12$(SUM(Z$PVZ("SIND",Z), R12Z(R12,Z)) ), REP95(T1,A,R12) ) ;
4417    REP95(T1,A,"M-PAKISTAN") = SUM(R11, REP95(T1,A,R11) ) ;
4418    REP95(T1,A,"A-PAKISTAN") = SUM(R12, REP95(T1,A,R12) ) ;
4419
4420    OPTION   REP90:3:2:1, REP95:3:2:1;
4421    OPTION EJECT ;
4422    DISPLAY " ANIMALS ARE IN THOUSAND, MEET IN MILLION KGS AND "
4423            " MILK IN MILLION LITERS", REP95, REP90 ;
4424    DISPLAY FSALEP, PP,SEEDP, MISC, ESALEP, ESEEDP, EMISC, EPP,IMPORTP
4425            EXPORTP, EXPLIMIT ;
4426
4427    PARAMETER REP100 ARTIFICIAL WATER USED AT THE NODES (MAF)
4428              REP110 COST OF ARTIFICIAL WATER AT NODES (MILLION RUPEES)
4429              REP120 COST OF IMPORTS ARTIFICIAL WATER AND FODDER (MILLION RUPEES)
4430              ;
4431    REP100(N,M)        = ARTWATERND.L(N,M) ;
4432    REP100(N,SEA)      = SUM(M$SEAM(SEA,M), REP100(N,M)) ;
4433    REP100(N,"ANNUAL") = SUM(SEA, REP100(N,SEA) );
4434    REP100("TOTAL",M1) = SUM(N, REP100(N,M1) );
4435    REP110(N)          = REP100(N,"ANNUAL")*PAWAT;
4436    REP110("TOTAL")    = REP100("TOTAL","ANNUAL")*PAWAT ;
4437    REP120("TOTAL","IMP-COST")  = SUM(CQ ,
4438                            REP49D("IMPORTS","PAKISTAN",CQ )*IMPORTP(CQ ));
4439    REP120(G,"AFOD-COST") = SUM((Z1,SEA), ARTFOD.L(Z1,G,SEA))*PAFOD ;
4440    REP120(G,"ARTWATER")  = SUM((M,Z1), ARTWATER.L(Z1,G,M)*PAWAT   )/1000 ;
4441    OPTION   REP100:3:1:1, REP110:3;
4442    OPTION EJECT ;
4443    DISPLAY REP100, REP110, REP120;
4444
4445      DISPLAY ENDPRICE ;
4446    * SYSTEM WATER BALANCE SUMMARY
4447
4448    SET R19/RIVER-INF, TRIB-INF,  TOT-INFLOW, RES-EVAP, LINK-LOSS,
4449            RIVER-LOSS, CANAL-DIV, TO-SEA,     BALNCE /
4450        R20/RIVER-INF, TRIB-INF/
4451        R21/RES-EVAP, LINK-LOSS, RIVER-LOSS, CANAL-DIV, TO-SEA    /
4452
4453    PARAMETER REP130  REPORT ON SYSTEM INFLOWS AND OUTFLOWS(MAF)
4454              ;
4455    REP130("RIVER-INF",M) = SUM(I, INFLOW(I,M) );
4456    REP130("TRIB-INF" ,M) = SUM( (N1,N), TRIB(N1,N,M)) ;
4457    REP130("TOT-INFLOW",M) = SUM(R20, REP130(R20,M) ) ;
4458
4459    REP130("RIVER-LOSS",M) = REP77A("TOTAL",M) ;
4460    REP130("LINK-LOSS",M)  = LINKLOSS("TOTAL","TOTAL",M) ;
4461    REP130("RES-EVAP",M)   = SUM(N$RCAP(N), REVAPL(N,M) )/1000 ;
4462    REP130("CANAL-DIV",M)  = REP72("PAKISTAN",M) ;
4463    REP130("TO-SEA",M)     = F.L("A-SEA","KOTRI-B",M) ;
4464    REP130("BALNCE",M)     = REP130("TOT-INFLOW",M) -
4465                                         SUM(R21, REP130(R21,M) ) ;
4466    REP130(R19,SEA1)       = SUM(M$SEA1M(SEA1,M), REP130(R19,M)) ;
4467    OPTION   REP130:3:1:1;
4468    OPTION EJECT ;
4469    DISPLAY "THIS REPORT IS MEANINGFUL WITH THE COMPLETE INDUS MODEL ONLY";
4470    DISPLAY REP130 ;
4471    DISPLAY PRODA.L;
4472    OPTION EJECT ;
4473    DISPLAY ARTWATER.L, ARTWATERND.L ,ARTFOD.L ;
```

```
INDUS BASIN MODEL REVISED (IBMR) FILENAME=WSISR41                    07/25/91 17:18:18
REPORT ON THE SOLUTION                                               GAMS 2.21 IBM CMS


*** FILE SUMMARY FOR USER KDAM

    RESTART    M41      WORK*     A
    INPUT      R41      GAMS      A
    OUTPUT     R41      LISTING   A
    SAVE       R41      WORK*     A

    COMPILATION TIME    =        1.790 SECONDS        VER: IBM-TB-003
```

Appendix A.4:

Groundwater Simulation Program

INDUS BASIN MODEL REVISED (IBMR) GROUNDWATER MODEL FILE=WSIRG41   07/25/91 17:18:26

GAMS 2.21 IBM CMS

```
4476   * MIXVOL = SOIL VOLUME + SEE
4477     SETS Y   YEAR  /2000*2010/
4478          Y1(Y)    /2002*2005, 2008*2009/
4479          Y2(Y)    /2002*2005/
4480          X1 /1*5/
4481          D1       /DRAINED, UNDRAINED, TOTAL/
4482          D(D1)    /DRAINED, UNDRAINED/
4483          UD(D1)   /UNDRAINED/
4484          R31 /RAIN,RAIN-RUNOF, CANAL,LINK, RIVER, DRAINAGE/
4485          R34 /RAIN, CANAL, LINK,RIVER,GW-EVAP,TUBEWELL, DRAINAGE,TOTAL/
4486          R35 /WOS-GW, WS-SOIL, WS-GW, WS-TOTAL, RATIO/
4487          R36 /RAIN, RAIN-RUNOF, CANAL,LINK, RIVER, SURF-WATER, TW, EVAP, TOT-SOIL,
4488             SEEP-MAF, IMP-PPM, SOIL-VOL, APPWAT, MIX-VOL,
4489                MIX-TONS,SEEP-PPM, TO-SOIL, TO-GW, SOIL-TONS/
4490          R32 /RAIN, LINK, CNL+WCFLD, TW, EVAP, TO-RUNOFF/
4491
4492          R37 /FROM-SEEP, TO-DRAIN, TO-EVAP, TO-PUMPS, FROM-RIVER, ANN-ADD, M-TONS, MAF,PPM/
4493          R33 /RAIN, RAIN-RUNOF, CANAL, LINK, RIVER, DRAINAGE, ANN-ADD, M-TONS, MAF, PPM/
4494          R91 /GW-EVAP, DRAINAGE,OUTFLOW/
4495          R92 /TOT-INF, TOT-OUTF,DRAINAGE, GW-EVAP, I-O/
4496          R94 /IN, OUT, MOVED /
4497          IND1(Z,G)
4498          IND2(Z)
4499           ;
4500          R92(R7)= YES;
4501          R37(R6) = YES;
4502     PARAMETERS
4503       REP81    INFLOWS AND OUTFLOWS TO THE GW (MAF)
4504       REP81A   DRAINAGE EFFLUENT (MAF)
4505       REP84    ANNUAL INFLOWS AND OUTFLOWS TO GW (MAF)
4506       REP84A   ANNUAL INFLOWS AND OUTFLOWS TO GW (MAF PER ACRE)
4507       REP86(*,D1,R9,*)  ANUAL INFLOWS AND OUTFLOWS TO GW (MAF)
4508       REP86A(*,*,D1,*)   ANNUAL SURFACE WATER FLOWS TO THE ZONE (MAF)
4509       APPWAT(*,*,D1)     TOTAL WATER APPLIED TO SOIL BY ZONE (MAF)
4510       MIXVOL(*,*,D1)     VOLUME OF MIXTURE IN THESOIL BY ZONE (MAF)
4511       REP86B   GW PPM
4512       REP86E(*,*,D1,Y,*,R94)  SALT MOVED FROM SOIL TO GW (MILLION TONS)
4513       REP86F   SALT MOVED FROM SOIL TO GW  (TONS PER ACRE OF GCA)
4514       REP86G   ANNUAL SALT ADDITION TO GW   (TONS PER ACRE)
4515       REP86H   PPM SUMMARY
4516       REP86C(*,*,*,*,Y)   SOIL SALT BALANCE (MILLION TONS)
4517       REP86D(*,D1,*,Y)   SALT BALANCE OF GW WITH SOIL (M TONS)
4518       REP86F
4519       REP862(*,*,D1,*,Y)   GW SALT BALANCE WITHOUT SOIL
4520       REP862A
4521       REP864   WITH AND WITHOUT SOIL COMPRISON
4522       REP87               GW REPORT FOR SEPTEMBER
4523       REP87A              GW REPORT FOR SEPTEMBER
4524       REP87B              GW REPORT FOR SEPTEMBER (PER ACRE)
4525       REP87C              GW REPORT FOR SEPTEMBER(PER ACRE)
4526       EVAPZ(Z,G,* )       PAN EVAPORATION BY ZONE (FEET)
4527       DEPTHZ1(Z,G,M)      AVERAGE GW DEPTH IN ACZ (YEAR 1988 FEET)
4528       DEPTHZ2(Z,G,D,*,M)  SIMULATED GW DEPTH IN ACZ (FEET)
4529       GWEVAPZ(*,*,D1,Y,* ) GW EVAPORATION (MAF )
4530       RIVSALT             SALTS TO GCA BY RIVER SEE
4531       TOTSEEP             TOTAL SEE
4532       STORCOEFF(Z)        GW STORAGE COEFFICIENT /(NWFP, PMW,PSW,PCW,PRW) .25
4533                                                 (SCWN,SCWS,SRWN,SRWS)    .30/
4534       SALTPPM(Z1)         SURRFACE WATER SALT CONCENTRATION(PPM TDS)
4535                                /NWFP  150
4536                                 (PMW,PSW,PRW,PCW) 200
4537                                 (SCWN,SCWS,SRWN,SRWS) 250/
```

```
4538     GWPPM(Z,G)          GW SALT CONCENTRATION
4539
4540                           /(NWFP,PMW,PCW,PSW,PRW).FRESH   900
4541                            (PMW,PCW,PSW)          .SALINE 3000
4542                                           SCWN.FRESH 1200
4543                                           SRWN.FRESH 1500
4544                                           SCWS.FRESH 1000
4545                            (SCWN,SRWN,SCWS,SRWS) .SALINE 4000/
4546
4547     SWPPM(Z,G)          SOIL WATER SALT CONCENTRATION
4548                           /(NWFP,PMW,PCW,PSW,PRW).FRESH  1280
4549                            (PMW,PCW,PSW)          .SALINE 4800
4550                                           SCWN.FRESH 1280
4551                                           SRWN.FRESH 1280
4552                                           SCWS.FRESH 1280
4553                            (SCWN,SRWN,SCWS,SRWS) .SALINE 5620/
4554
4555     GWAQDEPTH(Z,G)  ACTIVE GW AQUIFER DEPTH (FEET)
4556                       /NWFP.FRESH 300
4557                        (PMW,PCW,PSW,PRW).FRESH    400
4558                        (PMW,PCW,PSW,PRW).SALINE   300
4559                        (SCWN,SCWS,SRWN,SRWS).FRESH  275
4560                        (SCWN,SCWS,SRWN,SRWS).SALINE 185/
4561
4562     DAREA(*,*,D1)    CCA BY GW AND  DRAINAGE TYPE (MA)
4563     GDAREA(*,*,D1)   GCA BY GW AND  DRAINAGE TYPE (MA)
4564     GAREA(*,*)       GCA BY GW TYPE (MA)
4565     GCAF(Z)          GROSS AREA FACTOR (PROPORTION OF CCA)
4566                        /NWFP 1.174, PCW 1.122, PMW 1.277, PSW 1.149, PRW 1.130
4567                         SCWN 1.122, SRWN 1.149, SCWS 1.097, SRWS 1.074/
4568     SEEPF(G) SALT CONCENTRATION OF SEE
4569                         /FRESH .80, SALINE .90/
4570     SOILVOL   VOLUME OF SOIL WATER
4571     SOILSALT INITIAL SOIL SALT (MILLION TONS)
4572   SCALARS
4573   XX, EE, DD DRAINED DEPTH (FEET) /6/
4574   RAINPPM   SALT CONCENTRATION FOR THE RAIN WATER/ 0/
4575   SALTCONV CONVERSION FROM PPM TO TONS PER AF/.00136/
4576   SALTRIVF RIVER SEE
4577   RRSALTF  RAIN RUNOFF PPM FROM GCA      /2./
4578   SOILSC   STORAGE COEFF. FOR SOIL    /.2/
4579   SOILDEPTH SOIL DEPTH   /6./;
4580
4581     TABLE DAREAI(Z,G,D,IS)  SALINE AREAS WITH DRAINAGE TYPE (MA)
4582                               1988   2000
4583                PCW.SALINE. DRAINED .396   .735
4584                PSW.SALINE. DRAINED .080   .085
4585                SCWN.SALINE.DRAINED .123   .327
4586                SCWS.SALINE.DRAINED .206   .688
4587                 ;
4588 DAREA(Z1,G,"DRAINED") = SUM(ISR, DAREAI(Z1,G,"DRAINED",ISR));
4589 DAREA(Z1,G,"UNDRAINED")=RESOURCE(Z1,G,"CCA")-DAREA(Z1,G,"DRAINED");
4590 DAREA(Z1,"SALINE","TOTAL") = SUM(D, DAREA(Z1,"SALINE",D));
4591 DAREA(Z1,"TOTAL","TOTAL") = SUM( (G,D), DAREA(Z1,G,D));
4592 DAREA(PV,T1,D1)  = SUM(Z1$PVZ(PV,Z1), DAREA(Z1,T1,D1));
4593
4594 GAREA(Z1,G) = RESOURCE(Z1,G,"CCA")*GCAF(Z1);
4595 GAREA(Z1,"TOTAL") = SUM(G, GAREA(Z1,G));
4596 GAREA(PV,T1)  = SUM(Z1$PVZ(PV,Z1), GAREA(Z1,T1));
4597
4598 GDAREA(Z1,T1,D1) = DAREA(Z1,T1,D1)*GCAF(Z1);
4599 GDAREA(PV,T1,D1) = SUM(Z1$PVZ(PV,Z1), GDAREA(Z1,T1,D1));
4600 SOILVOL(Z1,T1,D1) = GDAREA(Z1,T1,D1)*SOILSC*SOILDEPTH;
4601 SOILVOL(PV,T1,D1) = GDAREA(PV,T1,D1)*SOILSC*SOILDEPTH;
4602 SOILSALT(Z1,G,D1) = SOILVOL(Z1,G,D1)*SWPPM(Z1,G)*SALTCONV;
```

INDUS BASIN MODEL REVISED (IBMR) GROUNDWATER MODEL FILE=WSIRG41   07/25/91 17:18:26

```
4603    SOILSALT(Z1,"TOTAL",D1) = SUM(G, SOILSALT(Z1,G,D1));
4604    SOILSALT(PV,T1,D1) = SUM(Z1$PVZ(PV,Z1), SOILSALT(Z1,T1,D1));
4605
4606    IND1(Z1,G)$(GDAREA(Z1,G,"DRAINED")*GDAREA(Z1,G,"UNDRAINED"))=YES;
4607    IND2(Z1)$(GAREA(Z1,"FRESH")*GAREA(Z1,"SALINE")) = YES;
4608
          loop(isr,
          EVAPZ(Z,G,M)$resource(z,g,"CCA") =
                 SUM( (CNL,SA)$(ZSA(Z,CNL,SA)$gwfg(cnl,sa,g)),
                    EVAP(CNL,M)*COMDEF(ISR,"CCA",CNL)*subdef(sa,cnl))/
                          resource(z,g,"cca");
          evapz(z,g,sea1)= sum(m$sea1m(sea1,m), evapz(z,g,m));
          depthz1(Z,G,M)$resource(z,g,"cca")=
                 SUM( (CNL,SA)$(ZSA(Z,CNL,SA)$gwfg(cnl,sa,g)),
                    depth(CNL,M)*COMDEF(ISR,"CCA",CNL)*subdef(sa,cnl))/
                          resource(z,g,"cca")  );
4621    EVAPZ(Z1,G,M1) = EVAPZ1(Z1,G,M1);
4622    DEPTHZ1(Z1,G,M) = DEPTH1(Z1,G,M);
4623    *- REPORT ON GROUNDWATER INFLOWS AND OUTFLOWS (MAF)
4624
4625    REP81(Z1,G,"UNDRAINED","SEEP-RAIN",M)  = SUM( (CNL,SA), ZWT(Z1,CNL,SA)*RAIN(CNL,M)) /12*GAREA(Z1,G)*(1-DRC)*
4626                                                                          (1-FLEFFZ(Z1))*SPFWF ;
4627    REP81(Z1,"FRESH","UNDRAINED","SEEP-PTW",M) = TW.L(Z1,M)*(1-TWEFFZ(Z1,M))*SPFWF/1000 ;
4628    REP81(Z1,G,"UNDRAINED","SEEP-GTW",M)    = REP70(Z1,G,"G-TUBEWELL",M)*(1-WCEFFZ(Z1,M)) *SPFWF     ;
4629    REP81(Z1,G,"UNDRAINED","SEEP-CANAL",M) = SUM( (CNL,SA)$(ZSA(Z1,CNL,SA)$GWFG(CNL,SA,G) ), CANALDIV.L(CNL,M) *
4630                                                          (1-CNEFFG(CNL,G))*SUBDEF(SA,CNL) )*SPFC ;
4631    REP81(Z1,G,"UNDRAINED","SEEP-WCFLD",M) = SUM( (CNL,SA)$(ZSA(Z1,CNL,SA)$GWFG(CNL,SA,G) ),
4632                             CANALDIV.L(CNL,M)*CNEFFG(CNL,G)*
4633                              (1-WCEFFG(CNL,G,M)) *SUBDEF(SA,CNL) ) *SPFWF ;
4634
4635    REP81(Z1,G,"UNDRAINED","SEEP-LINK",M)  = SUM( (N,N1),
4636                             LINKLOSS(N1,N,M)*LNKSP(N1,N,Z1) )
4637                                              *RATIOFS(Z1,G)*SPFC ;
4638    REP81(Z1,G,"UNDRAINED","SEEP-RIVER",M) = SUM( (N,N1),
4639                             LOSSGAIN(N,N1,M)*RIVSP(N1,N,Z1))
4640                                              *SPFRV*RATIOFS(Z1,G) ;
4641    REP81(Z1,"FRESH","UNDRAINED","P-TUBWELL",M)  =  TW.L(Z1,M) /1000 ;
4642    REP81(Z1,G,"UNDRAINED","G-TUBWELL",M)$GF(G) = SUM( (CNL,SA)$ZSA(Z1,CNL,SA),  GWTSA(CNL,SA,M) )/1000 ;
4643    REP81(Z1,G,"UNDRAINED","TOT-INF",M)   = SUM(R6, REP81(Z1,G,"UNDRAINED",R6,M) )                    ;
4644    REP81(Z1,G,"UNDRAINED","TOT-OUTF",M)  = SUM(R7, REP81(Z1,G,"UNDRAINED",R7,M) )                    ;
4645    REP81(Z1,G,"UNDRAINED","INF-OUTF",M)  = REP81(Z1,G,"UNDRAINED","TOT-INF",M) - REP81(Z1,G,"UNDRAINED","TOT-OUTF",M);
4646
4647    REP81(Z1,G,"UNDRAINED",R9,SEA1) = SUM(M$SEA1M(SEA1,M), REP81(Z1,G,"UNDRAINED",R9,M));
4648    REP81(Z1,G,D,R9,M1)$EX(Z1,G) = REP81(Z1,G,"UNDRAINED",R9,M1)*GDAREA(Z1,G,D)/GAREA(Z1,G);
4649    REP81(Z1,"SALINE","TOTAL",R9,M1) = SUM(D, REP81(Z1,"SALINE",D,R9,M1));
4650    REP81(PV,G,D1,R9,M1) = SUM(Z1$PVZ(PV,Z1), REP81(Z1,G,D1,R9,M1) );
4651
4652    REP86(G,D1,R9,Z1) = REP81(Z1,G,D1,R9,"ANNUAL");
4653    REP86("TOTAL",D,R9,Z1) = SUM(G, REP86(G,D,R9,Z1));
4654    REP86("TOTAL","TOTAL",R9,Z1) = SUM( (G,D), REP86(G,D,R9,Z1));
4655    REP86(T1,D1,R9,PV) = SUM(Z1$PVZ(PV,Z1), REP86(T1,D1,R9,Z1));
4656
4657    RIVSALT(Z1,G,D) = SUM(M,  MAX(0, REP81(Z1,G,D,"SEEP-RIVER",M)*SALTPPM(Z1))
4658                              +MIN(0, REP81(Z1,G,D,"SEEP-RIVER",M)*GWPPM(Z1,G)*SALTRIVF)) ;
4659    RIVSALT(Z1,G,D) = RIVSALT(Z1,G,D)*SALTCONV;
4660    RIVSALT(Z1,"SALINE","TOTAL")$IND1(Z1,"SALINE") = SUM(D, RIVSALT(Z1,"SALINE",D));
4661    RIVSALT(Z1,"TOTAL" ,"TOTAL")$IND2(Z1) = SUM( (G,D), RIVSALT(Z1,G,D));
4662    RIVSALT(PV,G,D)              = SUM(Z1$PVZ(PV,Z1), RIVSALT(Z1,G,D));
4663    RIVSALT(PV,"SALINE","TOTAL") = SUM(D, RIVSALT(PV,"SALINE",D));
4664    RIVSALT(PV,"TOTAL" ,"TOTAL") = SUM( (G,D), RIVSALT(PV,G,D));
4665
4666    TOTSEEP(Z1,T1,D1)  =  REP86(T1,D1,"TOT-INF",Z1) - REP86(T1,D1,"SEEP-RIVER",Z1);
4667    TOTSEEP(PV,T1,D1)  =  REP86(T1,D1,"TOT-INF",PV) - REP86(T1,D1,"SEEP-RIVER",PV);
4668
4669    DEPTHZ2(Z1,G,"UNDRAINED","2000",M)$GDAREA(Z1,G,"UNDRAINED") = DEPTHZ1(Z1,G,M);
```

```
4670
4671    LOOP(Y,
4672       LOOP(M,
4673         LOOP((Z1,G,D)$(GDAREA(Z1,G,D)$UD(D)),
4674            XX= DEPTHZ2(Z1,G,D,Y,M--1) ;
4675          LOOP(X1,
4676            XX = (DEPTHZ2(Z1,G,D,Y,M--1) + XX)/2;
4677            EE =MIN(1, 10.637/XX**2.558)*EVAPZ(Z1,G,M);
4678            XX = DEPTHZ2(Z1,G,D,Y,M--1) -
4679                (REP81(Z1,G,D,"INF-OUTF",M)/GDAREA(Z1,G,D)-EE)/
4680                                          STORCOEFF(Z1)
4681             );
4682            GWEVAPZ(Z1,G,D,Y,M) = EE;
4683            DEPTHZ2(Z1,G,D,Y,M) = XX ));
4684         DEPTHZ2(Z1,G,D,Y+1,M)= DEPTHZ2(Z1,G,D,Y,M)
4685         );
4686  *___ CALCULATE THE EVAPORATION AND DRAINAGE EFFLUENT IN THE DRAINED AREAS
4687
4688
4689     DEPTHZ2(Z1,G,"DRAINED",Y,M)$GDAREA(Z1,G,"DRAINED") =DD;
4690     GWEVAPZ(Z1,G,"DRAINED",Y,M)$GDAREA(Z1,G,"DRAINED") =
4691                          MIN(1, 10.637/DD**2.558)*EVAPZ(Z1,G,M);
4692
4693     GWEVAPZ(Z1,G,D,Y,M) = GWEVAPZ(Z1,G,D,Y,M)*GDAREA(Z1,G,D);
4694     GWEVAPZ(Z1,G,D,Y,M)$(GWEVAPZ(Z1,G,D,Y,M) LT .0001)= 0;
4695     GWEVAPZ(Z1,G,D,Y,SEA1)=SUM(M$SEA1M(SEA1,M), GWEVAPZ(Z1,G,D,Y,M));
4696
4697     GWEVAPZ(PV,G,D,Y,M1) =SUM(Z1$PVZ(PV,Z1), GWEVAPZ(Z1,G,D,Y,M1));
4698     GWEVAPZ(PV,"SALINE","TOTAL",Y,M1) = SUM(D, GWEVAPZ(PV,"SALINE",D,Y,M1));
4699
4700
4701
4702     REP81A(Z1,G,Y,M)$GDAREA(Z1,G,"DRAINED")   = MAX(0,
4703                (REP81(Z1,G,"DRAINED","INF-OUTF",M)-GWEVAPZ(Z1,G,"DRAINED",Y,M)) ) ;
4704
4705     REP81A(Z1,G,Y,SEA1) = SUM(M$SEA1M(SEA1,M),  REP81A(Z1,G,Y,M));
4706     REP81A(PV,G,Y,M1)   = SUM(Z1$PVZ(PV,Z1), REP81A(Z1,G,Y,M1));
4707
4708
4709     REP84(Z1,Y,G,D,"TOT-INF" ) = REP81(Z1,G,D,"TOT-INF" ,"ANNUAL");
4710     REP84(Z1,Y,G,D,"TOT-OUTF") = -REP81(Z1,G,D,"TOT-OUTF","ANNUAL");
4711     REP84(Z1,Y,G,D,"GW-EVAP") = -GWEVAPZ(Z1,G,D,Y,"ANNUAL");
4712     REP84(Z1,Y,G,"DRAINED","DRAINAGE") = -REP81A(Z1,G,Y,"ANNUAL");
4713     REP84(Z1,Y,G,D,"I-O") = REP84(Z1,Y,G,D,"TOT-INF")+REP84(Z1,Y,G,D,"TOT-OUTF")
4714                          +REP84(Z1,Y,G,D,"GW-EVAP")+REP84(Z1,Y,G,D,"DRAINAGE");
4715     REP84(Z1,Y,G,"TOTAL",R92)$IND1(Z1,G)    = SUM(D, REP84(Z1,Y,G,D,R92));
4716     REP84(Z1,Y,"TOTAL","TOTAL",R92)$IND2(Z1) = SUM( (G,D), REP84(Z1,Y,G,D,R92));
4717
4718     REP84(PV,Y,G ,D ,R92) = SUM(Z1$PVZ(PV,Z1), REP84(Z1,Y,G,D,R92));
4719     REP84(PV,Y,"SALINE","TOTAL",R92) = SUM(D,      REP84(PV,Y,"SALINE",D,R92));
4720     REP84(PV,Y,"TOTAL","TOTAL",R92)  = SUM((G,D), REP84(PV,Y,G,D,R92));
4721
4722     REP84A(Z1,Y,T1,D1,R92)$GDAREA(Z1,T1,D1) = REP84(Z1,Y,T1,D1,R92)/GDAREA(Z1,T1,D1);
4723     REP84A(PV,Y,T1,D1,R92)$GDAREA(PV,T1,D1) = REP84(PV,Y,T1,D1,R92)/GDAREA(PV,T1,D1);
4724  *___
4725     REP86A(G,Z1,D,"RAIN") = SUM( (CNL,SA,M), ZWT(Z1,CNL,SA)*RAIN(CNL,M))/12*(1-DRC)*GDAREA(Z1,G,D);
4726     REP86A(G,Z1,D,"RAIN-RUNOF") = SUM( (CNL,SA,M), ZWT(Z1,CNL,SA)*RAIN(CNL,M))/12*DRC*GDAREA(Z1,G,D);
4727
4728     REP86A(G,Z1,D,"CANAL")$GDAREA(Z1,G,D)=SUM((CNL,SA)$(ZSA(Z1,CNL,SA)$GWFG(CNL,SA,G)),
4729                          SUM(M, CANALDIV.L(CNL,M)*SUBDEF(SA,CNL)))*GDAREA(Z1,G,D)/GAREA(Z1,G);
4730     REP86A(G,Z1,D,"LINK")  = REP81(Z1,G,D,"SEEP-LINK","ANNUAL")/SPFC;
4731     REP86A(G,Z1,D,"RIVER") = REP81(Z1,G,D,"SEEP-RIVER","ANNUAL");
4732
4733     REP86A("SALINE",Z1,"TOTAL",R31) = SUM(D, REP86A("SALINE",Z1,D,R31));
4734     REP86A("TOTAL",Z1,D,R31)        = SUM(G, REP86A(G,Z1,D,R31));
```

```
4735    REP86A("TOTAL",Z1,"TOTAL",R31)  = SUM( (G,D), REP86A(G,Z1,D,R31));
4736    REP86A(T1,Z1,D1,"TOTAL")        = SUM(R31, REP86A(T1,Z1,D1,R31));
4737
4738    REP86A(T1,PV,D1,R31)    = SUM(Z1$PVZ(PV,Z1), REP86A(T1,Z1,D1,R31));
4739    REP86A(T1,PV,D1,"TOTAL") = SUM(R31,           REP86A(T1,PV,D1,R31));
4740
4741    APPWAT(Z1,G,D)   = REP86A(G,Z1,D,"RAIN")+REP86A(G,Z1,D,"CANAL")+
4742                REP86A(G,Z1,D,"LINK")+REP81(Z1,G,D,"TOT-OUTF","ANNUAL");
4743    APPWAT(Z1,"SALINE","TOTAL") = SUM(D, APPWAT(Z1,"SALINE",D));
4744    APPWAT(Z1,"TOTAL","TOTAL")  = SUM((G,D), APPWAT(Z1,G,D));
4745    APPWAT(PV,T1,D1) = SUM(Z1$PVZ(PV,Z1), APPWAT(Z1,T1,D1));
4746    MIXVOL(Z1,T1,D1) = TOTSEEP(Z1,T1,D1)+SOILVOL(Z1,T1,D1);
4747    MIXVOL(PV,T1,D1) = SUM(Z1$PVZ(PV,Z1), MIXVOL(Z1,T1,D1));
4748
4749    REP862(Z1,G,D,"RAIN"    ,Y)   = REP86A(G,Z1,D,"RAIN") *RAINPPM*SALTCONV;
4750    REP862(Z1,G,D,"RAIN-RUNOF",Y) = -REP86A(G,Z1,D,"RAIN-RUNOF")*SALTPPM(Z1)*RRSALTF*SALTCONV;
4751    REP862(Z1,G,D,"CANAL",Y)   = REP86A(G,Z1,D,"CANAL")*SALTPPM(Z1)*SALTCONV;
4752    REP862(Z1,G,D,"LINK" ,Y)   = REP86A(G,Z1,D,"LINK") *SALTPPM(Z1)*SALTCONV;
4753    REP862(Z1,G,D,"RIVER",Y)   = RIVSALT(Z1,G,D);
4754
4755
4756    REP86C(Z1,G,D,"RAIN",Y)   = REP86A(G,Z1,D,"RAIN")*RAINPPM*SALTCONV;
4757    REP86C(Z1,G,D,"RAIN-RUNOF",Y) = -REP86A(G,Z1,D,"RAIN-RUNOF")*SALTPPM(Z1)*RRSALTF*SALTCONV;
4758    REP86C(Z1,G,D,"CANAL",Y) = REP86A(G,Z1,D,"CANAL")*SALTPPM(Z1)*SALTCONV;
4759    REP86C(Z1,G,D,"LINK",Y ) = REP86A(G,Z1,D,"LINK")*SALTPPM(Z1)*SALTCONV;
4760    REP86C(Z1,G,D,"SURF-WATER",Y) = REP86C(Z1,G,D,"RAIN",Y)
4761                                  + REP86C(Z1,G,D,"RAIN-RUNOF",Y)
4762                                  + REP86C(Z1,G,D,"CANAL",Y)
4763                                  + REP86C(Z1,G,D,"LINK",Y);
4764    REP86D(Z1,G,D,"FROM-RIVER",Y) = RIVSALT(Z1,G,D) ;
4765
4766 *___ INITIAL CONDITIONS FOR THE SOIL
4767    REP86C(Z1,G,D,"MIX-TONS","2000") = SOILSALT(Z1,G,D);
4768    REP86C(Z1,G,D,"SOIL-TONS","2000") = SOILSALT(Z1,G,D);
4769
4770 *___ INITIAL CONDITIONS FOR THE GW
4771
4772    REP86D(Z1,G,D,"MAF","2000")=(GWAQDEPTH(Z1,G)-DEPTHZ1(Z1,G,"DEC"))*STORCOEFF(Z1)*GDAREA(Z1,G,D);
4773    REP86D(Z1,G,D,"M-TONS","2000")=REP86D(Z1,G,D,"MAF","2000")*GWPPM(Z1,G)*SALTCONV;
4774    REP86D(Z1,G,D,"PPM","2000")$REP86D(Z1,G,D,"MAF","2000") = GWPPM(Z1,G) ;
4775
4776    REP862(Z1,G,D,"M-TONS","2000")=REP86D(Z1,G,D,"MAF","2000")*GWPPM(Z1,G)*SALTCONV;
4777    REP862(Z1,G,D,"PPM","2000")$REP86D(Z1,G,D,"MAF","2000") = GWPPM(Z1,G) ;
4778    LOOP(Y,
4779
4780       REP86C(Z1,G,D,"TW",Y)    = REP81(Z1,G,D,"TOT-OUTF","ANNUAL")*REP86D(Z1,G,D,"PPM",Y)*SALTCONV;
4781       REP86C(Z1,G,D,"EVAP",Y) = GWEVAPZ(Z1,G,D,Y,"ANNUAL")*REP86D(Z1,G,D,"PPM",Y)*SALTCONV;
4782       REP86C(Z1,G,D,"TOT-SOIL",Y) = REP86C(Z1,G,D,"SURF-WATER",Y)
4783                                   +REP86C(Z1,G,D,"TW",Y)
4784                                   +REP86C(Z1,G,D,"EVAP",Y);
4785
4786       REP86C(Z1,G,D,"SEEP-MAF",Y) = TOTSEEP(Z1,G,D);
4787       REP86C(Z1,G,D,"IMP-PPM",Y)$REP86C(Z1,G,D,"SEEP-MAF",Y) =
4788                           REP86C(Z1,G,D,"TOT-SOIL",Y)/REP86C(Z1,G,D,"SEEP-MAF",Y)/SALTCONV;
4789       REP86C(Z1,G,D,"SOIL-VOL",Y) = SOILVOL(Z1,G,D);
4790       REP86C(Z1,G,D,"APPWAT",Y)  = APPWAT(Z1,G,D);
4791       REP86C(Z1,G,D,"MIX-VOL",Y) = MIXVOL(Z1,G,D);
4792
4793       REP86C(Z1,G,D,"MIX-TONS",Y) = REP86C(Z1,G,D,"TOT-SOIL",Y)+
4794                                     REP86C(Z1,G,D,"SOIL-TONS",Y);
4795
4796       REP86C(Z1,G,D,"SEEP-PPM",Y)$MIXVOL(Z1,G,D) = REP86C(Z1,G,D,"MIX-TONS",Y)
4797                                      /MIXVOL(Z1,G,D)/SALTCONV;
4798       REP86C(Z1,G,D,"TO-GW",Y)   = REP86C(Z1,G,D,"SEEP-PPM",Y)
4799                                    *REP86C(Z1,G,D,"SEEP-MAF",Y)*SALTCONV;
```

```
4800            REP86C(Z1,G,D,"TO-SOIL",Y) =  REP86C(Z1,G,D,"TOT-SOIL",Y)
4801                                          -REP86C(Z1,G,D,"TO-GW",Y);
4802            REP86C(Z1,G,D,"SOIL-TONS",Y) =  REP86C(Z1,G,D,"SOIL-TONS",Y)
4803                                          + REP86C(Z1,G,D,"TO-SOIL",Y);
4804
4805            REP86D(Z1,G,D,"FROM-SEEP",Y) = REP86C(Z1,G,D,"TO-GW",Y);
4806            REP86D(Z1,G,D,"TO-DRAIN",Y)  = REP84(Z1,Y,G,D,"DRAINAGE")*REP86D(Z1,G,D,"PPM",Y)*SALTCONV;
4807            REP86D(Z1,G,D,"TO-EVAP" ,Y)  = - GWEVAPZ(Z1,G,D,Y,"ANNUAL")*REP86D(Z1,G,D,"PPM",Y)*SALTCONV;
4808            REP86D(Z1,G,D,"TO-PUMPS",Y)  = - REP81(Z1,G,D,"TOT-OUTF","ANNUAL")*REP86D(Z1,G,D,"PPM",Y)*SALTCONV;
4809
4810            REP86D(Z1,G,D,"M-TONS",Y) = REP86D(Z1,G,D,"M-TONS",Y)
4811                                      + REP86C(Z1,G,D,"TO-GW",Y)
4812                                      + REP84(Z1,Y,G,D,"DRAINAGE")*REP86D(Z1,G,D,"PPM",Y)*SALTCONV
4813                                      - GWEVAPZ(Z1,G,D,Y,"ANNUAL")*REP86D(Z1,G,D,"PPM",Y)*SALTCONV
4814                                      - REP81(Z1,G,D,"TOT-OUTF","ANNUAL")*REP86D(Z1,G,D,"PPM",Y)*SALTCONV
4815                                      + REP86D(Z1,G,D,"FROM-RIVER",Y) ;
4816            REP862(Z1,G,D,"DRAINAGE",Y) = REP84(Z1,Y,G,D,"DRAINAGE")*REP862(Z1,G,D,"PPM",Y)*SALTCONV;
4817            REP862(Z1,G,D,"M-TONS",Y) = REP862(Z1,G,D,"M-TONS",Y)
4818                                       +REP862(Z1,G,D,"RAIN",Y)
4819                                       +REP862(Z1,G,D,"RAIN-RUNOF",Y)
4820                                       +REP862(Z1,G,D,"CANAL",Y)
4821                                       +REP862(Z1,G,D,"LINK",Y)
4822                                       +REP862(Z1,G,D,"RIVER",Y)
4823                                       +REP862(Z1,G,D,"DRAINAGE",Y);
4824
4825            REP86D(Z1,G,D,"MAF",Y) = REP86D(Z1,G,D,"MAF",Y)+REP84(Z1,Y,G,D,"I-O");
4826            REP86D(Z1,G,D,"PPM",Y)$REP86D(Z1,G,D,"MAF",Y) = REP86D(Z1,G,D,"M-TONS",Y)/REP86D(Z1,G,D,"MAF",Y)/SALTCONV;
4827            REP862(Z1,G,D,"PPM",Y)$REP86D(Z1,G,D,"MAF",Y) = REP862(Z1,G,D,"M-TONS",Y)/REP86D(Z1,G,D,"MAF",Y)/SALTCONV;
4828
4829            REP86C(Z1,G,D,"SOIL-TONS",Y+1) = REP86C(Z1,G,D,"SOIL-TONS",Y);
4830            REP86D(Z1,G,D,"MAF",Y+1)    = REP86D(Z1,G,D,"MAF",Y);
4831            REP86D(Z1,G,D,"PPM",Y+1)    = REP86D(Z1,G,D,"PPM",Y);
4832            REP86D(Z1,G,D,"M-TONS",Y+1) = REP86D(Z1,G,D,"M-TONS",Y);
4833            REP862(Z1,G,D,"PPM",Y+1)    = REP862(Z1,G,D,"PPM",Y);
4834            REP862(Z1,G,D,"M-TONS",Y+1) = REP862(Z1,G,D,"M-TONS",Y)
4835          );
4836
4837    REP862(Z1,G,D,"MAF",Y) = REP86D(Z1,G,D,"MAF",Y);
4838    REP862(Z1,G,D,"ANN-ADD",Y+1) = REP862(Z1,G,D,"M-TONS",Y+1) - REP862(Z1,G,D,"M-TONS",Y);
4839    REP862(Z1,G,D,"ANN-ADD","2000") = REP862(Z1,G,D,"M-TONS","2000")-(GWAQDEPTH(Z1,G)-DEPTHZ1(Z1,G,"DEC"))
4840                                      *STORCOEFF(Z1)*GDAREA(Z1,G,D)*GWPPM(Z1,G)*SALTCONV;
4841    REP862(Z1,"SALINE","TOTAL",R33,Y)$IND1(Z1,"SALINE") = SUM(D,  REP862(Z1,"SALINE",D,R33,Y));
4842    REP862(Z1,"SALINE","TOTAL","PPM",Y)$REP862(Z1,"SALINE","TOTAL","MAF",Y) = REP862(Z1,"SALINE","TOTAL","M-TONS",Y)
4843                                                                             /REP862(Z1,"SALINE","TOTAL","MAF",Y)/SALTCONV;
4844    REP862(Z1,"TOTAL","TOTAL",R33,Y)$IND2(Z1) = SUM( (G,D), REP862(Z1,G,D,R33,Y));
4845    REP862(Z1,"TOTAL","TOTAL","PPM",Y)$REP862(Z1,"TOTAL","TOTAL","MAF",Y) = REP862(Z1,"TOTAL","TOTAL","M-TONS",Y)
4846                                                                           /REP862(Z1,"TOTAL","TOTAL","MAF",Y)/SALTCONV;
4847
4848    REP862(PV,G,D,R33,Y) = SUM(Z1$PVZ(PV,Z1), REP862(Z1,G,D,R33,Y));
4849    REP862(PV,"SALINE","TOTAL",R33,Y) = SUM(D, REP862(PV,"SALINE",D,R33,Y) ) ;
4850    REP862(PV,"TOTAL","TOTAL",R33,Y) = SUM( (G,D), REP862(PV,G,D,R33,Y));
4851    REP862(PV,T1,D1,"PPM",Y)$REP862(PV,T1,D1,"MAF",Y) = REP862(PV,T1,D1,"M-TONS",Y)/REP862(PV,T1,D1,"MAF",Y)/SALTCONV;
4852
4853    REP86C(Z1,"SALINE","TOTAL",R36,Y)$IND1(Z1,"SALINE") = SUM(D, REP86C(Z1,"SALINE",D,R36,Y));
4854    REP86C(Z1,"SALINE","TOTAL","IMP-PPM",Y)$REP86C(Z1,"SALINE","TOTAL","SEEP-MAF",Y )=
4855                        REP86C(Z1,"SALINE","TOTAL","TOT-SOIL",Y)
4856                        /REP86C(Z1,"SALINE","TOTAL","SEEP-MAF",Y)/SALTCONV;
4857    REP86C(Z1,"SALINE","TOTAL","SEEP-PPM",Y)$IND1(Z1,"SALINE") =
4858                        REP86C(Z1,"SALINE","TOTAL","MIX-TONS",Y)/MIXVOL(Z1,"SALINE","TOTAL")/SALTCONV;
4859
4860    REP86C(Z1,"TOTAL","TOTAL",R36,Y)$IND2(Z1) = SUM( (G,D), REP86C(Z1,G,D,R36,Y));
4861    REP86C(Z1,"TOTAL","TOTAL","IMP-PPM",Y)$REP86C(Z1,"TOTAL","TOTAL","SEEP-MAF",Y)=
4862                        REP86C(Z1,"TOTAL","TOTAL","TOT-SOIL",Y)
4863                        /REP86C(Z1,"TOTAL","TOTAL","SEEP-MAF",Y)/SALTCONV;
4864
```

```
4865        REP86C(Z1,"TOTAL","TOTAL","SEEP-PPM",Y)$IND2(Z1) =
4866                             REP86C(Z1,"TOTAL","TOTAL","MIX-TONS",Y)
4867                             /MIXVOL(Z1,"TOTAL","TOTAL")/SALTCONV;
4868
4869        REP86C(PV,G,D,R36,Y) = SUM(Z1$PVZ(PV,Z1), REP86C(Z1,G,D,R36,Y));
4870        REP86C(PV,"SALINE","TOTAL",R36,Y) = SUM(D, REP86C(PV,"SALINE",D,R36,Y));
4871        REP86C(PV,"TOTAL","TOTAL",R36,Y) = SUM((G,D), REP86C(PV,G,D,R36,Y));
4872
4873        REP86C(PV,T1,D1,"IMP-PPM",Y)$REP86C(PV,T1,D1,"SEEP-MAF",Y) =
4874                             REP86C(PV,T1,D1,"TOT-SOIL",Y)/REP86C(PV,T1,D1,"SEEP-MAF",Y)/SALTCONV;
4875        REP86C(PV,T1,D1,"SEEP-PPM",Y)$MIXVOL(PV,T1,D1) =
4876                             REP86C(PV,T1,D1,"MIX-TONS",Y)
4877                             /MIXVOL(PV,T1,D1)/SALTCONV;
4878
4879   *____ GW SALT INFLOWS AND OUTFLOWS
4880        REP86D(Z1,G,D,R6,Y) = REP81(Z1,G,D,R6,"ANNUAL")*REP86C(Z1,G,D,"SEEP-PPM",Y)*SALTCONV;
4881        REP86D(Z1,G,D,"SEEP-RIVER",Y) = 0 ;
4882
4883        REP86D(Z1,G,D,"ANN-ADD",Y+1)   = REP86D(Z1,G,D,"M-TONS",Y+1) - REP86D(Z1,G,D,"M-TONS",Y);
4884        REP86D(Z1,G,D,"ANN-ADD","2000") = REP86D(Z1,G,D,"M-TONS","2000")-(GWAQDEPTH(Z1,G)-DEPTHZ1(Z1,G,"DEC"))
4885                                              *STORCOEFF(Z1)*GDAREA(Z1,G,D)*GWPPM(Z1,G)*SALTCONV;
4886        REP86D(Z1,"SALINE","TOTAL",R37,Y)$IND1(Z1,"SALINE") = SUM(D,  REP86D(Z1,"SALINE",D,R37,Y));
4887        REP86D(Z1,"SALINE","TOTAL","PPM",Y)$REP86D(Z1,"SALINE","TOTAL","MAF",Y) = REP86D(Z1,"SALINE","TOTAL","M-TONS",Y)
4888                                              /REP86D(Z1,"SALINE","TOTAL","MAF",Y)/SALTCONV;
4889        REP86D(Z1,"TOTAL","TOTAL",R37,Y)$IND2(Z1) = SUM( (G,D), REP86D(Z1,G,D,R37,Y));
4890        REP86D(Z1,"TOTAL","TOTAL","PPM",Y)$REP86D(Z1,"TOTAL","TOTAL","MAF",Y) =
4891                             REP86D(Z1,"TOTAL","TOTAL","M-TONS",Y)/REP86D(Z1,"TOTAL","TOTAL","MAF",Y)/SALTCONV;
4892        REP86D(PV,G,D,R37,Y) = SUM(Z1$PVZ(PV,Z1), REP86D(Z1,G,D,R37,Y));
4893        REP86D(PV,"SALINE","TOTAL",R37,Y) = SUM(D, REP86D(PV,"SALINE",D,R37,Y));
4894        REP86D(PV,"TOTAL","TOTAL",R37,Y) = SUM( (G,D), REP86D(PV,G,D,R37,Y));
4895
4896        REP86D(PV,T1,D1,"PPM",Y)$REP86D(PV,T1,D1,"MAF",Y) = REP86D(PV,T1,D1,"M-TONS",Y)/REP86D(PV,T1,D1,"MAF",Y)/SALTCONV;
4897
4898   *- ANNUAL SALT MOVED FROM SOIL TO GW
4899        REP86B(Z1,G,D,"2000")$GDAREA(Z1,G,D)=GWPPM(Z1,G);
4900        REP86B(Z1,G,D,Y+1)   =REP86D(Z1,G,D,"PPM",Y);
4901
4902        REP86E(Z1,G,D,Y,"CNL+WCFLD","IN" ) = REP86C(Z1,G,D,"CANAL",Y);
4903        REP86E(Z1,G,D,Y,"CNL+WCFLD","OUT") = REP86D(Z1,G,D,"SEEP-CANAL",Y) + REP86D(Z1,G,D,"SEEP-WCFLD",Y) ;
4904        REP86E(Z1,G,D,Y,"LINK", "IN" ) = REP86C(Z1,G,D,"LINK",Y );
4905        REP86E(Z1,G,D,Y,"LINK", "OUT") = REP86D(Z1,G,D,"SEEP-LINK ",Y);
4906        REP86E(Z1,G,D,Y,"TW"  , "IN" ) = REP86C(Z1,G,D,"TW",Y);
4907        REP86E(Z1,G,D,Y,"TW"  , "OUT") = REP86D(Z1,G,D,"SEEP-PTW",Y)
4908                                       + REP86D(Z1,G,D,"SEEP-GTW",Y) ;
4909
4910        REP86E(Z1,G,D,Y,R32,"MOVED")        = REP86E(Z1,G,D,Y,R32,"IN") - REP86E(Z1,G,D,Y,R32,"OUT");
4911        REP86E(Z1,G,D,Y,"RAIN","MOVED")     = - REP86D(Z1,G,D,"SEEP-RAIN",Y);
4912        REP86E(Z1,G,D,Y,"EVAP","MOVED")     = REP86C(Z1,G,D,"EVAP",Y);
4913        REP86E(Z1,G,D,Y,"TO-RUNOFF","MOVED") = REP86C(Z1,G,D,"RAIN-RUNOF",Y);
4914
4915
4916        REP86E(Z1,"SALINE","TOTAL",Y,R32,R94)$IND1(Z1,"SALINE") = SUM(D, REP86E(Z1,"SALINE",D,Y,R32,R94));
4917        REP86E(Z1,"TOTAL","TOTAL",Y,R32,R94)$IND2(Z1)  = SUM( (G,D), REP86E(Z1,G,D,Y,R32,R94) );
4918
4919        REP86E(PV,G,D,Y,R32,R94)   = SUM(Z1$PVZ(PV,Z1), REP86E(Z1,G,D,Y,R32,R94)  );
4920        REP86E(PV,"SALINE","TOTAL",Y,R32,R94) = SUM(D,     REP86E(PV,"SALINE",D,Y,R32,R94));
4921        REP86E(PV,"TOTAL" ,"TOTAL",Y,R32,R94) = SUM((G,D), REP86E(PV,G,D,Y,R32,R94));
4922
4923        REP86F(Z1,G,D,Y,R32,R94)$GDAREA(Z1,G,D) = REP86E(Z1,G,D,Y,R32,R94)/GDAREA(Z1,G,D);
4924
4925   *- SALT MOVE FROM GW
4926
4927        REP86G(Z1,Y,G,D,R37)$GDAREA(Z1,G,D) = REP86D(Z1,G,D,R37,Y)/GDAREA(Z1,G,D);
4928        REP86G(Z1,Y,G,D,"PPM")$REP86G(Z1,Y,G,D,"MAF") = REP86G(Z1,Y,G,D,"M-TONS")/REP86G(Z1,Y,G,D,"MAF")/SALTCONV;
4929
```

```
4930    REP86H(Z1,Y,T1,D1,"IMP-PPM")     = REP86C(Z1,T1,D1,"IMP-PPM",Y) ;
4931    REP86H(Z1,Y,T1,D1,"SEEP-PPM")    = REP86C(Z1,T1,D1,"SEEP-PPM",Y) ;
4932    REP86H(Z1,Y,T1,D1,"GW-PPM")      = REP86D(Z1,T1,D1,"PPM",Y) ;
4933
4934    REP86H(PV,Y,T1,D1,"IMP-PPM")     = REP86C(PV,T1,D1,"IMP-PPM",Y) ;
4935    REP86H(PV,Y,T1,D1,"SEEP-PPM")    = REP86C(PV,T1,D1,"SEEP-PPM",Y) ;
4936    REP86H(PV,Y,T1,D1,"GW-PPM")      = REP86D(PV,T1,D1,"PPM",Y) ;
4937
4938    REP864(Z1,T1,Y,D1,"WOS-GW")      = REP862(Z1,T1,D1,"ANN-ADD",Y);
4939    REP864(Z1,T1,Y,D1,"WS-SOIL")     = REP86C(Z1,T1,D1,"TO-SOIL",Y);
4940    REP864(Z1,T1,Y,D1,"WS-GW")       = REP86D(Z1,T1,D1,"ANN-ADD",Y);
4941    REP864(Z1,T1,Y,D1,"WS-TOTAL")    = REP864(Z1,T1,Y,D1,"WS-SOIL")+REP864(Z1,T1,Y,D1,"WS-GW");
4942    REP864(Z1,T1,Y,D1,"RATIO")$REP864(Z1,T1,Y,D1,"WS-TOTAL") = REP864(Z1,T1,Y,D1,"WOS-GW")/REP864(Z1,T1,Y,D1,"WS-TOTAL");
4943    REP864(PV,T1,Y,D1,R35) = SUM(Z1$PVZ(PV,Z1), REP864(Z1,T1,Y,D1,R35));
4944    REP864(PV,T1,Y,D1,"RATIO")$REP864(PV,T1,Y,D1,"WS-TOTAL") = REP864(PV,T1,Y,D1,"WOS-GW")/REP864(PV,T1,Y,D1,"WS-TOTAL");
4945
4946  * REPORT ON SEP.    SET R6 HAS ALL THE SEEPAGES AND R7 PUM
4947
4948    REP87(G,D,R6,Z1)  = REP81(Z1,G,D,R6,"SEP");
4949    REP87(G,D,R7,Z1)  = REP81(Z1,G,D,R7,"SEP");
4950    REP87(G,D,"TOT-INF",Z1)  = REP81(Z1,G,D,"TOT-INF","SEP");
4951
4952    REP87A(Z1,Y,G,D,"GW-EVAP")   = GWEVAPZ(Z1,G,D,Y,"SEP");
4953    REP87A(Z1,Y,G,"DRAINED","DRAINAGE") = REP81(Z1,G,Y,"SEP");
4954    REP87A(Z1,Y,G,D,"OUTFLOW") = REP81(Z1,G,D,"TOT-OUTF","SEP")+
4955                                 GWEVAPZ(Z1,G,D,Y,"SEP")+REP87A(Z1,Y,G,D,"DRAINAGE");
4956    REP87A(Z1,Y,G,D,"WTD")    = DEPTHZ2(Z1,G,D,Y,"SEP") ;
4957
4958    REP87B(G,D,R6,Z1)$GDAREA(Z1,G,D) = REP87(G,D,R6,Z1)/GDAREA(Z1,G,D);
4959    REP87B(G,D,R7,Z1)$GDAREA(Z1,G,D) = REP87(G,D,R7,Z1)/GDAREA(Z1,G,D);
4960
4961    REP87C(Z1,Y,G,D,R91)$GDAREA(Z1,G,D)  = REP87A(Z1,Y,G,D,R91)/GDAREA(Z1,G,D);
4962    REP87C(Z1,Y,G,D,"WTD")   = DEPTHZ2(Z1,G,D,Y,"SEP") ;
4963
4964    OPTION GDAREA:3:1:2, DAREA:3:1:2,  REP81A:3:1:1, REP81:3:1:1, GWEVAPZ:3:1:1, DEPTHZ2:2:1:1,
4965         REP84:3:1:3,   REP84A:3:1:3, REP86:3:1:1, REP86A:3:1:2, REP86C:3:1:1, REP86D:3:1:1, REP862:3:1:1,
4966         REP86E:3:1:2, REP86F:3:1:2, REP86G:3:1:3, REP86H:3:1:3,  REP87A:3:1:3,  REP87B:3:1:1, REP87C:3:1:3,
4967         REP864:3:1:2, RIVSALT:3:1:2,TOTSEEP:3:1:2
4968         SOILVOL:3:1:2,SOILSALT:3:1:2,APPWAT:3:1:2,MIXVOL:3:1:2;
        rep862(z1,t1,d1,r33,y)   $y2(y) = 0;
        rep86c(z1,t1,d1,r36,y)   $y2(y) = 0;
        rep86d(z1,t1,d1,r37,y)   $y2(y) = 0;
        rep86e(z1,g,d,y,r32,r94)$y2(y) = 0;
        rep86f(z1,g,d,y,r32,r94)$y2(y) = 0;
        rep86g(z1,y,g,d,r37)     $y2(y) = 0;
      * gwevapz(z1,g,d,y,m1)     $y2(y) = 0;
      * depthz2(z1,g,d,y,m)      $y2(y) = 0;
        rep84(z1,y,t1,d1,r92)    $y2(y) = 0;
        rep84a(z1,y,t1,d1,r92)   $y2(y) = 0;
        rep864(z1,t1,y,d1,r35)   $y2(y) = 0;
4982   DISPLAY IND1, IND2, SALTPPM, GWPPM, GWAQDEPTH,STORCOEFF,DD,SALTCONV,SALTRIVF, RRSALTF ;
4983   DISPLAY SOILSC, SOILDEPTH, SWPPM;
4984   DISPLAY GAREA, DAREA, GDAREA, SOILVOL,SOILSALT,APPWAT,MIXVOL
4985           DEPTHZ1,EVAPZ
4986   DISPLAY REP81, TOTSEEP, RIVSALT, REP81A,GWEVAPZ, DEPTHZ2;
4987
4988   DISPLAY REP86, REP84, REP86A, REP862;
4989  * DISPLAY REP86C, REP86D, REP86E, REP86H, REP864;
4990  * PER ACRE REPORTS ;
4991  * DISPLAY REP84A, REP86F, REP86G;
4992
4993  * SPREADSHEET METHOD
4994
4995    SET R95 /FLOW,SEEP-TOGW,FLOW-PPM,SALT-IN,SW-VOL,NEW-SALT,
4996                          SALT-SEEP,ADD-TOSOIL,ADD-TOGW,ADD-TOSGW/
```

INDUS BASIN MODEL REVISED (IBMR) GROUNDWATER MODEL FILE=WSIRG41   07/25/91 17:18:26

```
4997        R96 /RAIN-PERC,R-RUNOFF,CANALS,TUBWELLS,
4998                            RIVER-S,EVAPORATN,DRAINS,ROW-SUM/
4999        R97/SOILSALT, SOIL-MAF, SOIL-PPM, TTONS, GW-MAF, GWPPM/
5000        R98/ADD-TOSOIL,ADD-TOGW,ADD-TOSGW/
5001  PARAMETER REP865
5002            REP866
5003            REP867;
5004  REP866(Z1,G,D,"2000","TTONS")    = (GWAQDEPTH(Z1,G)-DEPTHZ1(Z1,G,"DEC"))*STORCOEFF(Z1)*GDAREA(Z1,G,D)
5005                                       *GWPPM(Z1,G)*SALTCONV;
5006  REP866(Z1,G,D,"2000","GWPPM")$SOILVOL(Z1,G,D)  = GWPPM(Z1,G) ;
5007  REP866(Z1,G,D,"2000","SOILSALT")$SOILVOL(Z1,G,D) = SOILSALT(Z1,G,D);
5008
5009  LOOP(Y,
5010  *- RAIN
5011  REP865(Z1,G,D,Y,"FLOW",     "RAIN-PERC") = REP86A(G,Z1,D,"RAIN") ;
5012  REP865(Z1,G,D,Y,"SEEP-TOGW","RAIN-PERC") = REP86(G,D,"SEEP-RAIN",Z1);
5013  REP865(Z1,G,D,Y,"SW-VOL",   "RAIN-PERC") = SOILVOL(Z1,G,D) + REP865(Z1,G,D,Y,"FLOW","RAIN-PERC");
5014  REP865(Z1,G,D,Y,"NEW-SALT", "RAIN-PERC") = REP866(Z1,G,D,Y,"SOILSALT");
5015  REP865(Z1,G,D,Y,"SALT-SEEP","RAIN-PERC")$REP865(Z1,G,D,Y,"SW-VOL","RAIN-PERC")
5016                           = REP865(Z1,G,D,Y,"NEW-SALT","RAIN-PERC")
5017                            /REP865(Z1,G,D,Y,"SW-VOL","RAIN-PERC")
5018                            *REP865(Z1,G,D,Y,"SEEP-TOGW","RAIN-PERC");
5019  REP865(Z1,G,D,Y,"ADD-TOSOIL","RAIN-PERC") = -REP865(Z1,G,D,Y,"SALT-SEEP","RAIN-PERC");
5020  REP865(Z1,G,D,Y,"ADD-TOGW"  ,"RAIN-PERC") =  REP865(Z1,G,D,Y,"SALT-SEEP","RAIN-PERC");
5021
5022  *-RAIN RUNOFF
5023  REP865(Z1,G,D,Y,"FLOW"      ,"R-RUNOFF") = REP86A(G,Z1,D,"RAIN-RUNOF") ;
5024  REP865(Z1,G,D,Y,"FLOW-PPM"  ,"R-RUNOFF") = SALTPPM(Z1)*RRSALTF ;
5025  REP865(Z1,G,D,Y,"SALT-IN "  ,"R-RUNOFF") = REP862(Z1,G,D,"RAIN-RUNOF",Y);
5026  REP865(Z1,G,D,Y,"ADD-TOSOIL","R-RUNOFF") = REP862(Z1,G,D,"RAIN-RUNOF",Y);
5027
5028  *- CANAL + WATER COURSES AND FILEDS.
5029  REP865(Z1,G,D,Y,"FLOW"      ,"CANALS" ) = REP86A(G,Z1,D,"CANAL") +
5030                                            REP86A(G,Z1,D,"LINK") ;
5031  REP865(Z1,G,D,Y,"SEEP-TOGW" ,"CANALS" ) = REP86(G,D,"SEEP-CANAL",Z1)+REP86(G,D,"SEEP-WCFLD",Z1) +
5032                                            REP86(G,D,"SEEP-LINK",Z1);
5033  REP865(Z1,G,D,Y,"FLOW-PPM"  ,"CANALS" ) = SALTPPM(Z1) ;
5034  REP865(Z1,G,D,Y,"SALT-IN"   ,"CANALS" ) = REP862(Z1,G,D,"CANAL",Y)+
5035                                            REP862(Z1,G,D,"LINK",Y);
5036  REP865(Z1,G,D,Y,"SW-VOL"    ,"CANALS" ) = SOILVOL(Z1,G,D) + REP865(Z1,G,D,Y,"FLOW","CANALS");
5037  REP865(Z1,G,D,Y,"NEW-SALT"  ,"CANALS" ) = REP866(Z1,G,D,Y,"SOILSALT") + REP865(Z1,G,D,Y,"SALT-IN","CANALS");
5038  REP865(Z1,G,D,Y,"SALT-SEEP" ,"CANALS" )$REP865(Z1,G,D,Y,"SW-VOL","CANALS")
5039                           = REP865(Z1,G,D,Y,"NEW-SALT","CANALS")
5040                            /REP865(Z1,G,D,Y,"SW-VOL","CANALS")
5041                            *REP865(Z1,G,D,Y,"SEEP-TOGW","CANALS");
5042  REP865(Z1,G,D,Y,"ADD-TOSOIL","CANALS"  ) = REP865(Z1,G,D,Y,"SALT-IN","CANALS")
5043                                            -REP865(Z1,G,D,Y,"SALT-SEEP","CANALS");
5044  REP865(Z1,G,D,Y,"ADD-TOGW"  ,"CANALS"  ) = REP865(Z1,G,D,Y,"SALT-SEEP","CANALS");
5045
      *- Link canals
      rep865(z1,g,d,y,"flow"      ,"links"  ) = rep86a(g,z1,d,"link") ;
      rep865(z1,g,d,y,"seep-togw" ,"links"  ) = rep86(g,d,"seep-link" ,z1);
      rep865(z1,g,d,y,"fiow-ppm"  ,"links"  ) = saltppm(z1) ;
      rep865(z1,g,d,y,"sw-vol"    ,"links"  ) = soilvol(z1,g,d) + rep865(z1,g,d,y,"flow","links");
      rep865(z1,g,d,y,"salt-in"   ,"links"  ) = rep862(z1,g,d,"link",y);
      rep865(z1,g,d,y,"new-salt"  ,"links"  ) = rep866(z1,g,d,y,"soilsalt") + rep865(z1,g,d,y,"salt-in","links");
      rep865(z1,g,d,y,"salt-seep" ,"links"  )$rep865(z1,g,d,y,"sw-vol","links")
                               = rep865(z1,g,d,y,"new-salt","links")
                                /rep865(z1,g,d,y,"sw-vol","links")
                                *rep865(z1,g,d,y,"seep-togw","links");
      rep865(z1,g,d,y,"add-tosoil","links"  ) = rep865(z1,g,d,y,"salt-in","links")
                                               -rep865(z1,g,d,y,"salt-seep","links");
      rep865(z1,g,d,y,"add-togw"  ,"links"  ) = rep865(z1,g,d,y,"salt-seep","links");
5062  *RIVER SEE
5063  REP865(Z1,G,D,Y,"FLOW"      ,"RIVER-S" ) = REP86(G,D,"SEEP-RIVER",Z1);
```

```
                                              193
INDUS BASIN MODEL REVISED (IBMR) GROUNDWATER MODEL FILE=WSIRG41                      07/25/91 17:18:26


5064  REP865(Z1,G,D,Y,"SEEP-TOGW" ,"RIVER-S"  ) = REP86(G,D,"SEEP-RIVER",Z1);
5065  REP865(Z1,G,D,Y,"SALT-SEEP" ,"RIVER-S"  ) = REP86D(Z1,G,D,"FROM-RIVER",Y);
5066  REP865(Z1,G,D,Y,"ADD-TOGW"  ,"RIVER-S"  ) = REP865(Z1,G,D,Y,"SALT-SEEP","RIVER-S");
5067
5068  *- EVAPORATION
5069
5070  REP865(Z1,G,D,Y,"FLOW"      ,"EVAPORATN") = -REP84(Z1,Y,G,D,"GW-EVAP");
5071  REP865(Z1,G,D,Y,"FLOW-PPM"  ,"EVAPORATN") = REP866(Z1,G,D,Y,"GWPPM") ;
5072  REP865(Z1,G,D,Y,"SALT-IN"   ,"EVAPORATN") = REP865(Z1,G,D,Y,"FLOW","EVAPORATN")*REP866(Z1,G,D,Y,"GWPPM")*SALTCONV;
5073  REP865(Z1,G,D,Y,"ADD-TOSOIL","EVAPORATN") = REP865(Z1,G,D,Y,"SALT-IN","EVAPORATN");
5074  REP865(Z1,G,D,Y,"ADD-TOGW"  ,"EVAPORATN") = -REP865(Z1,G,D,Y,"SALT-IN","EVAPORATN");
5075
5076  *- TUBEWELLS
5077  REP865(Z1,G,D,Y,"FLOW"      ,"TUBWELLS" ) = -REP84(Z1,Y,G,D,"TOT-OUTF");
5078  REP865(Z1,G,D,Y,"SEEP-TOGW" ,"TUBWELLS" ) = REP86(G,D,"SEEP-PTW" ,Z1)+REP86(G,D,"SEEP-GTW",Z1);
5079  REP865(Z1,G,D,Y,"FLOW-PPM"  ,"TUBWELLS" ) = REP866(Z1,G,D,Y,"GWPPM") ;
5080  REP865(Z1,G,D,Y,"SALT-IN"   ,"TUBWELLS" ) = REP865(Z1,G,D,Y,"FLOW","TUBWELLS" )*REP866(Z1,G,D,Y,"GWPPM");
5081  REP865(Z1,G,D,Y,"SW-VOL"    ,"TUBWELLS" )  = SOILVOL(Z1,G,D) + REP865(Z1,G,D,Y,"FLOW","TUBWELLS" );
5082  REP865(Z1,G,D,Y,"SALT-IN"   ,"TUBWELLS" ) = REP865(Z1,G,D,Y,"FLOW","TUBWELLS" )*REP866(Z1,G,D,Y,"GWPPM")*SALTCONV;
5083  REP865(Z1,G,D,Y,"NEW-SALT"  ,"TUBWELLS" ) = REP866(Z1,G,D,Y,"SOILSALT") + REP865(Z1,G,D,Y,"SALT-IN","TUBWELLS" );
5084  REP865(Z1,G,D,Y,"SALT-SEEP" ,"TUBWELLS" )$REP865(Z1,G,D,Y,"SW-VOL","TUBWELLS" )
5085                                           = REP865(Z1,G,D,Y,"NEW-SALT","TUBWELLS" )
5086                                            /REP865(Z1,G,D,Y,"SW-VOL","TUBWELLS" )
5087                                            *REP865(Z1,G,D,Y,"SEEP-TOGW","TUBWELLS" );
5088  REP865(Z1,G,D,Y,"ADD-TOSOIL","TUBWELLS" ) = REP865(Z1,G,D,Y,"SALT-IN","TUBWELLS" )
5089                                             -REP865(Z1,G,D,Y,"SALT-SEEP","TUBWELLS" );
5090  REP865(Z1,G,D,Y,"ADD-TOGW"  ,"TUBWELLS" ) = -REP865(Z1,G,D,Y,"ADD-TOSOIL","TUBWELLS" );
5091
5092  REP865(Z1,G,D,Y,"FLOW"      ,"DRAINS"   ) = -REP84(Z1,Y,G,D,"DRAINAGE");
5093  REP865(Z1,G,D,Y,"FLOW-PPM"  ,"DRAINS"   )$REP865(Z1,G,D,Y,"FLOW","DRAINS"   ) = REP866(Z1,G,D,Y,"GWPPM") ;
5094  REP865(Z1,G,D,Y,"SALT-IN"   ,"DRAINS"   ) = REP865(Z1,G,D,Y,"FLOW","DRAINS"   )*REP866(Z1,G,D,Y,"GWPPM")*SALTCONV;
5095  REP865(Z1,G,D,Y,"ADD-TOGW"  ,"DRAINS"   ) = -REP865(Z1,G,D,Y,"SALT-IN","DRAINS"   );
5096
5097  REP865(Z1,G,D,Y,R95,"ROW-SUM") = SUM(R96, REP865(Z1,G,D,Y,R95,R96));
5098  REP865(Z1,G,D,Y,"FLOW-PPM","ROW-SUM") = 0;
5099
5100  REP866(Z1,G,D,Y  ,"TTONS")   = REP865(Z1,G,D,Y,"ADD-TOGW","ROW-SUM") + REP866(Z1,G,D,Y,"TTONS");
5101  REP866(Z1,G,D,Y  ,"GWPPM")$REP86D(Z1,G,D,"MAF",Y) = REP866(Z1,G,D,Y,"TTONS")/
5102                             REP86D(Z1,G,D,"MAF",Y)/SALTCONV;
5103  REP866(Z1,G,D,Y  ,"SOILSALT") = REP866(Z1,G,D,Y,"SOILSALT") +
5104                                  REP865(Z1,G,D,Y,"ADD-TOSOIL","ROW-SUM");
5105  REP866(Z1,G,D,Y+1,"TTONS")    = REP866(Z1,G,D,Y,"TTONS");
5106  REP866(Z1,G,D,Y+1,"GWPPM")    = REP866(Z1,G,D,Y,"GWPPM");
5107  REP866(Z1,G,D,Y+1,"SOILSALT") = REP866(Z1,G,D,Y,"SOILSALT")
5108
5109  );
5110
5111  REP865(Z1,G,D,Y,R95,R96)$(NOT REP865(Z1,G,D,Y,"FLOW",R96) )=0;
5112  REP865(Z1,G,D,Y,"ADD-TOSGW",R96) = REP865(Z1,G,D,Y,"ADD-TOSOIL",R96)+
5113                                     REP865(Z1,G,D,Y,"ADD-TOGW",R96);
5114  REP865(Z1,G,D,Y,R95,R96)$Y1(Y) = 0 ;
5115
5116  REP865(Z1,"SALINE","TOTAL",Y,R95,R96) = SUM(D, REP865(Z1,"SALINE",D,Y,R95,R96));
5117  REP865(Z1,"TOTAL","TOTAL",Y,R95,R96)  = SUM((G,D), REP865(Z1,G,D,Y,R95,R96));
5118  REP865(PV,T1,D1,Y,R95,R96) = SUM(Z1$PVZ(PV,Z1), REP865(Z1,T1,D1,Y,R95,R96));
5119
5120
5121  REP866(Z1,G,D,Y,"GW-MAF")   = REP86D(Z1,G,D,"MAF",Y);
5122  REP866(Z1,G,D,Y,"SOIL-MAF") = SOILVOL(Z1,G,D);
5123  REP866(Z1,"SALINE","TOTAL",Y,R97) = SUM(D, REP866(Z1,"SALINE",D,Y,R97));
5124  REP866(Z1,"TOTAL","TOTAL",Y,R97)  = SUM((G,D), REP866(Z1,G,D,Y,R97));
5125
5126  REP866(Z1,T1,D1,Y,"SOIL-PPM")$SOILVOL(Z1,T1,D1)
5127           = REP866(Z1,T1,D1,Y,"SOILSALT")/SOILVOL(Z1,T1,D1)/SALTCONV;
5128
```

```
        rep866(z1,"Saline","total",y,"soil-ppm")$soilvol(z1,"saline","total")
                    = rep866(z1,"saline","total",y,"soilsalt")
                      /soilvol(z1,"saline","total",)/saltconv;

        rep866(z1,"total" ,"total",y,"soil-ppm")$soilvol(z1,"total" ,"total")
                    = rep866(z1,"Total" ,"total",y,"soilsalt")
                      /soilvol(z1,"Total" ,"total",)/saltconv;
5138
5139  REP866(Z1,T1,D1,Y,"GWPPM")$REP86D(Z1,T1,D1,"MAF",Y) =
5140                              REP866(Z1,T1,D1,Y,"TTONS")/
5141                              REP86D(Z1,T1,D1,"MAF",Y)/SALTCONV;
5142
5143  REP866(PV,T1,D1,Y,R97) = SUM(Z1$PVZ(PV,Z1), REP866(Z1,T1,D1,Y,R97));
5144
5145  REP866(PV,T1,D1,Y,"SOIL-PPM")$SOILVOL(PV,T1,D1)
5146            = REP866(PV,T1,D1,Y,"SOILSALT")/SOILVOL(PV,T1,D1)/SALTCONV;
5147
5148  REP866(PV,T1,D1,Y,"GWPPM")$REP86D(PV,T1,D1,"MAF",Y) =
5149                              REP866(PV,T1,D1,Y,"TTONS")/
5150                              REP86D(PV,T1,D1,"MAF",Y)/SALTCONV;
5151  REP867(Z1,T1,D1,Y,R98,R96) = REP865(Z1,T1,D1,Y,R98,R96);
5152  REP867(PV,T1,D1,Y,R98,R96) = REP865(PV,T1,D1,Y,R98,R96);
5153  REP866(Z1,T1,D1,Y,R97)$Y2(Y) = 0 ;
5154
5155  PARAMETER REP868 SALT ADDITION IN 2007;
5156
5157   REP868(T1,D1,R98,R96,Z1) =REP867(Z1,T1,D1,"2007",R98,R96);
5158   REP868(T1,D1,R98,R96,PV) =REP867(PV,T1,D1,"2007",R98,R96);
5159
5160  OPTION REP865:3:1:1,REP866:3:1:1, REP867:3:2:1, REP868:3:1:1;
5161
5162  DISPLAY SOILVOL, SOILSALT, REP865,REP866,REP867,REP868;


**** FILE SUMMARY FOR USER KDAM

RESTART    R41     WORK*     A
INPUT      RG41    GAMS      A
OUTPUT     RG41    LISTING   A
SAVE       RG41    WORK*     A

COMPILATION TIME    =       1.900 SECONDS        VER: IBM-TB-003
```

References

Ahmad, Masood, Anthony L. Brooke, and Gary P. Kutcher, "Guide to the Indus Basin Model Revised", World Bank Environment Department, January, 1990.

Ahmad, Masood, Gary P. Kutcher, and Alexander Meeraus, "The Agricultural Impact of Kalabagh Dam (as Simulated by the Indus Basin Model Revised)" (two volumes), World Bank Report No. 6884-PAK, Washington D.C., 1987.

Bresler, Eshel, "Irrigation and Soil Salinity", in Dan Yaron, ed., Salinity in Irrigation and Water Resources, New York, Dekker, 1981.

Carruthers, Ian, and Roy Stoner, "Economic Aspects and Policy Issues in Groundwater Development", World Bank Staff Working Paper No. 496, Washington, D.C., 1981.

Carruthers, Ian, "Protecting Irrigation Investment: the Drainage Factor", Ceres 106.

Duloy, John H. and Gerald T. O'Mara, "Issues of Efficiecny and Interdependence in Water Resource Investments: Lessons from the Indus Basin of Pakistan", World Bank Staff Working Paper No. 665, Washington, D.C., 1984.

Gardner, W. R., and Milton Fireman, "Laboratory Studies of Evaporation from Soil Columns in the Presence of a Water Table", Soil Science 85, 1958, pp. 244-249.

Government of Pakistan, Report of the National Commission on Agriculture, Rawalpindi, 1988.

Government of Pakistan, "Agricultural Statistics of Pakistan", Food and Agriculture Division, Ministry of Food, Agriculture and Cooperatives, Islamabad, various years.

Government of Pakistan, "Pakistan Agricultural Census, 1984" (five volumes), Lahore, 1985.

Government of Pakistan, Water and Power Development Authority, Revised Action Programme for Irrigated Agriculture (twelve volumes), Lahore, 1983.

Government of Pakistan, Federal Planning Cell, Water Sector Investment Planning Study (four volumes), Lahore, 1991.

Hanks, R. J., "Prediction of Crop Yield and Water Consumption under Saline Conditions", in I. Shainberg and J. Shalhevet (eds.), Soil Salinity Under Irrigation, New York, Springer-Verlag, 1984.

Jahania, Ch. Muhammad Hussain, Canal and Drainage Act of 1873, Lahore, Mansoor Bookhouse, 1973.

Kirmani, S. S., "Comprehensive Water Resources Management: a Prerequisite for Progess in Pakistan's Irrigated Agriculture", paper presented to the Consultative Meeting on Water Sector Investment Planning, Islamabad, March, 1991.

Maas, E. V. and G. J. Hoffman, "Crop Salt Tolerance - Current Assessment", <u>Journal of the Irrigation and Drainage Division</u>, June, 1977, pp. 115-134.

National Research Council, <u>Saline Agriculture: Salt-Tolerant Plants for Deveoping Countries</u>, Report of a Panel of the Board on Science and Technology for International Development, National Academy Press, Washington D.C., 1990.

Nijland, H. J., and S. El-Guindi, "Crop Yields, Watertable Depth, and Soil Salinity in the Nile Delta, Egypt", International Institute for Land Reclamation and Development, Annual Report, 1983.

World Bank, "Staff Appraisal Report Pakistan: Left Bank Outfall Drain Stage I Project, Report No. 5185-PAK, November 5, 1984. Restricted circulation.

World Bank, EMENA Region, "Pakistan: Medium Term Policy Agenda for Improving Natural Resources and Environmental Management", Washington D.C., 1991.

World Bank, "Staff Appraisal Report Pakistan: Third On-Farm Water Management Project", Report No. 9142-PAK, December 17, 1990. Restricted circulation.

Distributors of World Bank Publications

ARGENTINA
Carlos Hirsch, SRL
Galeria Guemes
Florida 165, 4th Floor-Ofc. 453/465
1333 Buenos Aires

AUSTRALIA, PAPUA NEW GUINEA, FIJI, SOLOMON ISLANDS, VANUATU, AND WESTERN SAMOA
D.A. Books & Journals
648 Whitehorse Road
Mitcham 3132
Victoria

AUSTRIA
Gerold and Co.
Graben 31
A-1011 Wien

BAHRAIN
Bahrain Research and Consultancy Associates Ltd.
Esterad Building No. 42
Diplomatic Area
P.O. Box 2750
Manama Town 317

BANGLADESH
Micro Industries Development Assistance Society (MIDAS)
House 5, Road 16
Dhanmondi R/Area
Dhaka 1209

Branch offices:
156, Nur Ahmed Sarak
Chittagong 4000

76, K.D.A. Avenue
Kulna

BELGIUM
Jean De Lannoy
Av. du Roi 202
1060 Brussels

CANADA
Le Diffuseur
C.P. 85, 1501B rue Ampère
Boucherville, Québec
J4B 5E6

CHINA
China Financial & Economic Publishing House
8, Da Fo Si Dong Jie
Beijing

COLOMBIA
Infoenlace Ltda.
Apartado Aereo 34270
Bogota D.E.

COTE D'IVOIRE
Centre d'Edition et de Diffusion Africaines (CEDA)
04 B.P. 541
Abidjan 04 Plateau

CYPRUS
MEMRB Information Services
P.O. Box 2098
Nicosia

DENMARK
SamfundsLitteratur
Rosenoerns Allé 11
DK-1970 Frederiksberg C

DOMINICAN REPUBLIC
Editora Taller, C. por A.
Restauración e Isabel la Católica 309
Apartado Postal 2190
Santo Domingo

EGYPT, ARAB REPUBLIC OF
Al Ahram
Al Galaa Street
Cairo

The Middle East Observer
41, Sherif Street
Cairo

EL SALVADOR
Fusades
Avenida Manuel Enrique Araujo #3530
Edificio SISA, ler. Piso
San Salvador

FINLAND
Akateeminen Kirjakauppa
P.O. Box 128
Helsinki 10
SF-00101

FRANCE
World Bank Publications
66, avenue d'Iéna
75116 Paris

GERMANY
UNO-Verlag
Poppelsdorfer Allee 55
D-5300 Bonn 1

GREECE
KEME
24, Ippodamou Street Platia Plastiras
Athens-11635

GUATEMALA
Librerias Piedra Santa
5a. Calle 7-55
Zona 1
Guatemala City

HONG KONG, MACAO
Asia 2000 Ltd.
48-48 Wyndham Street
Winning Centre
2nd Floor
Central Hong Kong

INDIA
Allied Publishers Private Ltd.
751 Mount Road
Madras - 600 002

Branch offices:
15 J.N. Heredia Marg
Ballard Estate
Bombay - 400 038

13/14 Asaf Ali Road
New Delhi - 110 002

17 Chittaranjan Avenue
Calcutta - 700 072

Jayadeva Hostel Building
5th Main Road Gandhinagar
Bangalore - 560 009

3-5-1129 Kachiguda Cross Road
Hyderabad - 500 027

Prarthana Flats, 2nd Floor
Near Thakore Baug, Navrangpura
Ahmedabad - 380 009

Patiala House
16-A Ashok Marg
Lucknow - 226 001

INDONESIA
Pt. Indira Limited
Jl. Sam Ratulangi 37
P.O. Box 181
Jakarta Pusat

ISRAEL
Yozmot Literature Ltd.
P.O. Box 56055
Tel Aviv 61560
Israel

ITALY
Licosa Commissionaria Sansoni SPA
Via Duca Di Calabria, 1/1
Casella Postale 552
50125 Florence

JAPAN
Eastern Book Service
37-3, Hongo 3-Chome, Bunkyo-ku 113
Tokyo

KENYA
Africa Book Service (E.A.) Ltd.
P.O. Box 45245
Nairobi

KOREA, REPUBLIC OF
Pan Korea Book Corporation
P.O. Box 101, Kwangwhamun
Seoul

KUWAIT
MEMRB Information Services
P.O. Box 5465

MALAYSIA
University of Malaya Cooperative Bookshop, Limited
P.O. Box 1127, Jalan Pantai Baru
Kuala Lumpur

MEXICO
INFOTEC
Apartado Postal 22-860
14060 Tlalpan, Mexico D.F.

MOROCCO
Société d'Etudes Marketing Marocaine
12 rue Mozart, Bd. d'Anfa
Casablanca

NETHERLANDS
De Lindeboom/InOr-Publikaties
P.O. Box 202
7480 AE Haaksbergen

NEW ZEALAND
Hills Library and Information Service
Private Bag
New Market
Auckland

NIGERIA
University Press Limited
Three Crowns Building Jericho
Private Mail Bag 5095
Ibadan

NORWAY
Narvesen Information Center
Book Department
P.O. Box 6125 Etterstad
N-0602 Oslo 6

OMAN
MEMRB Information Services
P.O. Box 1613, Seeb Airport
Muscat

PAKISTAN
Mirza Book Agency
65, Shahrah-e-Quaid-e-Azam
P.O. Box No. 729
Lahore 3

PERU
Editorial Desarrollo SA
Apartado 3824
Lima

PHILIPPINES
International Book Center
Fifth Floor, Filipinas Life Building
Ayala Avenue, Makati
Metro Manila

PORTUGAL
Livraria Portugal
Rua Do Carmo 70-74
1200 Lisbon

SAUDI ARABIA, QATAR
Jarir Book Store
P.O. Box 3196
Riyadh 11471

MEMRB Information Services
Branch offices:
Al Alsa Street
Al Dahna Center
First Floor
P.O. Box 7188
Riyadh

Haji Abdullah Alireza Building
King Khaled Street
P.O. Box 3969
Dammam

33, Mohammed Hassan Awad Street
P.O. Box 5978
Jeddah

SINGAPORE, TAIWAN, MYANMAR, BRUNEI
Information Publications
Private, Ltd.
02-06 1st Fl., Pei-Fu Industrial Bldg.
24 New Industrial Road
Singapore 1953

SOUTH AFRICA, BOTSWANA
For single titles:
Oxford University Press
Southern Africa
P.O. Box 1141
Cape Town 8000

For subscription orders:
International Subscription Service
P.O. Box 41095
Craighall
Johannesburg 2024

SPAIN
Mundi-Prensa Libros, S.A.
Castello 37
28001 Madrid

Librería Internacional AEDOS
Consell de Cent, 391
08009 Barcelona

SRI LANKA AND THE MALDIVES
Lake House Bookshop
P.O. Box 244
100, Sir Chittampalam A. Gardiner Mawatha
Colombo 2

SWEDEN
For single titles:
Fritzes Fackboksforetaget
Regeringsgatan 12, Box 16356
S-103 27 Stockholm

For subscription orders:
Wennergren-Williams AB
Box 30004
S-104 25 Stockholm

SWITZERLAND
For single titles:
Librairie Payot
6, rue Grenus
Case postale 381
CH 1211 Geneva 11

For subscription orders:
Librairie Payot
Service des Abonnements
Case postale 3312
CH 1002 Lausanne

TANZANIA
Oxford University Press
P.O. Box 5299
Dar es Salaam

THAILAND
Central Department Store
306 Silom Road
Bangkok

TRINIDAD & TOBAGO, ANTIGUA BARBUDA, BARBADOS, DOMINICA, GRENADA, GUYANA, JAMAICA, MONTSERRAT, ST. KITTS & NEVIS, ST. LUCIA, ST. VINCENT & GRENADINES
Systematics Studies Unit
#9 Watts Street
Curepe
Trinidad, West Indies

UNITED ARAB EMIRATES
MEMRB Gulf Co.
P.O. Box 6097
Sharjah

UNITED KINGDOM
Microinfo Ltd.
P.O. Box 3
Alton, Hampshire GU34 2PG
England

VENEZUELA
Libreria del Este
Aptdo. 60.337
Caracas 1060-A

WORLD BANK TECHNICAL PAPER NUMBER 166

Irrigation Planning with Environmental Considerations

A Case Study of Pakistan's Indus Basin

Masood Ahmad and Gary P. Kutcher

The World Bank
Washington, D.C.

Copyright © 1992
The International Bank for Reconstruction
and Development/THE WORLD BANK
1818 H Street, N.W.
Washington, D.C. 20433, U.S.A.

All rights reserved
Manufactured in the United States of America
First printing April 1992

Technical Papers are published to communicate the results of the Bank's work to the development community with the least possible delay. The typescript of this paper therefore has not been prepared in accordance with the procedures appropriate to formal printed texts, and the World Bank accepts no responsibility for errors.

The findings, interpretations, and conclusions expressed in this paper are entirely those of the author(s) and should not be attributed in any manner to the World Bank, to its affiliated organizations, or to members of its Board of Executive Directors or the countries they represent. The World Bank does not guarantee the accuracy of the data included in this publication and accepts no responsibility whatsoever for any consequence of their use. Any maps that accompany the text have been prepared solely for the convenience of readers; the designations and presentation of material in them do not imply the expression of any opinion whatsoever on the part of the World Bank, its affiliates, or its Board or member countries concerning the legal status of any country, territory, city, or area or of the authorities thereof or concerning the delimitation of its boundaries or its national affiliation.

The material in this publication is copyrighted. Requests for permission to reproduce portions of it should be sent to the Office of the Publisher at the address shown in the copyright notice above. The World Bank encourages dissemination of its work and will normally give permission promptly and, when the reproduction is for noncommercial purposes, without asking a fee. Permission to photocopy portions for classroom use is not required, though notification of such use having been made will be appreciated.

The complete backlist of publications from the World Bank is shown in the annual *Index of Publications*, which contains an alphabetical title list (with full ordering information) and indexes of subjects, authors, and countries and regions. The latest edition is available free of charge from the Distribution Unit, Office of the Publisher, Department F, The World Bank, 1818 H Street, N.W., Washington, D.C. 20433, U.S.A., or from Publications, The World Bank, 66, avenue d'Iéna, 75116 Paris, France.

ISSN: 0253-7494

Masood Ahmad is a water resources engineer in the World Bank's Europe, Middle East, and North Africa Region. This work was undertaken while he was with the Environment Department. Gary P. Kutcher is an agricultural and water resources economist. He is a former Bank staff member and currently a consultant to the Bank and other international organizations.

Library of Congress Cataloging-in-Publication Data

Ahmad, Masood.
   Irrigation planning with environmental considerations : a case
study of Pakistan's Indus Basin / Masood Ahmad and Gary P. Kutcher.
     p. cm. — (World Bank technical paper, ISSN 0253-7494 ; no.
166)
   Includes bibliographical references.
   ISBN 0-8213-2080-7
   1. Irrigation engineering—Environmental aspects—Pakistan—Case
studies. 2. Soil salinization—Pakistan—Case studies.
3. Waterlogging (Soils)—Pakistan—Case studies. 4. Irrigation-
-Pakistan—Management—Case studies. 5. Irrigation engineering-
-Environmental aspects—Indus River Watershed—Case studies.
I. Kutcher, Gary P., 1944-   . II. Title. III. Series.
TD195.H93A36 1992
627'.52'095491—dc20   92-6782
                                                                                CIP